Problems from the Book

Titu Andreescu
Gabriel Dospinescu

Problems from the Book

XYZ Press

Titu Andreescu
University of Texas at Dallas
School of Natural Sciences and Mathematics
800 W Campbell Road
Richardson, TX 75080
USA
titu.andreescu@utdallas.edu

Gabriel Dospinescu
École Normale Supérieure
Département de Mathématiques
45, rue d'Ulm
Paris, F-75230
France
gdospi2002@yahoo.com

Library of Congress Control Number: 2008924026

ISBN-13: 978-0-9799269-0-7

9 8 7 6 5 4 3 2 1

www.awesomemath.org

Cover design by Iury Ulzutuev

Math isn't the art of answering mathematical questions, it is the art of asking the right questions, the questions that give you insight, the ones that lead you in interesting directions, the ones that connect with lots of other interesting questions – the ones with beautiful answers.

– G. Chaitin

Preface

What can a new book of problems in elementary mathematics possibly contribute to the vast existing collection of journals, articles, and books? This was our main concern when we decided to write this book. The inevitability of this question does not facilitate the answer, because after five years of writing and rewriting we still had something to add. It could be a new problem, a comment we considered pertinent, or a solution that escaped our rationale until this predictive moment, when we were supposed to bring it under the scrutiny of a specialist in the field.

A mere perusal of this book should be sufficient to identify its target audience: students and coaches preparing for mathematical Olympiads, national or international. It takes more effort to realize that these are not the only potential beneficiaries of this work. While the book is rife with problems collected from various mathematical competitions and journals, one cannot neglect the classical results of mathematics, which naturally exceed the level of time-constrained competitions. And no, classical does not mean easy! These mathematical beauties are more than just proof that elementary mathematics can produce jewels. They serve as an invitation to mathematics beyond competitions, regarded by many to be the "true mathematics". In this context, the audience is more diverse than one might think.

Even so, as it will be easily discovered, many of the problems in this book are very difficult. Thus, the theoretical portions are short, while the emphasis is squarely placed on the problems. Certainly, more subtle results like quadratic reciprocity and existence of primitive roots are related to the basic results in linear algebra or mathematical analysis. Whenever their proofs are particularly useful, they are provided. We will assume of the reader a certain familiarity with classical theorems of elementary mathematics, which we will use freely. The selection of problems was made by weighing the need for routine exercises that engender familiarity with the joy of the difficult problems in which we find the truly beautiful ideas. We strove to select only those problems, easy and hard, that best illustrate the ideas we wanted to exhibit.

Allow us to discuss in brief the structure of the book. What will most likely surprise the reader when browsing just the table of contents is the absence of any chapters on geometry. This book was not intended to be an exhaustive treatment of elementary mathematics; if ever such a book appears, it will be a sad day for mathematics. Rather, we tried to assemble problems that enchanted us in order to give a sense of techniques and ideas that become leitmotifs not just in problem solving but in all of mathematics.

Furthermore, there are excellent books on geometry, and it was not hard to realize that it would be beyond our ability to create something new to add to this area of study. Thus, we preferred to elaborate more on three important fields of elementary mathematics: algebra, number theory, and combinatorics. Even after this narrowing of focus there are many topics that are simply left out, either in consideration of the available space or else because of the fine existing literature on the subject. This is, for example, the fate of functional equations, a field which can spawn extremely difficult, intriguing problems, but one which does not have obvious recurring themes that tie everything together.

Hoping that you have not abandoned the book because of these omissions, which might be considered major by many who do not keep in mind the stated objectives, we continue by elaborating on the contents of the chapters. To start out, we ordered the chapters in ascending order of difficulty of the mathematical tools used. Thus, the exposition starts out lightly with some classical substitution techniques in algebra, emphasizing a large number of examples and applications. These are followed by a topic dear to us: the Cauchy-Schwarz inequality and its variations. A sizable chapter presents applications of the Lagrange interpolation formula, which is known by most only through rôte, straightforward applications. The interested reader will find some genuine pearls in this chapter, which should be enough to change his or her opinion about this useful mathematical tool. Two rather difficult chapters, in which mathematical analysis mixes with algebra, are given at the end of the book. One of them is quite original, showing how simple consideration of integral calculus can solve very difficult inequalities. The other discusses properties of equidistribution and dense numerical series. Too many books consider the Weyl equidistribution theorem to be "much too difficult" to include, and we cannot resist contradicting them by presenting an elementary proof. Furthermore, the reader will quickly realize that for elementary problems we have not shied away from presenting the so-called non-elementary solutions which use mathematical analysis or advanced algebra. It would be a crime to consider these two types of mathematics as two different entities, and it would be even worse to present laborious elementary solutions without admitting the possibility of generalization for problems that have conceptual and easy non-elementary solutions. In the end we devote a whole chapter to discussing criteria for polynomial irreducibility. We observe that some extremely efficient criteria (like those of Peron and Capelli) are virtually unknown, even though they are more efficient than the well-known Eisenstein criterion.

The section dedicated to number theory is the largest. Some introductory chapters related to prime numbers of the form $4k + 3$ and to the order of an element are included to provide a better understanding of fundamental results which are used later in the book. A large chapter develops a tool which is as simple as it is useful: the exponent of a prime in the factorization of an integer. Some mathematical diamonds belonging to Paul Erdős and others appear within. And even though quadratic reciprocity is brought up in many books, we included an entire chapter on this topic because the problems available to us were too ingenious to exclude. Next come some difficult chapters concerning arithmetic properties of polynomials, the geometry of numbers (in which we present some arithmetic applications of the famous Minkowski's theorem), and the properties of algebraic numbers. A special chapter studies some applications of the extremely simple idea that a convergent series of integers is eventually stationary! The reader will have the chance to realize that in mathematics even simple ideas have great impact: consider, for example, the fundamental idea that in the interval $(-1, 1)$ the only integer is 0. But how many fantastic results concerning irrational numbers follow simply from that! Another chapter dear to us concerns the sum of digits, a subject that always yields unexpected and fascinating problems, but for which we could not find a unique approach.

Finally, some words about the combinatorics section. The reader will immediately observe that our presentation of this topic takes an algebraic slant, which was, in fact, our intention. In this way we tried to present some unexpected applications of complex numbers in combinatorics, and a whole chapter is dedicated to useful formal series. Another chapter shows how useful linear algebra can be when solving problems on set combinatorics. Of course, we are traditional in presenting applications of Turan's theorem and of graph theory in general, and the pigeonhole principle could not be omitted. We faced difficulties here, because this topic is covered extensively in other books, though rarely in a satisfying way. For this reason, we tried to present lesser-known problems, because this topic is so dear to elementary mathematics lovers. At the end, we included a chapter on special applications of polynomials in number theory and combinatorics, emphasizing the Combinatorial Nullstellensatz, a recent and extremely useful theorem by Noga Alon.

We end our description with some remarks on the structure of the chapters. In general, the main theoretical results are stated, and if they are sufficiently profound or obscure, a proof is given. Following the theoretical part, we present between ten and fifteen examples, most from mathematical contests or from journals such as Kvant, Komal, and American Mathematical Monthly. Others are new problems or classical results. Each chapter ends with a series of problems, the majority of which stem from the theoretical results.

Finally, a change that will please some and scare others: the end-of-chapter problems do not have solutions! We had several reasons for this. The first and most practical consideration was minimizing the mass of the book. But the second and more important factor was this: we consider solving problems to necessarily include the inevitably lengthy process of trial and research to which the inclusion of solutions provides perhaps too tempting of a shortcut. Keeping this in mind, the selection of the problems was made with the goal that the diligent reader could solve about a third of them, make some progress in the second third and have at least the satisfaction of looking for a solution in the remainder.

We come now to the most delicate moment, the one of saying thank you. First and foremost, we thank Marin Tetiva and Paul Stanford, whose close reading of the manuscript uncovered many errors that we would not have liked in this final version. We thank them for the great effort they put into reviewing the book. All of the remaining mistakes are the responsibility of the authors, who would be grateful for reports of errors so that in a future edition they will disappear. Many thanks to Radu Sorici for giving the book the look it has now and for the numerous suggestions for improvement. We thank Adrian Zahariuc for his help in writing the sections on the sums of digits and graph theory. Several solutions are either his own or the fruit of his experience. Special thanks are due to Valentin Vornicu for creating Mathlinks, which has generated many of the problems we have included. His website, mathlinks.ro, hosts a treasure trove of problems, and we invite every passionate mathematician to avail themselves of this fact. We would also like to thank Ravi Boppana, Vesselin Dimitrov, and Richard Stong for the excellent problems, solutions, and comments they provided. Lastly, we have surely forgotten many others who helped throughout the writing process; our thanks and apologies go out to them.

Titu Andreescu
titu.andreescu@utdallas.edu

Gabriel Dospinescu
gdospi2002@yahoo.com

Contents

Chapter 1

1.1 Theory and examples

We know that in most inequalities with a constraint such as $abc = 1$ the substitution $a = \frac{x}{y}$, $b = \frac{y}{z}$, $c = \frac{z}{x}$ simplifies the solution (don't kid yourself, not all problems of this type become easier!). The use of substitutions is far from being specific to inequalities; there are many other similar substitutions that usually make life easier. For instance, have you ever thought of other conditions such as

$$xyz = x + y + z + 2; \quad xy + yz + zx + 2xyz = 1; \quad x^2 + y^2 + z^2 + 2xyz = 1$$

or $x^2 + y^2 + z^2 = xyz + 4$? The purpose of this chapter is to present some of the most classical substitutions of this kind and their applications.

You will be probably surprised (unless you already know it...) when finding out that the condition $xyz = x + y + z + 2$ together with $x, y, z > 0$ implies the existence of positive real numbers a, b, c such that

$$x = \frac{b+c}{a}, \quad y = \frac{c+a}{b}, \quad z = \frac{a+b}{c}.$$

Let us explain why. The condition $xyz = x + y + z + 2$ can be written in the following equivalent way:

$$\frac{1}{1+x} + \frac{1}{1+y} + \frac{1}{1+z} = 1.$$

Proving this is just a matter of simple computations. Now take

$$a = \frac{1}{1+x}, \quad b = \frac{1}{1+y}, \quad c = \frac{1}{1+z}.$$

Then $a + b + c = 1$ and $x = \frac{1-a}{a} = \frac{b+c}{a}$. Of course, in the same way we find $y = \frac{c+a}{b}$, $z = \frac{a+b}{c}$. The converse (that is, $\frac{b+c}{a}, \frac{c+a}{b}, \frac{a+b}{c}$ satisfy $xyz = x + y + z + 2$) is much easier and is settled again by basic computations. Now, what about the second set of conditions, that is $x, y, z > 0$

and $xy+yz+zx+2xyz=1$? If you look carefully, you will see that it is closely related to the first one. Indeed, $x,y,z>0$ satisfy $xy+yz+zx+2xyz=1$ if and only if $\frac{1}{x},\frac{1}{y},\frac{1}{z}$ verify $\frac{1}{xyz}=\frac{1}{x}+\frac{1}{y}+\frac{1}{z}+2$, so the substitution here is

$$x=\frac{a}{b+c},\qquad y=\frac{b}{c+a},\qquad z=\frac{c}{a+b}.$$

Now, let us take a closer look at the other substitutions mentioned at the beginning of the chapter, namely $x^2+y^2+z^2+2xyz=1$ and $x^2+y^2+z^2=xyz+4$. Let us begin with the following question, which can be considered an exercise, too: consider three real numbers a,b,c such that $abc=1$ and let

$$x=a+\frac{1}{a},\qquad y=b+\frac{1}{b},\qquad z=c+\frac{1}{c} \tag{1.1}$$

The question is to find an algebraic relation between x,y,z, independent of a,b,c. An efficient way to answer this question (that is, without horrible computations that result from solving the quadratic equations) is to observe that

$$\begin{aligned} xyz &= \left(a+\frac{1}{a}\right)\left(b+\frac{1}{b}\right)\left(c+\frac{1}{c}\right)\\ &= \left(a^2+\frac{1}{a^2}\right)+\left(b^2+\frac{1}{b^2}\right)+\left(c^2+\frac{1}{c^2}\right)+2\\ &= (x^2-2)+(y^2-2)+(z^2-2)+2. \end{aligned}$$

Thus

$$x^2+y^2+z^2-xyz=4. \tag{1.2}$$

Because $\left|a + \frac{1}{a}\right| \geq 2$ for all real numbers a, it is clear that not every triple (x, y, z) satisfying (1.2) is of the form (1.1). However, with the extra-assumption $\min\{|x|, |y|, |z|\} \geq 2$ things get better and we do have the converse, that is if x, y, z are real numbers with $\min\{|x|, |y|, |z|\} \geq 2$ and satisfying (1.2), then there exist real numbers a, b, c with $abc = 1$ satisfying (1.1). Actually, it suffices to assume only that $\max(|x|, |y|, |z|) > 2$. Indeed, we may assume that $|x| > 2$. Thus there exists a nonzero real number u such that $x = u + \dfrac{1}{u}$. Now, let us regard (1.2) as a quadratic equation with respect to z. Because the discriminant is nonnegative, it follows that $(x^2 - 4)(y^2 - 4) \geq 0$. But since $|x| > 2$, we find that $y^2 \geq 4$ and so there exist a non-zero real number v for which $y = v + \dfrac{1}{v}$. How do we find the corresponding z? Simply by solving the second degree equation. We find two solutions:

$$z_1 = uv + \frac{1}{uv}, \quad z_2 = \frac{u}{v} + \frac{v}{u}$$

and now we are almost done. If $z = uv + \dfrac{1}{uv}$ we take $(a, b, c) = \left(u, v, \dfrac{1}{uv}\right)$ and if $z = \dfrac{u}{v} + \dfrac{v}{u}$, then we take $(a, b, c) = \left(\dfrac{1}{u}, v, \dfrac{u}{v}\right)$.

Inspired by the previous equation, let us consider another one,

$$x^2 + y^2 + z^2 + xyz = 4 \tag{1.3}$$

where $x, y, z > 0$. We will prove that the set of solutions of this equation is the set of triples $(2\cos A, 2\cos B, 2\cos C)$, where A, B, C are the angles of an acute triangle. First, let us prove that all these triples are solutions. This reduces to the identity

$$\cos^2 A + \cos^2 B + \cos^2 C + 2\cos A \cos B \cos C = 1.$$

This identity can be proved readily by using the sum-to-product formulas. For the converse, we see first that $0 < x, y, z < 2$, hence there are numbers $A, B \in \left(0, \frac{\pi}{2}\right)$ such that $x = 2\cos A$, $y = 2\cos B$. Solving the equation with respect to z and taking into account that $z \in (0, 2)$ we obtain $z = -2\cos(A + B)$. Thus we can take $C = \pi - A - B$ and we will have $(x, y, z) = (2\cos A, 2\cos B, 2\cos C)$.

Let us summarize: we have seen some nice substitutions, with even nicer proofs, but we still have not seen any applications. We will see them in a moment... and there are quite a few problems that can be solved by using these "tricks". First, an easy and classical problem, due to Nesbitt . It has so many extensions and generalizations that we must discuss it first.

Example 1. Prove that

$$\frac{a}{b+c} + \frac{b}{c+a} + \frac{c}{a+b} \geq \frac{3}{2}$$

for all $a, b, c > 0$.

Solution. With the "magical" substitution, it suffices to prove that if $x, y, z > 0$ satisfy $xy + yz + zx + 2xyz = 1$, then $x + y + z \geq \frac{3}{2}$. Let us suppose that this is not the case, i.e. $x + y + z < \frac{3}{2}$. Because $xy + yz + zx \leq \frac{(x+y+z)^2}{3}$, we must have $xy + yz + zx < \frac{3}{4}$ and since $xyz \leq \left(\frac{x+y+z}{3}\right)^3$, we also have $2xyz < \frac{1}{4}$. It follows that $1 = xy+yz+zx+2xyz < \frac{3}{4}+\frac{1}{4} = 1$, a contradiction, so we are done.

Let us now increase the level of difficulty and make an experiment: imagine that you did not know about these substitutions and try to solve the following problem. Then look at the solution provided and you will see that sometimes a good substitution can solve a problem almost alone.

Example 2. Let $x, y, z > 0$ be such that $xy + yz + zx + 2xyz = 1$. Prove that

$$\frac{1}{x} + \frac{1}{y} + \frac{1}{z} \geq 4(x + y + z).$$

[Mircea Lascu]

Solution. With our substitution the inequality becomes

$$\frac{b+c}{a} + \frac{c+a}{b} + \frac{a+b}{c} \geq 4\left(\frac{a}{b+c} + \frac{b}{c+a} + \frac{c}{a+b}\right).$$

But this follows from

$$\frac{4a}{b+c} \leq \frac{a}{b} + \frac{a}{c}, \quad \frac{4b}{c+a} \leq \frac{b}{c} + \frac{b}{a}, \quad \frac{4c}{a+b} \leq \frac{c}{a} + \frac{c}{b}.$$

Simple and efficient, these are the words that characterize this substitution. Here is a geometric application of the previous problem.

Example 3. Prove that in any acute triangle ABC the following inequality holds

$$\cos^2 A \cos^2 B + \cos^2 B \cos^2 C + \cos^2 C \cos^2 A$$
$$\leq \frac{1}{4}(\cos^2 A + \cos^2 B + \cos^2 C).$$

[Titu Andreescu]

Solution. We observe that the desired inequality is equivalent to

$$\frac{\cos A \cos B}{\cos C} + \frac{\cos B \cos C}{\cos A} + \frac{\cos A \cos C}{\cos B} \leq$$
$$\leq \frac{1}{4}\left(\frac{\cos A}{\cos B \cos C} + \frac{\cos B}{\cos C \cos A} + \frac{\cos C}{\cos A \cos B}\right).$$

Setting

$$x = \frac{\cos B \cos C}{\cos A}, \quad y = \frac{\cos A \cos C}{\cos B}, \quad z = \frac{\cos A \cos B}{\cos C},$$

the inequality reduces to

$$4(x + y + z) \leq \frac{1}{x} + \frac{1}{y} + \frac{1}{z}.$$

But this is precisely the inequality in the previous example. All that remains is to show that $xy + yz + zx + 2xyz = 1$. This is equivalent to

$$\cos^2 A + \cos^2 B + \cos^2 C + 2 \cos A \cos B \cos C = 1,$$

which we have already discussed.

The following problem is a nice characterization of the equation (1.2) by polynomials and also teaches us some things about polynomials, in two or three variables.

Example 4. Find all polynomials $f(x, y, z)$ with real coefficients such that

$$f\left(a + \frac{1}{a}, b + \frac{1}{b}, c + \frac{1}{c}\right) = 0$$

whenever $abc = 1$.

[Gabriel Dospinescu]

Solution. From the introduction, it is now clear that the polynomials divisible by $x^2 + y^2 + z^2 - xyz - 4$ are solutions to the problem. But it is not obvious why any desired polynomial should be of this form. To show this, we use the classical polynomial long division. There are polynomials $g(x, y, z)$, $h(y, z)$, $k(y, z)$ with real coefficients such that

$$f(x, y, z) = (x^2 + y^2 + z^2 - xyz - 4)g(x, y, z) + xh(y, z) + k(y, z)$$

Using the hypothesis, we deduce that

$$0 = \left(a + \frac{1}{a}\right) h\left(b + \frac{1}{b}, c + \frac{1}{c}\right) + k\left(b + \frac{1}{b}, c + \frac{1}{c}\right)$$

whenever $abc = 1$. Well, it seems that this is a dead end. Not exactly. Now we take two numbers x, y such that $\min\{|x|, |y|\} > 2$ and we write $x = b + \frac{1}{b}$, $y = c + \frac{1}{c}$ with $b = \frac{x + \sqrt{x^2 - 4}}{2}$, $c = \frac{y + \sqrt{y^2 - 4}}{2}$.
Then it is easy to compute $a + \frac{1}{a}$. It is exactly

$$xy + \sqrt{(x^2 - 4)(y^2 - 4)}.$$

So, we have found that

$$(xy + \sqrt{(x^2 - 4)(y^2 - 4)})h(x, y) + k(x, y) = 0$$

whenever $\min\{|x|, |y|\} > 2$. And now? The last relation suggests that we should prove that for each y with $|y| > 2$, the function $x \to \sqrt{x^2 - 4}$ is not rational, that is, there are not polynomials p, q such that $\sqrt{x^2 - 4} = \frac{p(x)}{q(x)}$. But this is easy because if such polynomials existed, than each zero of $x^2 - 4$ should have even multiplicity, which is not the case. Consequently, for each y with $|y| > 2$ we have $h(x, y) = k(x, y) = 0$ for all x. But this means that $h(x, y) = k(x, y) = 0$ for all x, y, that is our polynomial is divisible by $x^2 + y^2 + z^2 - xyz - 4$.

The level of difficulty continues to increase. When we say this, we refer again to the proposed experiment. The reader who will try first to solve the problems discussed without using the above substitutions will certainly understand why we consider these problems hard.

Example 5. Prove that if $x, y, z > 0$ and $xyz = x + y + z + 2$, then

$$2(\sqrt{xy} + \sqrt{yz} + \sqrt{zx}) \leq x + y + z + 6.$$

Solution. This is tricky, even with the substitution. There are two main ideas: using some identities that transform the inequality into an easier one and then using the substitution. Let us see. What does $2(\sqrt{xy}+\sqrt{yz}+\sqrt{zx})$ suggest? Clearly, it is related to

$$(\sqrt{x}+\sqrt{y}+\sqrt{z})^2-(x+y+z).$$

Consequently, our inequality can be written as

$$\sqrt{x}+\sqrt{y}+\sqrt{z}\leq\sqrt{2(x+y+z+3)}.$$

The first idea that comes to mind (that is using the Cauchy-Schwarz inequality in the form $\sqrt{x}+\sqrt{y}+\sqrt{z}\leq\sqrt{3(x+y+z)}\leq\sqrt{2(x+y+z+3)}$) does not lead to a solution. Indeed, the last inequality is not true: setting $x+y+z=s$, we have $3s\leq 2(s+3)$. This is because the AM-GM inequality implies that $xyz\leq\frac{s^3}{27}$, so $\frac{s^3}{27}\geq s+2$, which is equivalent to $(s-6)(s+3)^2\geq 0$, implying $s\geq 6$.

Let us see how the substitution helps. The inequality becomes

$$\sqrt{\frac{b+c}{a}}+\sqrt{\frac{c+a}{b}}+\sqrt{\frac{a+b}{c}}\leq\sqrt{2\left(\frac{b+c}{a}+\frac{c+a}{b}+\frac{a+b}{c}+3\right)}$$

The last step is probably the most important. We have to change the expression $\frac{b+c}{a}+\frac{c+a}{b}+\frac{a+b}{c}+3$ a little bit.

We see that if we add 1 to each fraction, then $a+b+c$ will appear as a common factor, so in fact

$$\frac{b+c}{a}+\frac{c+a}{b}+\frac{a+b}{c}+3=(a+b+c)\left(\frac{1}{a}+\frac{1}{b}+\frac{1}{c}\right).$$

And now we have finally solved the problem, amusingly, by employing again the Cauchy-Schwarz inequality:

$$\sqrt{\frac{b+c}{a}}+\sqrt{\frac{c+a}{b}}+\sqrt{\frac{a+b}{c}}\leq\sqrt{(b+c+c+a+a+b)\left(\frac{1}{a}+\frac{1}{b}+\frac{1}{c}\right)}.$$

We continue with a difficult 2003 USAMO problem. There are numerous proofs for this inequality, none of them easy. The following solution is again not simple, but seems natural for someone familiar with such a substitution.

Example 6. Prove that for any positive real numbers a, b, c the following inequality holds

$$\frac{(2a+b+c)^2}{2a^2+(b+c)^2} + \frac{(2b+c+a)^2}{2b^2+(c+a)^2} + \frac{(2c+a+b)^2}{2c^2+(a+b)^2} \le 8.$$

[Titu Andreescu, Zuming Feng] USAMO 2003

Solution. The desired inequality is equivalent to

$$\frac{\left(2+\frac{b+c}{a}\right)^2}{2+\left(\frac{b+c}{a}\right)^2} + \frac{\left(2+\frac{c+a}{b}\right)^2}{2+\left(\frac{c+a}{b}\right)^2} + \frac{\left(2+\frac{a+b}{c}\right)^2}{2+\left(\frac{a+b}{c}\right)^2} \le 8.$$

Taking our substitution into account, it suffices to prove that if $xyz = x+y+z+2$, then

$$\frac{(2+x)^2}{2+x^2} + \frac{(2+y)^2}{2+y^2} + \frac{(2+z)^2}{2+z^2} \le 8.$$

This is in fact the same as

$$\frac{2x+1}{x^2+2} + \frac{2y+1}{y^2+2} + \frac{2z+1}{z^2+2} \le \frac{5}{2}.$$

Now, we transform this inequality into

$$\frac{(x-1)^2}{x^2+2} + \frac{(y-1)^2}{y^2+2} + \frac{(z-1)^2}{z^2+2} \ge \frac{1}{2}.$$

This last form suggests using the Cauchy-Schwarz inequality to prove that

$$\frac{(x-1)^2}{x^2+2} + \frac{(y-1)^2}{y^2+2} + \frac{(z-1)^2}{z^2+2} \ge \frac{(x+y+z-3)^2}{x^2+y^2+z^2+6}.$$

So, we are left with proving that $2(x+y+z-3)^2 \geq x^2+y^2+z^2+6$. But this is not difficult. Indeed, this inequality is equivalent to

$$2(x+y+z-3)^2 \geq (x+y+z)^2 - 2(xy+yz+zx) + 6.$$

Now, from $xyz \geq 8$ (recall who x, y, z are and use the AM-GM inequality three times), we find that $xy+yz+zx \geq 12$ and $x+y+z \geq 6$ (by the same AM-GM inequality). This shows that it suffices to prove that $2(s-3)^2 \geq s^2 - 18$ for all $s \geq 6$, which is equivalent to $(s-3)(s-6) \geq 0$, clearly true. And this difficult problem is solved!

The following problem is also hard. Yet there is an easy solution using the substitutions described in this chapter.

Example 7. Prove that if $x, y, z \geq 0$ satisfy $xy + yz + zx + xyz = 4$ then $x + y + z \geq xy + yz + zx$.

India 1998

Solution. Let us write the given condition as

$$\frac{x}{2} \cdot \frac{y}{2} + \frac{y}{2} \cdot \frac{z}{2} + \frac{z}{2} \cdot \frac{x}{2} + 2\frac{x}{2} \cdot \frac{y}{2} \cdot \frac{z}{2} = 1.$$

Hence there are positive real numbers a, b, c such that

$$x = \frac{2a}{b+c}, \quad y = \frac{2b}{c+a}, \quad z = \frac{2c}{a+b}.$$

But now the solution is almost over, since the inequality

$$x + y + z \geq xy + yz + zx$$

is equivalent to

$$\frac{a}{b+c} + \frac{b}{c+a} + \frac{c}{a+b} \geq \frac{2ab}{(c+a)(c+b)} + \frac{2bc}{(a+b)(a+c)} + \frac{2ca}{(b+a)(b+c)}.$$

After clearing denominators, the inequality becomes

$$a(a+b)(a+c)+b(b+a)(b+c)+c(c+a)(c+b) \geq$$

$$\geq 2ab(a+b)+2bc(b+c)+2ca(c+a).$$

After basic computations, it reduces to

$$a(a-b)(a-c)+b(b-a)(b-c)+c(c-a)(c-b) \geq 0.$$

But this is Schur's inequality!

Here is a difficult problem, in which the substitution described plays a key role, but cannot solve the problem alone.

Example 8. Prove that if $x, y, z > 0$ satisfy $xyz = x+y+z+2$, then

$$xyz(x-1)(y-1)(z-1) \leq 8.$$

[Gabriel Dospinescu]

Solution. Using the substitution

$$x = \frac{b+c}{a}, \quad y = \frac{c+a}{b}, \quad z = \frac{a+b}{c},$$

the inequality becomes

$$(a+b)(b+c)(c+a)(a+b-c)(b+c-a)(c+a-b) \leq 8a^2b^2c^2 \quad (1.4)$$

for any positive real numbers a, b, c. It is readily seen that this form is stronger than Schur's inequality

$$(a+b-c)(b+c-a)(c+a-b) \leq abc.$$

First, we may assume that a, b, c are the sides of a triangle ABC, since otherwise the left-hand side in (1.4) is negative. This is true because no more than one of the numbers $a+b-c$, $b+c-a$, $c+a-b$ can be negative. Let R be the center of the circumcircle of triangle ABC. It is not difficult to deduce the following identity

$$(a+b-c)(b+c-a)(c+a-b) = \frac{a^2b^2c^2}{(a+b+c)R^2}.$$

Consequently, the desired inequality can be written as

$$(a+b+c)R^2 \geq \frac{(a+b)(b+c)(c+a)}{8}.$$

But we know that in each triangle ABC, $9R^2 \geq a^2+b^2+c^2$. Hence it suffices to prove that

$$8(a+b+c)(a^2+b^2+c^2) \geq 9(a+b)(b+c)(c+a).$$

This inequality is implied by the following ones:

$$8(a+b+c)(a^2+b^2+c^2) \geq \frac{8}{3}(a+b+c)^3$$

and

$$9(a+b)(b+c)(c+a) \leq \frac{8}{3}(a+b+c)^3.$$

The first inequality reduces to

$$a^2+b^2+c^2 \geq \frac{1}{3}(a+b+c)^2,$$

while the second is a consequence of the AM-GM inequality. By combining these two results, the desired inequality follows.

Of a different kind, the following problem and the featured solution prove that sometimes an efficient substitution can help more than ten complicated ideas.

Example 9. Let $a, b, c > 0$. Find all triples (x, y, z) of positive real numbers such that

$$\begin{cases} x + y + z = a + b + c \\ a^2x + b^2y + c^2z + abc = 4xyz \end{cases}$$

[Titu Andreescu] IMO Shortlist 1995

Solution. We try to use the information given by the second equation. This equation can be written as

$$\frac{a^2}{yz} + \frac{b^2}{zx} + \frac{c^2}{xy} + \frac{abc}{xyz} = 4$$

and we already recognize the relation

$$u^2 + v^2 + w^2 + uvw = 4$$

where $u = \frac{a}{\sqrt{yz}}$, $v = \frac{b}{\sqrt{zx}}$, $w = \frac{c}{\sqrt{xy}}$. According to example 3, we can find an acute-angled triangle ABC such that

$$u = 2\cos A, \quad v = 2\cos B, \quad w = 2\cos C.$$

We have made use of the second condition, so we use the first one to deduce that

$$x + y + z = 2\sqrt{xy}\cos C + 2\sqrt{yz}\cos A + 2\sqrt{zx}\cos B.$$

Trying to solve this as a second degree equation in $\sqrt{x}$, we find the discriminant

$$-4(\sqrt{y}\sin C - \sqrt{z}\sin B)^2.$$

Because this discriminant is nonnegative, we infer that

$$\sqrt{y}\sin C = \sqrt{z}\sin B \text{ and } \sqrt{x} = \sqrt{y}\cos C + \sqrt{z}\cos B.$$

Combining the last two relations, we find that

$$\frac{\sqrt{x}}{\sin A} = \frac{\sqrt{y}}{\sin B} = \frac{\sqrt{z}}{\sin C}$$

Now we square these relations and we use the fact that

$$\cos A = \frac{a}{2\sqrt{yz}}, \quad \cos B = \frac{b}{2\sqrt{zx}}, \quad \cos C = \frac{c}{2\sqrt{xy}}.$$

The conclusion is:

$$x = \frac{b+c}{2}, \quad y = \frac{c+a}{2}, \quad z = \frac{a+b}{2},$$

and it is immediate to see that this triple satisfies both conditions. Hence there is a unique triple that is solution to the given system.

And now, we come back to an earlier problem, this time with a solution based on geometric arguments.

Example 10. Prove that if the positive real numbers x, y, z satisfy $xy + yz + zx + xyz = 4$, then

$$x + y + z \geq xy + yz + zx.$$

India 1998

Solution. The relation given in the hypothesis of the problem is not an immediate analogue of the equation (1.3) Let us write the condition $xy + yz + zx + xyz = 4$ in the form

$$\sqrt{xy}^2 + \sqrt{yz}^2 + \sqrt{zx}^2 + \sqrt{xy} \cdot \sqrt{yz} \cdot \sqrt{zx} = 4.$$

Now, we can use the result from example 3 and we deduce the existence of an acute-angled triangle ABC such that

$$\begin{cases} \sqrt{yz} = 2\cos A \\ \sqrt{zx} = 2\cos B \\ \sqrt{xy} = 2\cos C. \end{cases}$$

We solve the system and we find the triplet

$$(x, y, z) = \left(\frac{2\cos B\cos C}{\cos A}, \frac{2\cos A\cos C}{\cos B}, \frac{2\cos A\cos B}{\cos C}\right).$$

Hence we need to prove that

$$\frac{\cos B\cos C}{\cos A} + \frac{\cos A\cos C}{\cos B} + \frac{\cos A\cos B}{\cos C} \geq 2(\cos^2 A + \cos^2 B + \cos^2 C).$$

This one is a hard inequality and it follows from a more general result.

Lemma 1.1. *If ABC is a triangle and x, y, z are arbitrary real numbers, then*

$$x^2 + y^2 + z^2 \geq 2yz\cos A + 2zx\cos B + 2xy\cos C.$$

Proof. Let us consider points P, Q, R on the lines AB, BC, CA, respectively, such that $AP = BQ = CR = 1$ and P, Q, R do not lie on the sides of the triangle. Then we see that the inequality is equivalent to

$$(x \cdot \overrightarrow{AP} + y \cdot \overrightarrow{BQ} + z \cdot \overrightarrow{CR})^2 \geq 0,$$

which is obviously true.

□

Note that the condition

$$x + y + z = 2\sqrt{xy}\cos C + 2\sqrt{yz}\cos A + 2\sqrt{zx}\cos B$$

is the equality case in the lemma. It offers another approach to Example 9.

The lemma being proved, we just have to take

$$x = \sqrt{\frac{2\cos B\cos C}{\cos A}} \quad y = \sqrt{\frac{2\cos A\cos C}{\cos B}}, \quad z = \sqrt{\frac{2\cos A\cos B}{\cos C}}$$

in the above lemma and the problem will be solved.

And finally, an apparently intricate recursive relation.

Example 11. Let $(a_n)_{n\geq 0}$ be a non-decreasing sequence of positive integers such that $a_0 = a_1 = 47$ and $a_{n-1}^2 + a_n^2 + a_{n+1}^2 - a_{n-1}a_n a_{n+1} = 4$ for $n \geq 1$. Prove that $2 + a_n$ and $2 + \sqrt{2 + a_n}$ are perfect squares for all $n \geq 0$.

[Titu Andreescu]

Solution. Let us write $a_n = x_n + \dfrac{1}{x_n}$, with $x_n > 1$. Then the given condition becomes $x_{n+1} = x_n x_{n-1}$ (we have used here explicitly that $x_n > 1$), which shows that $(\ln x_n)_{n\geq 0}$ is a Fibonacci-type sequence. Since $x_0 = x_1$, we deduce that $x_n = x_0^{F_n}$, where $F_0 = F_1 = 1$, $F_{n+1} = F_n + F_{n-1}$. Now, we have to do more: what is x_0? And the answer $x_0 = \dfrac{47 + \sqrt{47^2 - 1}}{2}$ won't suffice. Let us remark that

$$\left(\sqrt{x_0} + \frac{1}{\sqrt{x_0}}\right)^2 = 49$$

from where we find that

$$\sqrt{x_0} + \frac{1}{\sqrt{x_0}} = 7.$$

Similarly, we obtain that

$$\sqrt[4]{x_0} + \frac{1}{\sqrt[4]{x_0}} = 3.$$

Solving the equation, we obtain

$$\sqrt[4]{x_0} = \left(\frac{1 + \sqrt{5}}{2}\right)^2 = \lambda^2,$$

that is $x_0 = \lambda^8$. And so we have found the general formula $a_n = \lambda^{8F_n} + \lambda^{-8F_n}$. And now the problem becomes easy, since

$$a_n + 2 = (\lambda^{4F_n} + \lambda^{-4F_n})^2 \text{ and } 2 + \sqrt{2 + a_n} = (\lambda^{2F_n} + \lambda^{-2F_n})^2.$$

All we are left to prove is that $\lambda^{2k} + \dfrac{1}{\lambda^{2k}} \in \mathbf{N}$ for all $k \in \mathbf{N}$. But this is not difficult, since

$$\lambda^2 + \frac{1}{\lambda^2} \in \mathbf{N}, \quad \lambda^4 + \frac{1}{\lambda^4} \in \mathbf{N}$$

and

$$\lambda^{2(k+1)} + \frac{1}{\lambda^{2(k+1)}} = \left(\lambda^2 + \frac{1}{\lambda^2}\right)\left(\lambda^{2k} + \frac{1}{\lambda^{2k}}\right) - \left(\lambda^{2(k-1)} + \frac{1}{\lambda^{2(k-1)}}\right).$$

1.2 Practice Problems

1. Find all triples (x, y, z) of positive real numbers such that

$$\begin{cases} x^2 + y^2 + z^2 = xyz + 4 \\ xy + yz + zx = 2(x + y + z) \end{cases}$$

2. Prove that if $a, b, c \geq 0$ satisfy the condition $|a^2 + b^2 + c^2 - 4| = abc$, then

$$(a - 2)(b - 2) + (b - 2)(c - 2) + (c - 2)(a - 2) \geq 0.$$

Titu Andreescu, Gazeta Matematică

3. Prove that if $x, y, z > 0$ and $xyz = x + y + z + 2$, then

$$xy + yz + zx \geq 2(x + y + z) \text{ and } \sqrt{x} + \sqrt{y} + \sqrt{z} \leq \frac{3}{2}\sqrt{xyz}.$$

4. Let $x, y, z > 0$ such that $xy + yz + zx = 2(x + y + z)$. Prove that $xyz \leq x + y + z + 2$.

Gabriel Dospinescu, Mircea Lascu

5. Find all triples of positive integers (k, l, m) with sum 2002 and for which the system

$$\begin{cases} \dfrac{x}{y} + \dfrac{y}{x} = k \\[2ex] \dfrac{y}{z} + \dfrac{z}{y} = l \\[2ex] \dfrac{z}{x} + \dfrac{x}{z} = m \end{cases}$$

has real solutions.

Titu Andreescu, proposed for IMO 2002

6. Prove that if $a, b, c \geq 2$ satisfy the condition $a^2 + b^2 + c^2 = abc + 4$, then

$$a + b + c + ab + ac + bc \geq 2\sqrt{(a + b + c + 3)(a^2 + b^2 + c^2 - 3)}.$$

Marian Tetiva

7. Let $x, y, z > 0$ such that $xy + yz + zx + xyz = 4$. Prove that

$$3\left(\frac{1}{\sqrt{x}} + \frac{1}{\sqrt{y}} + \frac{1}{\sqrt{z}}\right)^2 \geq (x + 2)(y + 2)(z + 2).$$

Gabriel Dospinescu

8. Prove that in any acute triangle,

$$\left(\frac{\cos A}{\cos B}\right)^2 + \left(\frac{\cos B}{\cos C}\right)^2 + \left(\frac{\cos C}{\cos A}\right)^2 + 8\cos A \cos B \cos C \geq 4.$$

Titu Andreescu, MOSP 2000

9. Prove that in every acute triangle ABC,

$$(\cos A + \cos B)^2 + (\cos B + \cos C)^2 + (\cos C + \cos A)^2 \leq 3.$$

10. Prove that if $a, b, c \geq 0$ satisfy $a^2 + b^2 + c^2 + abc = 4$ then

$$0 \leq ab + bc + ca - abc \leq 2.$$

Titu Andreescu, USAMO 2001

11. Solve in positive integers the equation

$$(x + 2)(y + 2)(z + 2) = (x + y + z + 2)^2.$$

Titu Andreescu

12. Let $u, v, w > 0$ be real numbers such that $u + v + w + \sqrt{uvw} = 4$. Prove that

$$\sqrt{\frac{uv}{w}} + \sqrt{\frac{vw}{u}} + \sqrt{\frac{wu}{v}} \geq u + v + w.$$

Chinese TST 2007

13. Consider the sequence $(a_n)_{n\geq 0}$, where $a_0 = a_1 = 97$ and

$$a_{n+1} = a_n a_{n-1} + \sqrt{(a_n^2 - 1)(a_{n-1}^2 - 1)}$$

for all $n \geq 1$. Prove that $2 + \sqrt{2 + 2a_n}$ is a perfect square for all n.

Titu Andreescu

14. Prove that if $a, b, c > 0$ and $x = a + \frac{1}{b}$, $y = b + \frac{1}{c}$, $z = c + \frac{1}{a}$, then

$$xy + yz + zx \geq 2(x + y + z).$$

Vasile Cartoaje

15. Prove that for all $a, b, c > 0$,

$$\frac{(b+c-a)^2}{(b+c)^2 + a^2} + \frac{(c+a-b)^2}{(c+a)^2 + b^2} + \frac{(a+b-c)^2}{(a+b)^2 + c^2} \geq \frac{3}{5}.$$

Japan 1997

16. Prove that in any acute triangle,

$$\frac{a^2b^2}{c^2} + \frac{a^2c^2}{b^2} + \frac{b^2c^2}{a^2} \geq 9R^2.$$

Nguyen Son Ha

17. Find all real numbers k with the following property: for all positive numbers a, b, c the following inequality holds

$$\left(k+\frac{a}{b+c}\right)\left(k+\frac{b}{c+a}\right)\left(k+\frac{c}{a+b}\right) \geq \left(k+\frac{1}{2}\right)^3.$$

Vietnamese TST 2009

18. Let $a_1, a_2, \ldots, a_5$ be positive real numbers such that

$$a_1 a_2 \cdots a_5 = a_1(1+a_2) + a_2(1+a_3) + \cdots + a_5(1+a_1) + 2.$$

What is the least possible value of $\frac{1}{a_1} + \frac{1}{a_2} + \cdots + \frac{1}{a_5}$?

Gabriel Dospinescu, Mathematical Reflections

Chapter 2

2.1 Theory and examples

In recent years the Cauchy-Schwarz inequality has become one of the most used results in contest mathematics, an indispensable tool of any serious problem solver. There are countless problems that reduce readily to this inequality and even more problems in which the Cauchy-Schwarz inequality is the key idea of the solution. In this unit we will not focus on the theoretical results, since they are too well-known. Yet, examples that show the Cauchy-Schwarz inequality at work are not as readily available. This is the reason why we will see this inequality in action in several simple examples first, gradually leading to uses of the Cauchy-Schwarz inequality in some of the most difficult problems.
Let us begin with a very simple problem. Though it is a direct application of the inequality, it underlines something less emphasized: the analysis of the equality case.

Example 1. Prove that the finite sequence $a_0, a_1, \ldots, a_n$ of positive real numbers is a geometrical progression if and only if

$$(a_0^2+a_1^2+\cdots+a_{n-1}^2)(a_1^2+a_2^2+\cdots+a_n^2) = (a_0a_1+\cdots+a_{n-1}a_n)^2.$$

Solution. We see that the relation given in the problem is in fact the equality case in the Cauchy-Schwarz inequality. This is equivalent to the proportionality of the n-tuples $(a_0, a_1, \ldots, a_{n-1})$ and $(a_1, a_2, \ldots, a_n)$, that is

$$\frac{a_0}{a_1} = \frac{a_1}{a_2} = \cdots = \frac{a_{n-1}}{a_n}.$$

But this is just actually the definition of a geometrical progression. Hence the problem is solved. Note that Lagrange's identity allowed us to work with equivalences.

Another easy application of the Cauchy-Schwarz inequality is the following problem. This time the inequality is hidden in a closed form, which suggests using calculus. There exists a solution that uses derivatives, but it is not as elegant as the one featured:

Example 2. Let p be a polynomial with positive real coefficients. Prove that $p(x^2)p(y^2) \geq p^2(xy)$ for any positive real numbers x, y.

Russian Mathematical Olympiad

Solution. If we work only with the closed expression $p(x^2)p(y^2) \geq p^2(xy)$, the chances of seeing a way to proceed are small. So, let us write $p(x) = a_0 + a_1x + \cdots + a_nx^n$. The desired inequality becomes

$$(a_0 + a_1x^2 + \cdots + a_nx^{2n})(a_0 + a_1y^2 + \cdots + a_ny^{2n})$$
$$\geq (a_0 + a_1xy + \cdots + a_nx^ny^n)^2.$$

And now the Cauchy-Schwarz inequality comes into the picture:

$$(a_0 + a_1xy + \cdots + a_nx^ny^n)^2$$
$$= (\sqrt{a_0} \cdot \sqrt{a_0} + \sqrt{a_1x^2} \cdot \sqrt{a_1y^2} + \cdots + \sqrt{a_nx^n} \cdot \sqrt{a_ny^n})^2$$
$$\leq (a_0 + a_1x^2 + \cdots + a_nx^{2n})(a_0 + a_1y^2 + \cdots + a_ny^{2n}).$$

And the problem is solved. Moreover, we see that the conditions $x, y > 0$ are redundant, since we have of course $p^2(xy) \leq p^2(|xy|)$. Additionally, note an interesting consequence of the problem: the function $f : (0, \infty) \to (0, \infty)$, $f(x) = \ln p(e^x)$ is convex, that is why we said in the introduction to this problem that it has a solution based on calculus. The idea of that solution is to prove that the second derivative of this function is nonnegative. We will not prove this here, but we note a simple consequence: the more general inequality

$$p(x_1^k)p(x_2^k)\dots p(x_k^k) \geq p^k(x_1x_2\dots x_k),$$

which follows from Jensen's inequality for the convex function $f(x) = \ln p(e^x)$.

Here is another application of the Cauchy-Schwarz inequality, though this time you might be surprised why the "trick" fails at a first approach:

Example 3. Prove that if $x, y, z > 0$ satisfy $\frac{1}{x}+\frac{1}{y}+\frac{1}{z}=2$, then

$$\sqrt{x-1}+\sqrt{y-1}+\sqrt{z-1} \leq \sqrt{x+y+z}.$$

Iran 1998

Solution. The obvious and most natural approach is to apply the Cauchy-Schwarz inequality in the form

$$\sqrt{x-1}+\sqrt{y-1}+\sqrt{z-1} \leq \sqrt{3(x+y+z-3)}$$

and then to try to prove the inequality $\sqrt{3(x+y+z-3)} \leq \sqrt{x+y+z}$, which is equivalent to $x+y+z \leq \frac{9}{2}$. Unfortunately, this inequality is not true. In fact, the reversed inequality holds, that is $x+y+z \geq \frac{9}{2}$, since $2=\frac{1}{x}+\frac{1}{y}+\frac{1}{z} \geq \frac{9}{x+y+z}$. Thus this approach fails, so we try another, using again the Cauchy-Schwarz inequality, but this time in the form

$$\sqrt{x-1}+\sqrt{y-1}+\sqrt{z-1} = \sqrt{a}\cdot\sqrt{\frac{x-1}{a}}+\sqrt{b}\cdot\sqrt{\frac{y-1}{b}}+\sqrt{c}\cdot\sqrt{\frac{z-1}{c}}$$

$$\leq \sqrt{(a+b+c)\left(\frac{x-1}{a}+\frac{y-1}{b}+\frac{z-1}{c}\right)}.$$

We would like to have the last expression equal to $\sqrt{x+y+z}$. This encourages us to take $a=x$, $b=y$, $c=z$, since in this case

$$\frac{x-1}{a}+\frac{y-1}{b}+\frac{z-1}{c}=1 \text{ and } a+b+c=x+y+z.$$

Hence this idea works and the problem is solved.

We continue with a classical result, the not so well-known inequality of Aczel. We will also see during our trip through the world of the Cauchy-Schwarz inequality a nice application of Aczel's inequality.

Example 4. Let $a_1, a_2, \ldots, a_n, b_1, b_2, \ldots, b_n$ be real numbers and let $A, B > 0$ such that

$$A^2 \geq a_1^2 + a_2^2 + \cdots + a_n^2 \text{ or } B^2 \geq b_1^2 + b_2^2 + \cdots + b_n^2.$$

Then

$$(A^2 - a_1^2 - a_2^2 - \cdots - a_n^2)(B^2 - b_1^2 - b_2^2 - \cdots - b_n^2)$$
$$\leq (AB - a_1b_1 - a_2b_2 - \cdots - a_nb_n)^2.$$

[Aczel]

Solution. We observe first that we may assume that

$$A^2 > a_1^2 + a_2^2 + \cdots + a_n^2 \text{ and } B^2 > b_1^2 + b_2^2 + \cdots + b_n^2.$$

Otherwise the left-hand side of the desired inequality is less than or equal to 0 and the inequality becomes trivial. From our assumption and the Cauchy-Schwarz inequality, we infer that

$$a_1b_1 + a_2b_2 + \cdots + a_nb_n \leq \sqrt{a_1^2 + a_2^2 + \cdots + a_n^2} \cdot \sqrt{b_1^2 + b_2^2 + \cdots + b_n^2} < AB$$

Hence we can rewrite the inequality in the more appropriate form

$$a_1b_1 + a_2b_2 + \cdots + a_nb_n + \sqrt{(A^2 - a)(B^2 - b)} \leq AB,$$

where $a = a_1^2 + a_2^2 + \cdots + a_n^2$ and $b = b_1^2 + b_2^2 + \cdots + b_n^2$. Now, we can apply the Cauchy-Schwarz inequality, first in the form

$$a_1b_1 + a_2b_2 + \cdots + a_nb_n + \sqrt{(A^2 - a)(B^2 - b)} \leq \sqrt{ab} + \sqrt{(A^2 - a)(B^2 - b)}$$

and then in the form

$$\sqrt{ab} + \sqrt{(A^2 - a)(B^2 - b)} \leq \sqrt{(a + A^2 - a)(b + B^2 - b)} = AB.$$

And by combining the last two inequalities the desired inequality follows.

As a consequence of this inequality we discuss the following problem, in which the condition seems to be redundant. In fact, it is the key that suggests using Aczel's inequality.

Example 5. Let $a_1, a_2, \ldots, a_n, b_1, b_2, \ldots, b_n$ be real numbers such that

$$(a_1^2 + \cdots + a_n^2 - 1)(b_1^2 + \cdots + b_n^2 - 1) > (a_1b_1 + \cdots + a_nb_n - 1)^2.$$

Prove that $a_1^2 + a_2^2 + \cdots + a_n^2 > 1$ and $b_1^2 + b_2^2 + \cdots + b_n^2 > 1$.

[Titu Andreescu, Dorin Andrica] USA TST 2004

Solution. First of all, it is not difficult to observe that an indirect approach is more efficient. Moreover, we may even assume that both numbers $a_1^2 + a_2^2 + \cdots + a_n^2 - 1$ and $b_1^2 + b_2^2 + \cdots + b_n^2 - 1$ are negative, since they have the same sign (this follows immediately from the hypothesis of the problem). Now, we want to prove that

$$(a_1^2 + \cdots + a_n^2 - 1)(b_1^2 + \cdots + b_n^2 - 1) \leq (a_1b_1 + \cdots + a_nb_n - 1)^2 \quad (2.1)$$

in order to obtain the desired contradiction. And all of a sudden we arrived at the result in the previous problem. Indeed, we have now the conditions $1 > a_1^2 + a_2^2 + \cdots + a_n^2$ and $1 > b_1^2 + b_2^2 + \cdots + b_n^2$, while the conclusion is (2.1). But this is exactly Aczel's inequality, with $A = 1$ and $B = 1$.The conclusion follows.

The Cauchy-Schwarz inequality is extremely well hidden in the next problem. It is also a refinement of the Cauchy-Schwarz inequality, as we can see from the solution.

Example 6. For given $n > k > 1$ find in closed form the best constant $T(n, k)$ such that for any real numbers $x_1, x_2, \ldots, x_n$ the following inequality holds:

$$\sum_{1\le i<j\le n}(x_i-x_j)^2 \ge T(n,k)\sum_{1\le i<j\le k}(x_i-x_j)^2.$$

[Gabriel Dospinescu]

Solution. In this form, we cannot make any reasonable conjecture about $T(n,k)$, so we need an efficient transformation. We observe that

$$\sum_{1\le i<j\le n}(x_i-x_j)^2$$

is nothing else than

$$n\sum_{i=1}^{n}x_i^2-\left(\sum_{i=1}^{n}x_i\right)^2$$

and also

$$\sum_{1\le i<j\le k}(x_i-x_j)^2 = k\sum_{i=1}^{k}x_i^2-\left(\sum_{i=1}^{k}x_i\right)^2,$$

according to Lagrange's identity. Consequently, the inequality can be written in the equivalent form

$$n\sum_{i=1}^{n}x_i^2-\left(\sum_{i=1}^{n}x_i\right)^2 \ge T(n,k)\left[k\sum_{i=1}^{k}x_i^2-\left(\sum_{i=1}^{k}x_i\right)^2\right].$$

And now we see that it is indeed a refinement of the Cauchy-Schwarz inequality, only if in the end it turns out that $T(n,k)>0$. We also observe that in the left-hand side there are $n-k$ variables that do not appear in the right-hand side and that the left-hand side is minimal when these variables are equal. So, let us take them all to be zero. The result is

$$n\sum_{i=1}^{k}x_i^2-\left(\sum_{i=1}^{k}x_i\right)^2 \ge T(n,k)\left[k\sum_{i=1}^{k}x_i^2-\left(\sum_{i=1}^{k}x_i\right)^2\right],$$

which is equivalent to

$$(T(n,k)-1)\left(\sum_{i=1}^{k} x_i\right)^2 \geq (kT(n,k)-n)\sum_{i=1}^{k} x_i^2 \qquad (2.2)$$

Now, if $kT(n,k)-n>0$, we can take a k-tuple $(x_1, x_2, \ldots, x_k)$ such that $\sum_{i=1}^{k} x_i = 0$ and $\sum_{i=1}^{k} x_i^2 \neq 0$ and we contradict the inequality (2.2). Hence we must have $kT(n,k)-n \leq 0$ that is $T(n,k) \leq \frac{n}{k}$. Now, let us proceed with the converse, that is showing that

$$n\sum_{i=1}^{n} x_i^2 - \left(\sum_{i=1}^{n} x_i\right)^2 \geq \frac{n}{k}\left[k\sum_{i=1}^{k} x_i^2 - \left(\sum_{i=1}^{k} x_i\right)^2\right] \qquad (2.3)$$

for all real numbers $x_1, x_2, \ldots, x_n$. If we manage to prove this inequality, then it will follow that $T(n,k) = \frac{n}{k}$. But (2.3) is of course equivalent to

$$n\sum_{i=k+1}^{n} x_i^2 \geq \left(\sum_{i=1}^{n} x_i\right)^2 - \frac{n}{k}\left(\sum_{i=1}^{k} x_i\right)^2.$$

Now, we have to apply the Cauchy-Schwarz inequality, because we need $\sum_{i=k+1}^{n} x_i$.

We find that

$$n\sum_{i=k+1}^{n} x_i^2 \geq \frac{n}{n-k}\left(\sum_{i=k+1}^{n} x_i\right)^2$$

and so it suffices to prove that

$$\frac{n}{n-k}A^2 \geq (A+B)^2 - \frac{n}{k}B^2, \tag{2.4}$$

where we have taken $A = \sum_{i=k+1}^{n} x_i$ and $B = \sum_{i=1}^{k} x_i$. But (2.4) is straightforward, since it is equivalent to

$$(kA - (n-k)B)^2 + k(n-k)B^2 \geq 0,$$

which is clear. Finally, the conclusion is settled: $T(n,k) = \frac{n}{k}$ is the best constant.

We continue the series of difficult inequalities with a very nice problem of Murray Klamkin. This time, one part of the problem is obvious from the Cauchy-Schwarz inequality, but the second one is not immediate. Let us see.

Example 7. Let a, b, c be positive real numbers. Find the extreme values of the expression

$$\sqrt{a^2x^2 + b^2y^2 + c^2z^2} + \sqrt{b^2x^2 + c^2y^2 + a^2z^2} + \sqrt{c^2x^2 + a^2y^2 + b^2z^2}$$

where x, y, z are real numbers such that $x^2 + y^2 + z^2 = 1$.

[Murray Klamkin] Crux Mathematicorum

Solution. Finding the upper bound does not seem to be too difficult, since from the Cauchy-Schwarz inequality it follows that

$$\sqrt{a^2x^2 + b^2y^2 + c^2z^2} + \sqrt{b^2x^2 + c^2y^2 + a^2z^2} + \sqrt{c^2x^2 + a^2y^2 + b^2z^2} \leq$$

$$\leq \sqrt{3(a^2x^2+b^2y^2+c^2z^2+c^2y^2+a^2z^2+c^2x^2+a^2y^2+b^2z^2)}$$
$$=\sqrt{3(a^2+b^2+c^2)}.$$

We have used here the hypothesis $x^2+y^2+z^2=1$. Thus, $\sqrt{3(a^2+b^2+c^2)}$ is the upper bound and this value if attained for $x=y=z=\frac{\sqrt{3}}{3}$.
But for the lower bound things are not so easy. Investigating what happens when $xyz=0$, we conclude that the minimal value should be $a+b+c$, attained when two variables are zero and the third one is 1 or -1. Hence, we should try to prove the inequality

$$\sqrt{a^2x^2+b^2y^2+c^2z^2}+\sqrt{b^2x^2+c^2y^2+a^2z^2}$$
$$+\sqrt{c^2x^2+a^2y^2+b^2z^2}\geq a+b+c.$$

Why not square it? After all, we observe that

$$a^2x^2+b^2y^2+c^2z^2+b^2x^2+c^2y^2+a^2z^2+c^2x^2+a^2y^2+b^2z^2=a^2+b^2+c^2,$$

so the new inequality cannot have a very complicated form. It becomes

$$\sqrt{a^2x^2+b^2y^2+c^2z^2}\cdot\sqrt{b^2x^2+c^2y^2+a^2z^2}$$
$$+\sqrt{b^2x^2+c^2y^2+a^2z^2}\cdot\sqrt{c^2x^2+a^2y^2+b^2z^2}$$
$$+\sqrt{c^2x^2+a^2y^2+b^2z^2}\cdot\sqrt{a^2x^2+b^2y^2+c^2z^2}\geq ab+bc+ca$$

which has great chances to be true. And indeed, it is true and it follows from – what else, the Cauchy-Schwarz inequality:

$$\sqrt{a^2x^2+b^2y^2+c^2z^2}\cdot\sqrt{b^2x^2+c^2y^2+a^2z^2}\geq abx^2+bcy^2+caz^2$$

and the other two similar inequalities. This shows that the minimal value is indeed $a+b+c$, attained for example when $(x,y,z)=(1,0,0)$.

It is now time for the champion inequalities. Do not worry if the time you spend on them is much longer than the time spent for the other examples: these problems are difficult! There are inequalities where you can immediately see that you should apply the Cauchy-Schwarz inequality. Yet, applying it incorrectly can be very annoying. This is the case with the following example, where there is only one possibility to solve the problem using Cauchy-Schwarz:

Example 8. Prove that for any real numbers a, b, c, x, y, z the following inequality holds:

$$ax + by + cz + \sqrt{(a^2+b^2+c^2)(x^2+y^2+z^2)} \geq \frac{2}{3}(a+b+c)(x+y+z).$$

[Vasile Cartoaje] Kvant

Solution. It is quite clear that a direct application of the Cauchy-Schwarz inequality for

$$\sqrt{(a^2+b^2+c^2)(x^2+y^2+z^2)}$$

has no chance to work. Instead, if we develop $\frac{2}{3}(a+b+c)(x+y+z)$ we may group a, b, c and therefore try again the same method. Let us see:

$$\frac{2}{3}(a+b+c)(x+y+z) - (ax+by+cz) = a \cdot \frac{2y+2z-x}{3} + b \cdot \frac{2x+2z-y}{3} + c \cdot \frac{2x+2y-z}{3}$$

and the latter can be bounded by $\sqrt{a^2+b^2+c^2} \cdot \sqrt{\sum \left(\frac{2x+2y-z}{3}\right)^2}$. All we have to do now is to prove the easy inequality $\sum \left(\frac{2x+2y-z}{3}\right)^2 \leq x^2+y^2+z^2$, which is actually an equality!

Example 9. Prove that for any nonnegative numbers $a_1, a_2, \ldots, a_n$ such that $\sum_{i=1}^{n} a_i = \frac{1}{2}$, the following inequality holds:

$$\sum_{1\le i<j\le n}\frac{a_ia_j}{(1-a_i)(1-a_j)}\le\frac{n(n-1)}{2(2n-1)^2}.$$

[Vasile Cartoaje]

Solution. This is a very hard problem, in which intuition is better than technique. We will concoct a solution using a combination of the Cauchy-Schwarz inequality and Jensen's inequality, but we warn the reader that such a solution cannot be invented easily. Fasten your seat belts! Let us write the inequality in the form

$$\left(\sum_{i=1}^n\frac{a_i}{1-a_i}\right)^2\le\sum_{i=1}^n\frac{a_i^2}{(1-a_i)^2}+\frac{n(n-1)}{(2n-1)^2}.$$

We apply now the Cauchy-Schwarz inequality to find that

$$\left(\sum_{i=1}^n\frac{a_i}{1-a_i}\right)^2\le\left(\sum_{i=1}^n a_i\right)\left(\sum_{i=1}^n\frac{a_i}{(1-a_i)^2}\right)=\sum_{i=1}^n\frac{a_i/2}{(1-a_i)^2}.$$

Thus, it remains to prove the inequality

$$\sum_{i=1}^n\frac{a_i/2}{(1-a_i)^2}\le\sum_{i=1}^n\frac{a_i^2}{(1-a_i)^2}+\frac{n(n-1)}{(2n-1)^2}.$$

The latter can be written of course in the following form:

$$\sum_{i=1}^n\frac{a_i(1-2a_i)}{(1-a_i)^2}\le\frac{2n(n-1)}{(2n-1)^2}.$$

This encourages us to study the function

$$f:\left[0,\frac{1}{2}\right]\to\mathbb{R},\quad f(x)=\frac{x(1-2x)}{(1-x)^2}$$

and to see if it is concave. This is not difficult, for a short computation shows that $f''(x)=\dfrac{-6x}{(1-x)^4}\le 0$. Hence we can apply Jensen's inequality to complete the solution.

We continue this discussion with a remarkable solution, found by Claudiu Raicu, a member of the Romanian Mathematical Olympiad Committee, to the difficult problem given in 2004 in one of the Romanian Team Selection Tests.

Example 10. Let $a_1, a_2, \ldots, a_n$ be real numbers and let S be a non-empty subset of $\{1, 2, \ldots, n\}$. Prove that

$$\left(\sum_{i \in S} a_i\right)^2 \leq \sum_{1 \leq i \leq j \leq n} (a_i + \cdots + a_j)^2.$$

[Gabriel Dospinescu] Romanian TST 2004

Solution. Let us define $s_i = a_1 + a_2 + \cdots + a_i$ for $i \geq 1$ and $s_0 = 0$. Now, partition S into groups of consecutive numbers. Then $\sum\limits_{i \in S} a_i$ is of the form $s_{j_1} - s_{i_1} + s_{j_2} - s_{i_2} + \cdots + s_{j_k} - s_{i_k}$, with $0 \leq i_1 < i_2 < \cdots < i_k \leq n$, $j_1 < j_2 < \cdots < j_k$ and also $i_1 < j_1, \ldots, i_k < j_k$. Now, let us observe that the left-hand side is nothing other than

$$\sum_{i=1}^{n} s_i^2 + \sum_{1 \leq i < j \leq n} (s_j - s_i)^2 = \sum_{1 \leq i < j \leq n} (s_j - s_i)^2.$$

Hence we need to prove that

$$(s_{j_1} - s_{i_1} + s_{j_2} - s_{i_2} + \cdots + s_{j_k} - s_{i_k})^2 \leq \sum_{0 \leq i < j \leq n+1} (s_j - s_i)^2.$$

Let us take $a_1 = s_{i_1}$, $a_2 = s_{j_1}, \ldots,$ $a_{2k-1} = s_{i_k}$, $a_{2k} = s_{j_k}$ and observe the obvious (but important) inequality

$$\sum_{0 \leq i < j \leq n} (s_j - s_i)^2 \geq \sum_{1 \leq i < j \leq 2k} (a_i - a_j)^2.$$

And this is how we arrived at the inequality

$$(a_1 - a_2 + a_3 - \cdots + a_{2k-1} - a_{2k})^2 \leq \sum_{1 \leq i < j \leq 2k} (a_i - a_j)^2. \qquad (2.5)$$

The latter inequality can be proved by using the Cauchy-Schwarz inequality k-times:

$$\begin{cases} (a_1 - a_2 + a_3 - \cdots + a_{2k-1} - a_{2k})^2 \\ \quad \leq k((a_1 - a_2)^2 + (a_3 - a_4)^2 + \cdots + (a_{2k-1} - a_{2k})^2) \\ (a_1 - a_2 + a_3 - \cdots + a_{2k-1} - a_{2k})^2 \\ \quad \leq k((a_1 - a_4)^2 + (a_3 - a_6)^2 + \cdots + (a_{2k-1} - a_2)^2) \\ \ldots \\ (a_1 - a_2 + a_3 - \cdots + a_{2k-1} - a_{2k})^2 \\ \quad \leq k((a_1 - a_{2k})^2 + (a_3 - a_2)^2 + \cdots + (a_{2k-1} - a_{2k-2})^2) \end{cases}$$

and by summing up all these inequalities. In the right-hand side we obtain an even smaller quantity than $\sum\limits_{1 \leq i < j \leq 2k} (a_i - a_j)^2$, which proves that (2.5) is correct. The solution ends here.

The following is a remarkable inequality in which the Cauchy-Schwarz inequality is extremely well hidden. We must confess that the following solution was found after several weeks of trial and error:

Example 11. Prove that for any positive real numbers a, b, c, x, y, z such that $xy + yz + zx = 3$,

$$\frac{a}{b+c}(y+z) + \frac{b}{c+a}(x+z) + \frac{c}{a+b}(x+y) \geq 3.$$

[Titu Andreescu, Gabriel Dospinescu]

Solution. This is probably the best example of how finding the good homogeneous inequality simplifies the solution. In our case, it suffices to prove the homogeneous inequality

$$\frac{a}{b+c}(y+z)+\frac{b}{c+a}(x+z)+\frac{c}{a+b}(x+y) \geq \sqrt{3(xy+yz+zx)}.$$

And now we can assume that $x+y+z=1$! Let us apply then the Cauchy-Schwarz inequality:

$$\frac{a}{b+c}x+\frac{b}{c+a}y+\frac{c}{a+b}z+\sqrt{3(xy+yz+zx)} \leq \sqrt{\sum\left(\frac{a}{b+c}\right)^2}\cdot\sqrt{\sum x^2+}$$

$$\sqrt{\frac{3}{4}(xy+yz+zx)}+\sqrt{\frac{3}{4}(xy+yz+zx)} \leq \sqrt{\frac{3}{2}+\sum\left(\frac{a}{b+c}\right)^2}\cdot\sqrt{(x+y+z)^2}$$

Therefore, the problem will be solved if we manage to prove that

$$\sqrt{\frac{3}{2}+\sum\left(\frac{a}{b+c}\right)^2} \leq \sum\frac{a}{b+c},$$

which is the same as

$$\sum\frac{ab}{(a+c)(b+c)} \geq \frac{3}{4}.$$

This reduces to $(a+b+c)(ab+bc+ca) \geq 9abc$ which is clearly true.

Finally, two classical inequalities show the power of a clever application of the Cauchy-Schwarz inequality combined with some analytic tools:

Example 12. Prove that for any real numbers $a_1, a_2, ..., a_n$ the following inequality holds:

$$\sum_{i=1}^{n}\sum_{j=1}^{n}\frac{a_i a_j}{i+j} \leq \pi\cdot\sum_{i=1}^{n}a_i^2$$

[Hilbert]

Solution. Here is a unusual way to apply the Cauchy-Schwarz inequality:

$$\left(\sum_{i=1}^{n}\sum_{j=1}^{n}\frac{a_i a_j}{i+j}\right)^2 = \left(\sum_{i,j=1}^{n}\frac{\sqrt[4]{i}a_i}{\sqrt[4]{j}\sqrt{i+j}}\cdot\frac{\sqrt[4]{j}a_j}{\sqrt[4]{i}\sqrt{i+j}}\right)^2$$

$$\leq \left(\sum_{i,j=1}^{n}\frac{\sqrt{i}a_i^2}{\sqrt{j}(i+j)}\right)\cdot\left(\sum_{i,j=1}^{n}\frac{\sqrt{j}a_j^2}{\sqrt{i}(i+j)}\right).$$

By rearranging terms in both sums, it is enough to prove that for any positive integer m

$$\sum_{n\geq 1}\frac{\sqrt{m}}{(m+n)\sqrt{n}} \leq \pi.$$

Fortunately, this is not difficult, because the inequality

$$\frac{1}{(n+m+1)\sqrt{n+1}} \leq \int_n^{n+1}\frac{dx}{(x+m)\sqrt{x}}$$

holds as a consequence of the monotonicity of $f(x) = \frac{1}{(x+m)\sqrt{x}}$. By adding up these inequalities, we deduce that

$$\sum_{n\geq 0}\frac{1}{(n+m+1)\sqrt{n+1}} \leq \int_0^{\infty}\frac{dx}{(x+m)\sqrt{x}}.$$

With the change of variable $x = mu^2$, a simple computation shows that the last integral is $\frac{\pi}{\sqrt{m}}$ and this finishes the solution.

We end this chapter with a remarkable inequality due to Fritz Carlson . There are many analytic methods of proving this result, but undoubtedly the following one, due to Hardy, will make you say: always Cauchy-Schwarz!

Example 13. For any real numbers $a_1, a_2, ..., a_n$ we have

$$\pi^2 \cdot (a_1^2 + a_2^2 + \cdots + a_n^2)(a_1^2 + 4a_2^2 + \cdots + n^2 a_n^2) \geq (a_1 + \cdots + a_n)^4.$$

[Fritz Carlson]

Solution. Choose some arbitrary positive numbers x, y and use the Cauchy-Schwarz inequality in the form

$$(a_1 + a_2 + ... + a_n)^2 \leq \sum_{k=1}^{n} (x + yk^2) a_k^2 \cdot \sum_{k \geq 1} \frac{1}{x + yk^2}.$$

Because the function $f(z) = \frac{1}{x+yz^2}$ is decreasing, we have

$$\sum_{k \geq 1} \frac{1}{x + yk^2} \leq \int_0^{\infty} \frac{dz}{x + yz^2}.$$

It is immediate to check that the last integral equals $\frac{\pi}{2\sqrt{xy}}$. Therefore, if we let $S = a_1^2 + a_2^2 + ... + a_n^2$ and $T = a_1^2 + 2^2 a_2^2 + ... + n^2 a_n^2$, then we have for all positive numbers x, y the inequality

$$(a_1 + a_2 + ... + a_n)^2 \leq \frac{\pi}{2\sqrt{xy}} (Sx + Ty).$$

And now, we can make a choice for x, y, so as to minimize the last quantity. It is not difficult to see that a smart choice is $x = \sqrt{\frac{T}{S}}$ and $y = \frac{1}{x}$. All it remains is to insert these values in the previous inequality and to take the square of this relation.

2.2 Practice problems

1. Let a, b, c be nonnegative real numbers. Prove that

$$(ax^2 + bx + c)(cx^2 + bx + a) \geq (a + b + c)^2 x^2$$

for all nonnegative real numbers x.

Titu Andreescu, Gazeta Matematică

2. Let p be a polynomial with positive coefficients. Prove that if the inequality $p\left(\frac{1}{x}\right) \geq \frac{1}{p(x)}$ holds for $x = 1$, then it holds for all $x > 0$.

Titu Andreescu, Revista Matematică Timişoara

3. Prove that for any real numbers $a, b, c \geq 1$,

$$\sqrt{a-1} + \sqrt{b-1} + \sqrt{c-1} \leq \sqrt{a(bc+1)}.$$

4. For any positive integer n find the number of ordered n-tuples of integers $(a_1, a_2, \ldots, a_n)$ such that

$$a_1 + a_2 + \cdots + a_n \geq n^2 \text{ and } a_1^2 + a_2^2 + \cdots + a_n^2 \leq n^3 + 1.$$

China 2002

5. Prove that for any positive real numbers a, b, c,

$$\frac{1}{a+b} + \frac{1}{b+c} + \frac{1}{c+a} + \frac{1}{2\sqrt[3]{abc}} \geq \frac{(a+b+c+\sqrt[3]{abc})^2}{(a+b)(b+c)(c+a)}.$$

Titu Andreescu, MOSP 1999

6. Let $x_1, x_2, \ldots, x_{10}$ be real numbers between 0 and $\frac{\pi}{2}$ such that

$$\sin^2 x_1 + \sin^2 x_2 + \cdots + \sin^2 x_{10} = 1.$$

Prove that

$$3(\sin x_1 + \sin x_2 + \cdots + \sin x_{10}) \leq \cos x_1 + \cos x_2 + \cdots + \cos x_{10}.$$

Saint Petersburg 2001

7. Let ABC be a triangle such that

$$\left(\cot \frac{A}{2}\right)^2 + \left(2\cot \frac{B}{2}\right)^2 + \left(3\cot \frac{C}{2}\right)^2 = \left(\frac{6s}{7r}\right)^2.$$

Prove that ABC is similar to a triangle T whose side lengths are positive integers with no common divisor and determine these integers.

Titu Andreescu, USAMO 2002

8. Let $n \geq 2$ be an even integer. We consider all polynomials of the form $x^n + a_{n-1}x^{n-1} + \cdots + a_1 x + 1$, with real coefficients and having at least one real zero. Determine the least possible value of $a_1^2 + a_2^2 + \cdots + a_{n-1}^2$.

Czech-Polish-Slovak Competition 2002

9. Prove that for any positive real numbers x, y, z such that $xyz \geq 1$ the following inequality holds

$$\frac{x}{x^3 + y^2 + z} + \frac{y}{y^3 + z^2 + x} + \frac{z}{z^3 + x^2 + y} \leq 1.$$

Tuan Le, Komal

10. Let $n \geq 2$ and let $a_1, \ldots, a_n$ and $b_1, \ldots, b_n$ be real numbers such that

$$a_1^2 + a_2^2 + \cdots + a_n^2 = b_1^2 + b_2^2 + \cdots + b_n^2 = 1$$

and $a_1b_1 + a_2b_2 + \cdots + a_nb_n = 0$. Prove that

$$(a_1 + a_2 + \cdots + a_n)^2 + (b_1 + b_2 + \cdots + b_n)^2 \leq n.$$

Cezar and Tudorel Lupu, Romania TST 2007

11. Let $x_1, x_2, \ldots, x_n$ be positive real numbers such that

$$\frac{1}{1+x_1} + \frac{1}{1+x_2} + \cdots + \frac{1}{1+x_n} = 1.$$

Prove the inequality

$$\sqrt{x_1} + \sqrt{x_2} + \cdots + \sqrt{x_n} \geq (n-1)\left(\frac{1}{\sqrt{x_1}} + \frac{1}{\sqrt{x_2}} + \cdots + \frac{1}{\sqrt{x_n}}\right).$$

Vojtech Jarnik Competition 2002

12. Find the greatest real number T such that for any nonnegative real numbers a, b, c, d, e such that $a + b = c + d + e$ we have

$$\sqrt{a^2 + b^2 + c^2 + d^2 + e^2} \geq T(\sqrt{a} + \sqrt{b} + \sqrt{c} + \sqrt{d} + \sqrt{e})^2$$

Iran 2007

13. Prove that for any real numbers $x_1, x_2, \ldots, x_n$,

$$\frac{x_1}{1+x_1^2} + \frac{x_2}{1+x_1^2+x_2^2} + \cdots + \frac{x_n}{1+x_1^2+\cdots+x_n^2} < \sqrt{n}.$$

Bogdan Enescu, IMO Shortlist 2001

14. For $n \geq 2$ let $a_1, a_2, \ldots, a_n$ be positive real numbers such that

$$(a_1 + a_2 + \cdots + a_n)\left(\frac{1}{a_1} + \frac{1}{a_2} + \cdots + \frac{1}{a_n}\right) \leq \left(n + \frac{1}{2}\right)^2.$$

Prove that $\max(a_1, a_2, \ldots, a_n) \leq 4\min(a_1, a_2, \ldots, a_n)$.

Titu Andreescu, USAMO 2009

15. Let $n > 2$ and $x_1, x_2, \ldots, x_n$ be positive real numbers such that

$$(x_1 + x_2 + \cdots + x_n)\left(\frac{1}{x_1} + \frac{1}{x_2} + \cdots + \frac{1}{x_n}\right) = n^2 + 1.$$

Prove that

$$(x_1^2 + x_2^2 + \cdots + x_n^2)\left(\frac{1}{x_1^2} + \frac{1}{x_2^2} + \cdots + \frac{1}{x_n^2}\right) > n^2 + 4 + \frac{2}{n(n-1)}.$$

Gabriel Dospinescu

16. Prove that if a, b, c, d are positive real numbers, then

$$\frac{a}{b^2 + c^2 + d^2} + \frac{b}{a^2 + c^2 + d^2} + \frac{c}{a^2 + b^2 + d^2} + \frac{d}{a^2 + b^2 + c^2} > \frac{4}{a + b + c + d}.$$

P.K.Hung

17. Let $n \geq 2$ be an integer and let $x_1, x_2, \ldots, x_n$ be real numbers satisfying

$$x_1^2 + x_2^2 + \cdots + x_n^2 + x_1x_2 + x_2x_3 + \cdots + x_{n-1}x_n = 1.$$

For a fixed $1 \leq k \leq n$, find the maximum value that $|x_k|$ can take.

China 1998

18. Let a, b, c be positive real numbers. Prove that

$$\frac{1}{a^2}+\frac{1}{b^2}+\frac{1}{c^2}+\frac{1}{(a+b+c)^2} \geq \frac{7}{25}\left(\frac{1}{a}+\frac{1}{b}+\frac{1}{c}+\frac{1}{a+b+c}\right)^2$$

Iran 2010

19. Let x, y, z be real numbers and let A, B, C be the angles of a triangle. Prove that

$$x \sin A + y \sin B + z \sin C \leq \sqrt{(1+x^2)(1+y^2)(1+z^2)}.$$

20. Let a, b, c, x, y, z be real numbers and let

$$A = ax + by + cz, B = ay + bz + cx, C = az + bx + cy.$$

Assuming that

$$\min(|A-B|, |B-C|, |C-A|) \geq 1,$$

find the least possible value of $(a^2+b^2+c^2)(x^2+y^2+z^2)$.

Adrian Zahariuc, Mathematical Reflections

21. Let a, b, c, d, e be nonnegative real numbers such that $a^2+b^2+c^2 = d^2+e^2$ and $a^4+b^4+c^4 = d^4+e^4$. Prove that $a^3+b^3+c^3 \leq d^3+e^3$.

IMC

22. Prove that for any real numbers $x_1, x_2, \ldots, x_n$,

$$\left(\sum_{i=1}^{n}\sum_{i=1}^{n}|x_i - x_j|\right)^2 \leq \frac{2(n^2-1)}{3}\left(\sum_{i=1}^{n}\sum_{j=1}^{n}|x_i - x_j|^2\right).$$

IMO 2003

23. Let a, b, c, d be real numbers such that

$$(a^2+1)(b^2+1)(c^2+1)(d^2+1) = 16.$$

Prove that

$$-3 \le ab + bc + cd + da + ac + bd - abcd \le 5.$$

Titu Andreescu, Gabriel Dospinescu, Mathematical Reflections

24. Let $a_1, a_2, \dots, a_n$ be positive real numbers which add up to 1. Let n_i be the number of integers k such that $2^{1-i} \ge a_k > 2^{-i}$. Prove that

$$\sum_{i \ge 1} \sqrt{\frac{n_i}{2^i}} \le 4 + \sqrt{\log_2 n}.$$

L. Leindler, Miklos Schweitzer Competition

25. Let $n > 2$ be an integer. Find the largest real number k with the following property: if the positive real numbers $x_1, x_2, \dots, x_n$ satisfy

$$k > (x_1 + x_2 + \cdots + x_n)\left(\frac{1}{x_1} + \frac{1}{x_2} + \cdots + \frac{1}{x_n}\right),$$

then any three of them are sides of a triangle.

Adapted after IMO 2004

26. If a, b, c, d, e are real numbers such that $a + b + c + d + e = 0$, then

$$(a^2+b^2+c^2+d^2+e^2)^2 \le \frac{30}{7}(a^4+b^4+c^4+d^4+e^4).$$

Vasile Cartoaje

27. Prove that for any positive real numbers $a_1, a_2, \ldots, a_n, x_1, x_2, \ldots, x_n$ such that

$$\sum_{i \leq i < j \leq n} x_i x_j = \binom{n}{2},$$

the following inequality holds

$$\frac{a_1}{a_2 + \cdots + a_n}(x_2 + \cdots + x_n) + \cdots + \frac{a_n}{a_1 + \cdots + a_{n-1}}(x_1 + \cdots + x_{n-1}) \geq n.$$

Vasile Cartoaje, Gabriel Dospinescu

Chapter 3

3.1 Theory and examples

Most of the time, proving divisibility reduces to congruences or to the famous theorems such as those of Fermat, Euler, or Wilson. But what do we do when we have to prove, for example, that $\operatorname{lcm}(a,b,c)^2 \mid \operatorname{lcm}(a,b)\cdot\operatorname{lcm}(b,c)\cdot\operatorname{lcm}(c,a)$ for any positive integers a,b,c? One thing is sure: the above methods fail. Yet, another smart idea appears: if we have to prove that $a|b$, then it suffices to show that the exponent of any prime number in the prime factorization of a is at most the exponent of that prime in the prime factorization of b. For simplicity, let us denote by $v_p(a)$ the exponent of the prime number p in the prime factorization of a. Of course, if p does not divide a, then $v_p(a) = 0$. Also, it is easy to prove the following properties of $v_p(a)$:

- $v_p(a+b) \geq \min\{v_p(a), v_p(b)\}$
- $v_p(ab) = v_p(a) + v_p(b)$
  for any positive integers a and b. Now, let us rephrase the above idea in terms of $v_p(a)$: $a|b$ if and only if for any prime p we have $v_p(a) \leq v_p(b)$, and $a = b$ if and only if for any prime p, $v_p(a) = v_p(b)$.
- $v_p(\gcd(a_1, a_2, \ldots, a_n)) = \min\{v_p(a_1), v_p(a_2), \ldots, v_p(a_n)\}$,
- $v_p(\operatorname{lcm}(a_1, a_2, \ldots, a_n)) = \max\{v_p(a_1), v_p(a_2), \ldots, v_p(a_n)\}$
- $v_p(n!) = \left\lfloor \frac{n}{p} \right\rfloor + \left\lfloor \frac{n}{p^2} \right\rfloor + \left\lfloor \frac{n}{p^3} \right\rfloor + \cdots = \frac{n - s_p(n)}{p-1}$.

Here, $s_p(n)$ is the sum of the digits of n when written in base p. Observe that the third and fourth properties are simple consequences of the definitions. Less straightforward is the fifth property; it follows from the fact that among the numbers $1, \ldots, n$ there are $\left\lfloor \frac{n}{p} \right\rfloor$ multiples of p, $\left\lfloor \frac{n}{p^2} \right\rfloor$ multiples of p^2 and so on. The other equality is not difficult. Indeed, let us write $n = a_0 + a_1p + \cdots + a_kp^k$, where $a_0, a_1, \ldots, a_k \in \{0, 1, \ldots, p-1\}$ and $a_k \neq 0$. Then

$$\left\lfloor \frac{n}{p} \right\rfloor + \left\lfloor \frac{n}{p^2} \right\rfloor + \cdots = a_1 + a_2p + \cdots + a_kp^{k-1} + a_2 + a_3p + \cdots + a_kp^{k-2} + \cdots + a_k,$$

and now, using the formula

$$1 + p + \cdots + p^i = \frac{p^{i+1} - 1}{p - 1},$$

we find exactly the fifth property.

Example 1. Let a and b be positive integers such that $a|b^2$, $b^3|a^4$, $a^5|b^6$, $b^7|a^8, \ldots$. Prove that $a = b$.

Solution. We will prove that $v_p(a) = v_p(b)$ for any prime p. The hypothesis $a|b^2$, $b^3|a^4$, $a^5|b^6$, $b^7|a^8, \ldots$ is the same as $a^{4n+1}|b^{4n+2}$ and $b^{4n+3}|a^{4n+4}$ for all positive integers n. But the relation $a^{4n+1}|b^{4n+2}$ can be written as $(4n + 1)v_p(a) \leq (4n + 2)v_p(b)$ for all n, so that

$$v_p(a) \leq \lim_{n \to \infty} \frac{4n + 2}{4n + 1} v_p(b) = v_p(b).$$

Similarly, the condition $b^{4n+3}|a^{4n+4}$ implies $v_p(a) \geq v_p(b)$ and so $v_p(a) = v_p(b)$. The conclusion now follows.

We have mentioned at the beginning of the discussion a nice and easy problem, so probably it is time to solve it, although you might have already done this.

Example 2. Prove that $\operatorname{lcm}(a, b, c)^2 | \operatorname{lcm}(a, b) \cdot \operatorname{lcm}(b, c) \cdot \operatorname{lcm}(c, a)$ for any positive integers a, b, c.

Solution. Let p be an arbitrary prime number. We have $v_p(\operatorname{lcm}(a, b, c)^2) = 2 \max\{x, y, z\}$ and

$$v_p(\operatorname{lcm}(a, b) \cdot \operatorname{lcm}(b, c) \cdot \operatorname{lcm}(c, a)) = \max\{x, y\} + \max\{y, z\} + \max\{z, x\},$$

where $x = v_p(a)$, $y = v_p(b)$, $z = v_p(c)$. So we need to prove that

$$\max\{x, y\} + \max\{y, z\} + \max\{z, x\} \geq 2 \max\{x, y, z\}$$

for any nonnegative integers x, y, z. But this follows by symmetry: we may assume that $x \geq y \geq z$ and the inequality reduces to $2x + y \geq 2x$.

It is time for some difficult problems. The ones we chose to present are all based on the observations from the beginning of the chapter.

Example 3. Prove that there exists a constant c such that for any positive integers a, b, n that satisfy $a! \cdot b! | n!$ we have $a + b < n + c \ln n$.

[Paul Erdős]

Solution. Of course, there is no reasonable estimation of this constant, so we should better see what happens if $a! \cdot b! | n!$. Then $v_2(a!) + v_2(b!) \leq v_2(n!)$, which can be also written as $a - s_2(a) + b - s_2(b) \leq n - s_2(n) < n$. So we have found almost exactly what we needed: $a + b < n + s_2(a) + s_2(b)$. Now, we need another observation: the sum of digits of a number A when written in binary is at most the number of digits of A in base 2, which is $1 + [\log_2 A]$ (this follows from the fact that $2^{k-1} \leq A < 2^k$, where k is the number of digits of A in base 2). Hence we have the estimations

$$a + b < n + s_2(a) + s_2(b) \leq n + 2 + \log_2 ab \leq n + 2 + 2\log_2 n$$

(since we have of course $a, b \leq n$). And now the conclusion is immediate.

The following problem appeared in Kvant. It took quite a long time before an Olympian, S. Konyagin, found a simple solution. We will not present his solution here, but another one, even simpler.

Example 4. Is there an infinite set of positive integers such that no matter how we choose some elements of this set, their sum is not a perfect power?

Kvant

Solution. Let us take $A = \{2^n \cdot 3^{n+1} | n \geq 1\}$ If we consider some different numbers from this set, their sum will be of the form $2^x \cdot 3^{x+1} \cdot y$, where $(y, 6) = 1$. This is certainly not a perfect power, since otherwise the exponent should divide both x and $x + 1$. Thus this set is actually a good choice.

The following problem shows the beauty of elementary Number Theory. It combines diverse ideas and techniques, and the result we are about to present is truly beautiful. You might also want to try a combinatorial approach by counting the invertible matrices with entries in the field $\mathbb{Z}/2\mathbb{Z}$.

Example 5. Prove that for any positive integer n, $n!$ is a divisor of

$$\prod_{k=0}^{n-1}(2^n - 2^k).$$

Solution. Let us take a prime number p. We may assume that $p \leq n$. First, let us see what happens if $p = 2$. We have

$$v_2(n!) = n - s_2(n) \leq n - 1$$

and also

$$v_2\left(\prod_{k=0}^{n-1}(2^n - 2^k)\right) = \sum_{k=0}^{n-1} v_2(2^n - 2^k) \geq n - 1$$

(since $2^n - 2^k$ is even for $k \geq 1$). Now, let us assume that $p > 2$. From Fermat's theorem we have $p | 2^{p-1} - 1$, so $p | 2^{k(p-1)} - 1$ for all $k \geq 1$. Now,

$$\prod_{k=0}^{n-1}(2^n - 2^k) = 2^{\frac{n(n-1)}{2}} \prod_{k=1}^{n}(2^k - 1)$$

and from the above remarks we infer that

$$v_p\left(\prod_{k=0}^{n-1}(2^n - 2^k)\right) = \sum_{k=1}^{n} v_p(2^k - 1)$$

$$\geq \sum_{1\leq k(p-1)\leq n} v_p(2^{k(p-1)}-1) \geq \text{card}\{k|1\leq k(p-1)\leq n\}.$$

Because

$$\text{card}\{k|1\leq k(p-1)\leq n\} = \left\lfloor \frac{n}{p-1} \right\rfloor,$$

we find that

$$v_p\left(\prod_{k=0}^{n-1}(2^n-2^k)\right) \geq \left\lfloor \frac{n}{p-1} \right\rfloor.$$

But

$$v_p(n!) = \frac{n-s_p(n)}{p-1} \leq \frac{n-1}{p-1} < \frac{n}{p-1},$$

and since $v_p(n!) \in \mathbb{Z}$, we must have $v_p(n!) \leq \left\lfloor \frac{n}{p-1} \right\rfloor$.
From these two inequalities, we conclude that

$$v_p\left(\prod_{k=0}^{n-1}(2^n-2^k)\right) \geq v_p(n!)$$

and the problem is solved.
Diophantine equations can also be solved using the method described in this chapter. Here is a difficult one, given at a Russian Olympiad.

Example 6. Prove that the equation

$$\frac{1}{10^n} = \frac{1}{n_1!} + \frac{1}{n_2!} + \cdots + \frac{1}{n_k!}$$

does not have integer solutions such that $1 \leq n_1 < \cdots < n_k$.

Tuymaada Olympiad

Solution. We have

$$10^n((n_1+1)\ldots(n_k-1)n_k + \cdots + (n_{k-1}+1)\cdots(n_k-1)n_k+1) = n_k!$$

which shows that n_k divides 10^n. Let us write $n_k = 2^x \cdot 5^y$. Let

$$S = (n_1+1)\dots(n_k-1)n_k + \dots + (n_{k-1}+1)\dots(n_k-1)n_k + 1.$$

First of all, suppose that x, y are positive. Thus, S is relatively prime to 10. It follows that $v_2(n_k!) = v_5(n_k!)$. This implies $\left\lfloor \frac{n_k}{2^j} \right\rfloor = \left\lfloor \frac{n_k}{5^j} \right\rfloor$ for all j (because we clearly have $\left\lfloor \frac{n_k}{2^j} \right\rfloor \geq \left\lfloor \frac{n_k}{5^j} \right\rfloor$) and so $n_k \leq 3$. A simple verification shows that there is no solution in this case. Next, suppose that $y = 0$. Then S is odd and thus $v_2(n_k!) = n \leq v_5(n_k!)$. Again, this implies $v_2(n_k!) = v_5(n_k!)$ and we have seen that this yields no solution. Thus $x = 0$. A crucial observation is that if $n_k > n_{k-1} + 1$, then S is odd and thus we find again that $v_2(n_k!) = n \leq v_5(n_k!)$, impossible. Hence $n_k = n_{k-1}+1$. But then, taking into account that n_k is a power of 5, we deduce that S is congruent to 2 modulo 4 and thus $v_2(n_k!) = n + 1 \leq v_5(n_k!) + 1$. It follows that $\left\lfloor \frac{n_k}{2} \right\rfloor \leq 1 + \left\lfloor \frac{n_k}{5} \right\rfloor$ and thus $n_k \leq 6$. Because n_k is a power of 5, we find that $n_k = 5$, $n_{k-1} \leq 4$ and exhausting all of the possibilities shows that there are no solutions.

A tricky APMO 1997 problem asked to prove that there is a number $100 < n < 1997$ such that $n | 2^n + 2$. We will invite you to verify that $2 \cdot 11 \cdot 43$ is a solution, and especially to find out how we arrived at this number. Yet... small verifications show that all such numbers are even. Proving this turns out to be a difficult problem and this was proved for the first time by Schinzel.

Example 7. Prove that for any $n > 1$ we cannot have $n | 2^{n-1} + 1$.

[Schinzel]

Solution. Although very short, the proof is tricky. Suppose n is a solution. Let $n = \prod_{i=1}^{s} p_i^{k_i}$ where $p_1 < p_2 < \dots < p_s$ are prime numbers. The idea is to look at $v_2(p_i - 1)$. Choose that p_i which minimizes this quantity and write $p_i = 1 + 2^{r_i} m_i$ with m_i odd. Then $n \equiv 1 \pmod{2^{r_i}}$ and we can write

$n-1=2^{r_i}t$. We have $2^{2^{r_i}t} \equiv -1 \pmod{p_i}$, thus

$$-1 \equiv 2^{2^{r_i}tm_i} \equiv 2^{(p_i-1)t} \equiv 1 \pmod{p_i}$$

(the last congruence being derived from Fermat's little theorem). Thus $p_i = 2$, which is clearly impossible.

We continue with a very nice and difficult problem, in which the idea of looking at the exponents is really helpful. It seems to have appeared for the first time in AMM, but over the last few years, it has been proposed in various national and international contests.

Example 8. Prove that for any integers $a_1, a_2, \ldots, a_n$ the number

$$\prod_{1\le i<j\le n} \frac{a_i-a_j}{i-j}$$

is an integer.

[Armond E. Spencer] AMM E 2637

Solution. We consider a prime number p and prove that for each $k \ge 1$, there are more numbers divisible by p^k in the sequence of differences $(a_i-a_j)_{1\le i<j\le n}$ than in the sequence $(i-j)_{1\le i<j\le n}$. Because

$$v_p\left(\prod_{1\le i<j\le n}(a_i-a_j)\right) = \sum_{k\ge 1} N_{p^k}\left(\prod_{1\le i<j\le n}(a_i-a_j)\right),$$

where $N_{p^k}(\{(i,j)|1 \le i < j \le n\})$ is the number of terms in the sequence A that are multiples of x and

$$v_p\left(\prod_{1\le i<j\le n}(i-j)\right) = \sum_{k\ge 1} N_{p^k}\left(\prod_{1\le i<j\le n}(i-j)\right),$$

the problem will be solved if we prove our claim. Fix $k \geq 1$ and suppose that there are exactly b_i indices $j \in \{1, 2, \ldots, n\}$ such that $a_j \equiv i \pmod{p^k}$, for each $i \in \{0, 1, \ldots, p^k - 1\}$. Then

$$N_{p^k}\left(\prod_{1\leq i<j\leq n}(a_i - a_j)\right) = \sum_{i=0}^{p^k-1}\binom{b_i}{2}.$$

Let us see what happens for $a_i = i$. If $i = 0$, then the number of $1 \leq j \leq n$ for which $j = 0 \pmod{p^k}$ is $\left\lfloor \frac{n}{p^k} \right\rfloor$. If $i > 0$ then any $1 \leq j \leq n$ for which $j = i \pmod{p^k}$ has the form $rp^k + i$ for some $0 \leq r \leq \left\lfloor \frac{n-i}{p^k} \right\rfloor$. Thus we find $1 + \left\lfloor \frac{n-i}{p^k} \right\rfloor$ indices in this case. Hence

$$N_{p^k}\left(\prod_{1\leq i<j\leq n}(i - j)\right) = \sum_{i=1}^{p^k-1}\binom{1+\left\lfloor \frac{n-i}{p^k}\right\rfloor}{2} + \binom{\left\lfloor \frac{n}{p^k}\right\rfloor}{2} \qquad (3.1)$$

By changing $j = p^k - 1$ in (3.1), we infer that

$$N_{p^k}\left(\prod_{1\leq i<j\leq n}(i - j)\right) = \sum_{j=0}^{p^k-1}\binom{\left\lfloor \frac{n+j}{p^k}\right\rfloor}{2},$$

so it suffices to prove that

$$\sum_{i=0}^{p^k-1}\binom{b_i}{2} \geq \sum_{j=0}^{p^k-1}\binom{\left\lfloor \frac{n+j}{p^k}\right\rfloor}{2}.$$

Now, observe that we need to find the minimum of $\sum_{i=0}^{p^k-1} \binom{x_i}{2}$, when $\sum_{i=0}^{p^k-1} x_i = n$ (it is clear from the definition of b_i that

$$\sum_{i=0}^{p^k-1} b_i = n = \sum_{j=0}^{p^k-1} \left\lfloor \frac{n+j}{p^k} \right\rfloor$$

from the definition of b_i). For this, let us suppose that $x_0 \le x_1 \le x_2 \le \cdots \le x_{p^k-1}$ is the p^k-tuple for which the minimum is reached (such a p^k-tuple exists since the equation $\sum_{i=0}^{p^k-1} x_i = n$ has a finite number of solutions). If $x_{p^k-1} > x_0 + 1$, then we consider the n-tuple $(x_0+1, x_1, \ldots, x_{p^k-2}, x_{p^k-1}-1)$, where the sum of components is n, but for which

$$\binom{x_0+1}{2} + \binom{x_1}{2} + \cdots + \binom{x_{p^k-2}}{2} + \binom{x_{p^k-1}-1}{2}$$
$$< \binom{x_0}{2} + \binom{x_1}{2} + \cdots + \binom{x_{p^k-2}}{2} + \binom{x_{p^k-1}}{2}.$$

The last inequality is true, since it is equivalent to $x_{p^k-1} > x_0 + 1$. But this contradicts the minimality of $(x_0, x_1, \ldots, x_2, \ldots, x_{p^k-1})$. So, $x_{p^k-1} \le x_0 + 1$, and from here it follows that $x_i \in \{x_0, x_0+1\}$ for all $i \in \{0, 1, 2, \ldots, p^k-1\}$. Hence there is a $j \in \{0, 1, 2, \ldots, p^k-1\}$ such that $x_0 = x_1 = \cdots = x_j$ and $x_{j+1} = x_{j+2} = \cdots = x_{p^k-1} = x_0 + 1$. Because the variables x_r add up to n, we must have

$$(j+1)x_0 + (p^k - j - 1)(x_0+1) = n,$$

thus $p^k(x_0+1) = n + j + 1$. Therefore $\sum_{i=0}^{p^k-1} \binom{b_i}{2} \ge (j+1)\binom{x_0}{2} + (p^k - j - 1)\binom{x_0+1}{2}$. Finally, observe that for all $0 \le i \le p^k - 1$ we have $\left\lfloor \frac{n+i}{p^k} \right\rfloor = x_0 + 1 + \left\lfloor \frac{i-j-1}{p^k} \right\rfloor$, and this is equal to x_0+1 if $i \ge j+1$ and to x_0 otherwise. Therefore

$$\sum_{i=0}^{p^k-1} \binom{\left\lfloor \frac{n+i}{p^k} \right\rfloor}{2} = (j+1)\binom{x_0}{2} + (p^k - j - 1)\binom{x_0+1}{2}.$$

The next exercise is particularly difficult, but the ideas used in its solution are extremely useful when solving some other problems.

Example 9. Let a and b be two distinct positive rational numbers such that for infinitely many integers n, $a^n - b^n$ is an integer. Prove that a and b are also integers.

[Gabriel Dospinescu] Mathlinks Contest

<u>**Solution.**</u> Let us start by writing $a = \frac{x}{z}$, $b = \frac{y}{z}$, where x, y, z are distinct positive integers with no common factor, and $x \neq y$. We are given that $z^n | x^n - y^n$ for all positive integers n in an infinite set M. Assume that $z > 1$ and take p a prime divisor of z. If p does not divide x, it follows that it cannot divide y. Now, we have two cases:
i) If $p = 2$, then let n be such that $2^n | x^n - y^n$. Write $n = 2^{u_n} j_n$, where j_n is odd. From the identity

$$x^{2^{u_n} j_n} - y^{2^{u_n} j_n} = (x^{j_n} - y^{j_n})(x^{j_n} + y^{j_n}) \dots (x^{2^{u_n - 1} j_n} + y^{2^{u_n - 1} j_n})$$

it follows that

$$v_2(x^n - y^n) = v_2(x^{j_n} - y^{j_n}) + \sum_{k=0}^{u_n - 1} v_2(x^{2^k j_n} + y^{2^k j_n}).$$

But $x^{j_n - 1} + x^{j_n - 2} y + \dots + x y^{j_n - 2} + y^{j_n - 1}$ is clearly odd (since j_n, x, y are odd), hence

$$v_2(x^{j_n} - y^{j_n}) = v_2(x - y).$$

Similarly, we can prove that

$$v_2(x^{j_n} + y^{j_n}) = v_2(x + y).$$

Because

$$x^{2^k j_n} + y^{2^k j_n} \equiv 2 \pmod 4,$$

for $k > 0$, we finally deduce that

$$2^{u_n} j_n \leq v_2(x^n - y^n) \leq v_2(x+y) + v_2(x-y) + u_n - 1 \tag{3.2}$$

Consequently, $(2^{u_n})_{n \in M}$ is bounded, a simple reason being the inequality $2^{u_n} \leq v_2(x+y) + v_2(x-y) + u_n - 1$. Hence $(u_n)_{n \in M}$ takes only a finite number of values, and from (3.2) it follows that $(j_n)_{n \in M}$ also takes a finite number of values, that is M is finite, a contradiction.
ii) Suppose that p is odd and let d be the least positive integer k such that $p | x^k - y^k$. Then for any n in M we have $p | x^n - y^n$. Let $x = tu$, $y = tv$, where $(u, v) = 1$. Clearly, tuv is not a multiple of p. It follows that

$$p \mid (u^d - v^d, u^n - v^n) = u^{(n,d)} - v^{(n,d)} \mid x^{(n,d)} - y^{(n,d)}$$

and by the choice of d, we must have $d | n$. Therefore any element of M is a multiple of d. Take now n in M and write it in the form $n = md$, for some positive integer m. Let $A = x^d$ and $B = y^d$. Then

$$p^m \mid p^n \mid x^n - y^n = A^m - B^m,$$

and this happens for infinitely many m. Moreover, $p | A - B$. Let R be the infinite set of those m. We will prove now a very useful result in this type of problems:

Theorem 3.1. *Let p be an odd prime and let A, B be positive integers, not divisible by p and such that $p | A - B$. Then for all positive integers n we have*

$$v_p(A^n - B^n) = v_p(n) + v_p(A - B).$$

Proof. The proof of this theorem is natural, even though it is quite long and technical. Indeed, let us write $n = p^k \cdot l$ with $\gcd(l, p) = 1$. We will prove the result by induction on k. First, suppose that $k = 0$. Observe that $v_p(A^n - B^n) = v_p(A - B)$ if and only if p does not divide $A^{n-1} + A^{n-2}B + \cdots +$

$AB^{n-2}+B^{n-1}$. If the latter does not hold, because $A=B \pmod p$, we infer that $p|nA^{n-1}$ and this cannot hold because $k=0$ and $\gcd(A,p)=1$. Suppose now that the result holds for k and take $n=p^{k+1}l$ with $\gcd(l,p)=1$. Then, if $m=p^k l$ we can apply the inductive hypothesis and write:

$$v_p(A^n-B^n)=v_p(A^{mp}-B^{mp})=v_p(A^m-B^m)+$$

$$v_p(A^{m(p-1)}+A^{m(p-2)}B^m+\cdots+A^mB^{m(p-2)}+B^{m(p-1)})$$

$$=v_p(A-B)+k+v_p(A^{m(p-1)}+A^{m(p-2)}B^m+\cdots+A^mB^{m(p-2)}+B^{m(p-1)}).$$

So, we need to prove that

$$v_p(A^{m(p-1)}+A^{m(p-2)}B^m+\cdots+A^mB^{m(p-2)}+B^{m(p-1)})=1.$$

But this is not difficult. First, note that if we put $A^m=a, B^m=b$, it is enough to prove that if $v_p(a)=v_p(b)=v_p(a-b)-1=0$, then

$$v_p(a^{p-1}+a^{p-2}b+\cdots+ab^{p-2}+b^{p-1})=1.$$

Now, write $b=a+pc$ for some integer c and observe that using the binomial formula we can write

$$a^{p-1}+a^{p-2}b+\cdots+ab^{p-2}+b^{p-1}=a^{p-1}+a^{p-2}(a+pc)+a^{p-3}(a^2+2apc)+$$

$$\cdots+a^2(a^{p-3}+(p-3)a^{p-4}pc)+a(a^{p-2}+(p-2)a^{p-3}pc)+a^{p-1}+(p-1)a^{p-2}pc$$

$$=pa^{p-1}+ca^{p-2}p^2\cdot\frac{p-1}{2}=pa^{p-1} \pmod{p^2},$$

which proves the inductive step and finishes the proof of the theorem.

□

Let us come back to our problem. Using the theorem, we deduce that for infinitely many m we have

$$m\le v_p(A^m-B^m)=v_p(A-B)+v_p(m)\le v_p(A-B)+\lfloor \log_p m\rfloor,$$

which is clearly impossible. Hence $p|x$ and $p|y$, in contradiction with the fact that x, y, z are relatively prime. This shows that $z = 1$ and a, b are integers.

If you thought this is the last challenge on this chapter, you are wrong! The following problems can be called The Erdős Corner. They were especially kept for the end of the chapter, because of their beauty and difficulty.

Example 10. a) Prove that for any positive integer n there exist positive integers $a_1 < a_2 < \cdots < a_n$ such that $a_i - a_j | a_i$ for all $i \leq j$.

b) Prove that there exists a positive constant c such that for any n and any sequence $a_1 < a_2 < ... < a_n$ which satisfies the conditions of a), $a_1 > n^{cn}$.

[Paul Erdős] Miklos Schweitzer Competition

Solution. If a) is not so difficult, b) needs culture and ingenuity. The proof of a) is of course by induction on n. For $n = 1$ it is enough to take $a_1 = 1$. Suppose that $a_1 < a_2 < ... < a_n$ is a good sequence and let us take $b = a_1 a_2 \cdots a_n$. The sequence $b, b + a_1, b + a_2, ..., b + a_n$ is also good and shows how the inductive step works. Now, let us discuss b). Take any prime number $p \leq n$ and observe that if $a_i \equiv a_j \pmod{p}$ then $a_i \equiv a_j \equiv 0 \pmod{p}$. Therefore at most $p - 1$ among the numbers $a_1, a_2, ..., a_n$ are not multiples of p. Consider the multiples of p among $a_1, a_2, ..., a_n$ and divide them by p. We obtain another good sequence, and the previous argument shows that this new sequence has at most $p - 1$ terms not divisible by p. Repeating this argument yields

$$v_p(a_1 a_2 \cdots a_n) \geq (n - (p-1)) + (n - 2(p-1)) + \cdots + \left(n - \left\lfloor \frac{n}{p-1} \right\rfloor (p-1)\right).$$

A small computation shows that if $p \leq \sqrt{n}$, then the last quantity exceeds $\frac{n^2}{3p}$. Therefore $a_1 a_2 ... a_n \geq \prod_{p \leq \sqrt{n}} p^{\frac{n^2}{3p}}$. But it is clear that $a_1 \geq a_n - a_1$, so

$$a_1 \geq \frac{a_n}{2} \geq \frac{\sqrt[n]{a_1 a_2 \cdots a_n}}{2},$$

which shows that

$$a_1 \geq \frac{1}{2} \cdot e^{\frac{n}{3} \cdot \sum\limits_{p \leq \sqrt{n}} \frac{\ln p}{p}}.$$

So, all we need now is to prove that there exists a constant $c > 0$ such that $\sum\limits_{p \leq n} \frac{\ln p}{p} \geq c \cdot \ln n$. Actually, we will prove more, that

$$\sum_{p \leq n} \frac{\ln p}{p} = \ln n + O(1).$$

The tool will be again the factorization of $n!$. Indeed, this gives the identity

$$\ln(n!) = \sum_{p} v_p(n!) \cdot \ln p.$$

On the one hand, using Stirling's formula $n! \approx (\frac{n}{e})^n \sqrt{2\pi n}$, we deduce that $\ln(n!) = n(\ln n - 1) + O(\ln n)$. On the other hand, $\frac{n}{p} - 1 \leq v_p(n!) < \frac{n}{p-1}$. Therefore

$$n \cdot \sum_{p \leq n} \frac{\ln p}{p} + n \cdot \sum_{p \leq n} \frac{\ln p}{p(p-1)} \geq \sum_{p} v_p(n!) \cdot \ln p > n \cdot \sum_{p \leq n} \frac{\ln p}{p} - \ln \prod_{p \leq n} p. \tag{3.3}$$

Because the series $\sum\limits_{p} \frac{\ln p}{p(p-1)}$ is clearly convergent, it follows that $n \cdot \sum\limits_{p \leq n} \frac{\ln p}{p(p-1)} = O(n)$.

And now, we will prove the following result also due to Erdős: $\prod\limits_{p \leq n} p \leq 4^{n-1}$ if $n \geq 1$. The proof of this theorem is magnificent. We use induction. For small values of n it is clear. Now, assume the inequality true for all values smaller than n and let us prove that $\prod\limits_{p \leq n} p \leq 4^{n-1}$. If n is even, we have nothing to prove, since $\prod\limits_{p \leq n} p = \prod\limits_{p \leq n-1} p \leq 4^{n-2} < 4^{n-1}$.

Now, assume that $n = 2k+1$ and consider the binomial coefficient

$$\binom{2k+1}{k} = \frac{(k+2)\dots(2k+1)}{k!}.$$

An application of the identity $2^{2k+1} = \sum_{i\geq 0}\binom{2k+1}{i}$ shows that $\binom{2k+1}{k} \leq 4^k$. Thus, using the inductive hypothesis, we find

$$\prod_{p\leq n} p \leq \prod_{p\leq k+1} p \cdot \prod_{k+2\leq p\leq 2k+1} p \leq 4^k \cdot 4^k = 4^{n-1}.$$

This result shows that $\ln \prod\limits_{p\leq n} p = O(n)$, so using the previous estimations we can write

$$\sum_{p\leq n} \frac{\ln p}{p} = \ln n + O(1).$$

Here is a refinement and proof of the famous Bertrand's postulate, asserting that between n and $2n$ there is always a prime number if $n > 1$. Actually, the result proved in the next example shows that much more is true for sufficiently large n and also gives an effectively computable constant $c < 10000$ for the proof of Bertrand's postulate. Simple computations allow after that a complete proof of this result. However, we prefer the more quantitative result below:

Example 11. For any $\epsilon > 0$ there exists an n_0 such that for all $n > n_0$ there are at least $\left(\frac{2}{3} - \epsilon\right)\frac{n}{\log_2(n)}$ primes between n and $2n$.

[Paul Erdős]

Solution. A very good way of obtaining interesting bounds for the counting functions of prime numbers is to study the powers that divide the binomial coefficient $\binom{2n}{n}$. Why is this number so special? First of all, because it is quite easy to evaluate it asymptotically. One can easily prove, for instance using Stirling's formula that $\binom{2n}{n} \approx \frac{4^n}{\sqrt{\pi n}}$. There are, however, much more elementary estimations. For example, using the fact that $\binom{2n}{n}$ is the largest binomial coefficient and that the sum of these binomial coefficients is 4^n, we easily infer the inequality $\binom{2n}{n} \geq \frac{4^n}{2n+1}$, which is more than enough for our modest purposes. Now, another important fact about this binomial coefficient is that the prime powers dividing it do not have large exponents. Indeed,

$$v_p\left(\binom{2n}{n}\right) = \sum_{k\geq 1}\left(\left\lfloor \frac{2n}{p^k}\right\rfloor - 2\left\lfloor \frac{n}{p^k}\right\rfloor\right) \leq \lfloor \log_p 2n \rfloor,$$

which shows that the largest power of p dividing $\binom{2n}{n}$ does not exceed $2n$. This implies that the exponent of any prime $p > \sqrt{2n}$ is at most 1. But the remarkable observation that Erdős had is that actually this special binomial coefficient is not a multiple of any prime between $\frac{2n}{3}$ and n, as you can immediately establish using the fact that $v_p(\binom{2n}{n}) = \sum\limits_{k\geq 1}\left(\left\lfloor \frac{2n}{p^k}\right\rfloor - 2\left\lfloor \frac{n}{p^k}\right\rfloor\right)$. So, using all these observations, we infer that

$$\frac{4^n}{2n+1} \leq \binom{2n}{n} \leq \prod_{p\leq\sqrt{2n}} 2n \cdot \prod_{\sqrt{2n}\leq p\leq \frac{2n}{3}} p \cdot \prod_{n<p\leq 2n} p.$$

Using the result proved in the solution of the previous example, we deduce that $\prod\limits_{\sqrt{2n}\leq p\leq \frac{2n}{3}} p \leq 4^{\frac{2n}{3}-1}$. Also, it is clear that $\prod\limits_{p\leq\sqrt{2n}} 2n \leq (2n)^{\sqrt{2n}+1}$, so if $f(n)$ is the number of primes between n and $2n$, then

$$\frac{4^n}{2n+1} \leq (2n)^{1+\sqrt{2n}} \cdot 4^{\frac{2n}{3}-1} \cdot (2n)^{f(n)}.$$

By taking logarithms, we finally deduce that

$$f(n) \geq \frac{\frac{2n}{3} - O(\sqrt{n}\cdot \ln n)}{\log_2 n},$$

from which the conclusion follows immediately.

But the most subtle and difficult problem of this chapter (and probably of the whole book) is the following fascinating result, conjectured by Palfy and proved by Erdős using Sylvester's theorem on prime divisors of consecutive numbers. The following marvelous solution by M. Szegedi was taken from the note "$a \pmod p \le b \pmod p$ *for all primes* p *implies* $a = b$", published in the second issue of the American Mathematical Monthly, 1987:

Example 12. Let a, b be positive integers such that for all prime numbers p, $a \pmod p \le b \pmod p$. Then $a = b$.

[Erdős, Palfy] Miklos Schweitzer Competition 1984

Solution. This solution will not be short, but it has the merit of being completely elementary. It follows from a very subtle analysis of the prime powers dividing $\binom{b}{a}$ (for it is clear that by taking a prime $p > a + b$ we obtain $a \le b$). Hence suppose that $a < b$. Observe that if $\frac{b}{2} < a$ then by letting $c = b - a$ we have $0 < c < \frac{b}{2}$ and also $c \pmod p = b \pmod p - a \pmod p \le b \pmod p$ (because $0 \le b \pmod p - a \pmod p < p$). Therefore it is enough to prove that the case $0 < a \le \frac{b}{2}$ is impossible. Let $\binom{b}{a} = \frac{B}{A}$ where $A = a!$ and $B = b(b-1)\cdots(b-a+1)$. Also, let $A(p^k)$ and $B(p^k)$ be the number of factors of A and B respectively, that are multiples of p^k. It is clear that $A(p^k) = \left\lfloor \frac{a}{p^k} \right\rfloor$ and

$$B(p^k) = \left\lfloor \frac{b}{p^k} \right\rfloor - \left\lfloor \frac{b-a}{p^k} \right\rfloor.$$

Then, by using the fact that $0 \le \lfloor x + y \rfloor - \lfloor x \rfloor - \lfloor y \rfloor \le 1$ for all real numbers x, y, we infer that $B(p^k) - A(p^k)$ is 0 or 1. Now, the crucial observation is that $A(p) \ge B(p)$. Indeed, the first multiple of p that appears in the product $a \cdot (a-1) \cdots 2 \cdot 1$ is $a - a \pmod p$, while the first multiple of p in $b \cdot (b-1) \cdots (b-a+2) \cdot (b-a+1)$ is $b - b \pmod p$. Using this remark and the fact that the sequences $1, 2, \ldots, a$ and $b-a+1, b-a+2, \ldots, b$ have the same length, we infer that $A(p) \ge B(p)$. But, as we have already seen, this implies $A(p) = B(p)$. Therefore if $p > a$ then surely $A(p) = 0$, so $B(p) = 0$ and so $A(p^k) = B(p^k) = 0$ for all positive integers k and all $p > a$. Therefore

$$A = \prod_{p \le a} p^{A(p)+A(p^2)+\cdots} \text{ and } B = \prod_{p \le a} p^{B(p)+B(p^2)+\cdots},$$

so

$$\binom{b}{a} = \frac{B}{A} = \prod_{p \le a} p^{B(p)-A(p)+B(p^2)-A(p^2)+\cdots}$$

There is another crucial observation to be made: if $m(p)$ is the largest k such that $B(p^k)$ is not zero, then using the fact that $A(p) = B(p)$ we obtain

$$B(p) - A(p) + B(p^2) - A(p^2) + \cdots = \sum_{j=2}^{m(p)} (B(p^j) - A(p^j)),$$

so

$$B(p) - A(p) + B(p^2) - A(p^2) + \cdots \le m(p) - 1$$

(recall that we have established the inequality $B(p^k) - A(p^k) \le 1$). Therefore $\binom{b}{a}$ is a divisor of $\prod_{p \le a} p^{m(p)-1}$ and so

$$\frac{(b-a+1)\cdot(b-a+2)\cdots b}{\prod_{p \le a} p^{m(p)}}$$

is a divisor of $\frac{a!}{\prod_{p \le a} p}$. However, the last divisibility cannot hold for $b \ge 2a$. Indeed, it is clear that

$$\frac{a!}{\prod_{p \le a} p} \le a^{a-\pi(a)} < \frac{(b-a+1)\cdot(b-a+2)\cdots b}{\prod_{p \le a} p^{m(p)}},$$

because after cancellations are made in

$$\frac{(b-a+1)\cdot(b-a+2)\cdots b}{\prod_{p \le a} p^{m(p)}},$$

we obtain $a - \pi(a)$ factors all equal to at least $b - a + 1 > a$, a contradiction.

3.2 Practice problems

1. Prove the identity

$$\frac{\operatorname{lcm}(a,b,c)^2}{\operatorname{lcm}(a,b)\cdot\operatorname{lcm}(b,c)\cdot\operatorname{lcm}(c,a)}=\frac{\gcd(a,b,c)^2}{\gcd(a,b)\cdot\gcd(b,c)\cdot\gcd(c,a)}$$

for all positive integers a, b, c.

USAMO 1972

2. Let $a_1, a_2, \ldots, a_k, b_1, b_2, \ldots, b_k$ be positive integers such that $\gcd(a_i, b_i) = 1$ for all $i \in \{1, 2, \ldots, k\}$. Let $m = \operatorname{lcm}(b_1, b_2, \ldots, b_k)$. Prove that

$$\gcd\left(\frac{a_1m}{b_1}, \frac{a_2m}{b_2}, \ldots, \frac{a_km}{b_k}\right) = \gcd(a_1, a_2, \ldots, a_k).$$

IMO 1974 Shortlist

3. Prove that if n is a positive integer and a and b are integers, then $n!$ divides $a(a+b)(a+2b)\cdots(a+(n-1)b)b^{n-1}$.

IMO 1985 Shortlist

4. Let a, b, c be positive integers such that $c \mid a^c - b^c$. Prove that $c \mid \frac{a^c-b^c}{a-b}$.

I. Niven, AMM E 564

5. Prove that for all integers a, b with $b \neq 0$ there is a positive integer n such that $v_2(n!) = a \pmod{b}$.

Komal

6. Prove that for any positive integer n,

$$(n+1)\operatorname{lcm}\left(\binom{n}{0},\binom{n}{1},\dots,\binom{n}{n}\right)=\operatorname{lcm}(1,2,\dots,n+1).$$

Peter L. Montgomery, AMM E 2686

7. Let m be an integer greater than 1. Suppose that a positive integer n satisfies $n \mid a^m - 1$ for all integers a relatively prime to n. Prove that $n \le 4m(2^m - 1)$. Find all cases of equality.

Gabriel Dospinescu, Marian Andronache, Romanian TST 2004

8. Let n be a positive integer and let $a > b > 1$ be integers such that b is odd and $b^n \mid a^n - 1$. Prove that $a^b > \frac{3^n}{n}$.

Chinese TST 2009

9. Find all primes p and all positive integers n such that $n^{p-1} \mid (p-1)^n + 1$.

After IMO 1999

10. Let a be a positive integer. Prove that the set of prime divisors of $2^{2^n} + a$ for $n = 1, 2, \dots$ is infinite.

Iranian TST 2009

11. Let $p > 7$ be a prime. Prove that p^4 divides the numerator of the fraction

$$2 \cdot \sum_{k=1}^{p-1} \frac{1}{k} + p \cdot \sum_{k=1}^{p-1} \frac{1}{p^2}$$

when written in lowest terms.

Gabriel Dospinescu

12. Let $p_1, p_2, \ldots, p_k$ be distinct prime numbers and let S be the set of numbers all of whose prime factors are among $p_1, p_2, \ldots, p_k$. If A is a finite set of integers, let $G(A)$ be the graph whose set of vertices is A, two vertices $a, b \in A$ being connected if $a - b \in S$. Is it true that for any $m \geq 3$ we can find A with m elements such that

    (a) $G(A)$ is complete.

    (b) $G(A)$ is connected and all vertices have degree at most 2?

Miklos Schweitzer Competition 2009

13. Solve in positive integers $x^{2007} - y^{2007} = x! - y!$.

Romanian TST 2007

14. Prove that for all positive integers n different from 3 and 5, $n!$ is divisible by the number of its positive divisors.

Paul Erdős, Miklos Schweitzer Competition

15. Find all positive integers a, b, c such that $(2^a - 1)(3^b - 1) = c!$.

Gabriel Dospinescu, Mathematical Reflections

16. Let a be a fixed positive integer. Prove that the equation $n! = a^b - a^c$ has a finite number of solutions (n, b, c) in positive integers.

Chinese TST 2004

17. Let $m > n^{n-1}$ be positive integers such that $m+1, m+2, \ldots, m+n$ are composite numbers. Prove that we can find pairwise distinct prime numbers $p_1, p_2, \ldots, p_n$ such that p_i divides $m+i$ for all $1 \leq i \leq n$.

Tuymaada Olympiad 2004

18. Find all positive integers n with the following property: there exist natural numbers $b_1, b_2, \ldots, b_n$, not all equal and such that the number $(b_1 + k)(b_2 + k) \cdots (b_n + k)$ is a power of an integer for each natural number k. Here, a power means a number of the form x^y with $x, y > 1$.

Russia 2008

19. Let $(a_n)_{n \geq 1}$ be a sequence of positive integers such that $\gcd(a_m, a_n) = a_{\gcd(m,n)}$ for all positive integers m, n. Prove that there exists a unique sequence of positive integers $(b_n)_{n \geq 1}$ such that $a_n = \prod_{d|n} b_d$.

Marcel Ţena, Romanian TST

20. Let $n \geq 2$ and let $a_1, a_2, \ldots, a_n$ be positive integers, not all of them equal. Prove that there are infinitely many prime numbers p with the property: there exists a positive integer k such that

$$p \mid a_1^k + a_2^k + \cdots + a_n^k$$

Iran 2004

21. Let n, k be positive integers such that $n > 9^k$. Prove that $\binom{n}{k}$ has at least k distinct prime factors.

Paul Erdős, Miklos Schweitzer Competition

22. Let $f(n)$ be the maximum size of a subset A of $\{1, 2, \ldots, n\}$ which does not contain two elements i, j such that $i \mid 2j$. Prove that there exists a constant $C > 0$ such that for all n we have

$$\left| f(n) - \frac{4n}{9} \right| \leq C \ln n.$$

Paul Erdős, AMM E 3403

23. Prove that $\lim_{n\to\infty} x_n = \infty$, where x_n is the exponent of 2 in the numerator of $\frac{2}{1} + \frac{2^2}{2} + \cdots + \frac{2^n}{n}$. Moreover, prove that $x_{2^n} \geq 2^n - n + 1$.

Adapted from a Kvant problem

24. Find the exponent of 2 in the prime factorization of the number

$$\binom{2^{n+1}}{2^n} - \binom{2^n}{2^{n-1}}.$$

J. Desmong, W.R.Hastings, AMM E 2640

25. Let x, y be relatively prime positive integers. Prove that for infinitely many primes p, the exponent of p in $x^{p-1} - y^{p-1}$ is odd.

Barry Powell, AMM E 2948

26. Let p be a prime and let $n > s + 1$ be positive integers. Prove that p^d divides

$$\sum_{p|k, 0 \leq k \leq n} (-1)^k k^s \binom{n}{k},$$

where $d = \left\lfloor \frac{n-s-1}{p-1} \right\rfloor$.

Gabriel Dospinescu, Mathematical Reflections

27. Prove that for any $c > 0$ there are infinitely many n such that the largest prime divisor of $n^2 + 1$ is greater than cn.

Chebyshev, Nagell

28. (a) Let p be a prime and let $a_0, a_1, \dots$ be integers such that

$$\sum_{k=0}^{n} p^k \binom{n}{k} a_k = 0$$

for infinitely many positive integers n. Prove that $a_n = 0$ for all n.

(b) A sequence $(a_n)_n$ of integers satisfies

$$a_{n+d} = x_1 a_{n+d-1} + x_2 a_{n+d-2} + \dots + x_d a_n$$

for all $n \geq 0$, where $d \geq 1$ and $x_1, x_2, \dots, x_d$ are integers. Prove that there exists a finite set S and integers $c_1, c_2, \dots, c_N, d_1, d_2, \dots, d_N$ such that

$$\{n \geq 0 | a_n = 0\} = S \cup (c_1 + d_1\mathbb{N}) \cup \dots \cup (c_N + d_N\mathbb{N}).$$

Skolem-Mahler-Lech theorem

Chapter 4

4.1 Theory and examples

The study of the properties of prime numbers is very well-developed, yet many old conjectures and open questions are still waiting to be solved. In this chapter, we present properties of some classes of primes and also of some classical results related to representations as sum of two squares. At the end of the unit, we will discuss, as usual, some nonstandard and surprising problems. Because we will use some facts several times, we prefer to fix some notations before discussing the problems. So, we will consider the sets P_1 and P_3 of all prime numbers of the form $4k+1$ and $4k+3$, respectively. Also, Q_2 will be the set of all numbers that can be written as the sum of two perfect squares. Our purpose is to present some classical results related to P_1, P_3, Q_2. The most spectacular property of the set P_1 is the fact that any of its elements is the sum of the squares of two positive integers. This is not a trivial property and we will present a beautiful proof of it next.

Example 1. Prove that P_1 is a subset of Q_2.

[Fermat]

Solution. We need to prove that any prime number of the form $4k+1$ is the sum of two squares. We will use a very nice result:

Theorem 4.1 (Thue). *If n is a positive integer and a is relatively prime to n, then there exist integers $0 < x, y \leq \sqrt{n}$ such that $xa \equiv \pm y \pmod{n}$ for a suitable choice of the signs $+$ or $-$.*

Proof. The proof is simple, but the theorem itself is a diamond. Indeed, let us consider all the values $xa - y$, with $0 \leq x, y \leq \lfloor\sqrt{n}\rfloor$. So, we have a list of $(\lfloor\sqrt{n}\rfloor + 1)^2 > n$ numbers and it follows that two numbers among them give the same remainder when divided by n, let them be $ax_1 - y_1$ and $ax_2 - y_2$. It is not difficult to see that we may assume that $x_1 > x_2$ (we certainly cannot have $x_1 = x_2$ or $y_1 = y_2$). If we take $x = x_1 - x_2$ and $y = |y_1 - y_2|$, all the conditions will be satisfied, so the theorem is proved. □

We will use now Wilson's theorem to find an integer n such that $p|n^2+1$. Indeed, let us write $p = 4k+1$ and observe that we can take $n = (2k)!$. Why? Because from Wilson's theorem we have

$$-1 \equiv (p-1)! \equiv 1 \cdot 2 \cdot \ldots \cdot \left(\frac{p-1}{2}\right) \cdot \left(p - \frac{p-1}{2}\right) \cdot \ldots \cdot (p-1)$$

$$\equiv (-1)^{\frac{p-1}{2}} \left(\left(\frac{p-1}{2}\right)!\right)^2 \equiv ((2k)!)^2 \pmod{p}$$

and the claim is proved. Now, since $p|n^2+1$, it is clear that p and n are relatively prime. Hence we can apply in order to find positive integers $0 < x, y < \sqrt{p}$ (since $\sqrt{p} \notin \mathbb{Q}$) such that $p|n^2x^2 - y^2$. Because $p|n^2+1$, we find that $p|x^2+y^2$ and because $0 < x, y < \sqrt{p}$, we conclude that we have in fact $p = x^2+y^2$. The theorem is proved.

It is time now to study some properties of the set P_3. Because they are easier, we will discuss them in a single example.

Example 2. Let $p \in P_3$ and suppose that x and y are integers such that $p|x^2+y^2$. Show that $p|\gcd(x,y)$. Consequently, any number of the form n^2+1 has only prime factors that belong to P_1 or are equal to 2. Conclude that P_1 is infinite and then that P_3 is infinite.

Solution. Let us focus on the first question. Suppose that $p|\gcd(x,y)$ is not true. Then, it is obvious that xy is not a multiple of p. Because $p|x^2+y^2$, we can write $x^2 \equiv -y^2 \pmod{p}$. Combining this with the observation that $\gcd(x,p) = \gcd(y,p) = 1$ and with Fermat's little theorem, we find that $1 \equiv x^{p-1} \equiv (-1)^{\frac{p-1}{2}} y^{p-1} \equiv (-1)^{2k+1} \equiv -1 \pmod{p}$ (for $p = 4k+3$), which is impossible. This settles the first question. The second one follows clearly from the first one. Now, it remains to prove the third assertion. Proving that P_3 is infinite is almost identical with the proof that there are infinitely many primes. Indeed, suppose that $p_1, p_2, \ldots, p_n$ are all the elements of P_3 greater

than 3 and consider the odd number $N = 4p_1p_2\dots p_n + 3$. Because $N \equiv 3 \pmod 4$, N must have a prime factor that belongs to P_3. But since p_i is not a divisor of N for any $i = 1, 2, ..., n$, the contradiction is reached and thus P_3 is infinite. In the same manner we can prove that P_1 is infinite, but this time we must use the second question. Indeed, we consider this time the number $M = (q_1q_2\dots q_m)^2 + 1$, where $q_1, q_2, \dots, q_m$ are the elements of P_1 and then simply apply the result from the second question. The conclusion is clear.
It is not difficult now to characterize the elements of the set Q_2. A number is a sum of two squares if and only if any of its prime factors that also belongs to P_3 appears at an even exponent in the decomposition of that number. The proof is just a consequence of the first example and we will not insist on anything more.

Having presented some basic results that we will further use in this unit, it is time to see some applications that these two examples have. As a simple consequence of the first example, we consider the following problem, which is certainly easy for someone who knows Fermat's theorem regarding the elements of P_1 and difficult enough otherwise.

Example 3. Find the number of integers $x \in \{-1997, \dots, 1997\}$ for which $1997 | x^2 + (x+1)^2$.

India 1998

Solution. We know that any quadratic congruence reduces to the congruence $x^2 \equiv a \pmod p$. So, let us proceed and reduce the given congruence to this special form. This is not difficult, since $x^2+(x+1)^2 \equiv 0 \pmod{1997}$ is equivalent to $2x^2+2x+1 \equiv 0 \pmod{1997}$, which in turn becomes $(2x+1)^2+1 \equiv 0 \pmod{1997}$. Because $1997 \in P_1$, the congruence $n^2 \equiv -1 \pmod{1997}$ has at least one solution. More precisely, there are exactly two solutions that belong to $\{1, 2, \dots, 1996\}$, because if n_0 is a solution, then so is $1997 - n_0$ and it is clear that this equation has at most two noncongruent solutions mod 1997. Because $\gcd(2, 1997) = 1$, the function $x \mapsto 2x + 1$ is a permutation of $\mathbb{Z}/1997\mathbb{Z}$, and so the initial congruence has exactly two solutions with $x \in \{1, 2, \ldots, 1996\}$. In a similar way, we find that there are exactly two solutions with $x \in \{-1997, -1996, \ldots, -1\}$. Therefore there are exactly four numbers $x \in \{-1997, \ldots, 1997\}$ such that $1997 | x^2 + (x+1)^2$.

We continue with a much trickier problem, proposed by Romania for the 1996 IMO. Even though it uses only the elementary facts about P_3 proved before, this problem is fairly difficult:

Example 4. Let $\mathbb{N}_0$ denote the set of nonnegative integers. Is there a bijective function $f : \mathbb{N}_0 \to \mathbb{N}_0$ such that for all nonnegative integers m, n we have $f(3mn + m + n) = 4f(m)f(n) + f(m) + f(n)$?

IMO 1996 Shortlist

Solution. The first step is to notice that one can change the given relation into

$$f\left(\frac{(3m+1)(3n+1)-1}{3}\right) = \frac{(4f(m)+1)(4f(n)+1)-1}{4}.$$

This has the advantage that after introducing the function $g : 3 \cdot \mathbb{N}_0 + 1 \to 4 \cdot \mathbb{N}_0 + 1$, $g(n) = 4f\left(\frac{n-1}{3}\right) + 1$, it becomes $g(mn) = g(m)g(n)$, which is much easier than the initial relation. Because one can easily reconstruct f from g by $f(n) = \frac{g(3n+1)-1}{4}$, the question becomes: is there a bijective multiplicative function g between $3 \cdot \mathbb{N}_0 + 1$ and $4 \cdot \mathbb{N}_0 + 1$, that is are the monoids $3 \cdot \mathbb{N}_0 + 1$ and $4 \cdot \mathbb{N}_0 + 1$ isomorphic? Let us introduce the analogous sets T_1, T_2 of positive primes of the form $3k + 1$ and $3k + 2$. In the same way as we proved that P_1, P_3 are infinite, you can prove that T_1, T_2 are infinite. Because they are clearly countable, there exists a bijection between P_1 and T_1 and a bijection between P_3 and T_2. This gives us a bijection ψ between $P_1 \cup P_3$ and $T_1 \cup T_2$ which maps P_1 onto T_1 and P_3 onto T_2 bijectively. Now, it is not difficult to construct an isomorphism g: define $g(1) = 1$ and if $n > 1$ is in $3 \cdot \mathbb{N}_0 + 1$ write $n = p_1 p_2 \cdots p_k$ for some prime numbers $p_i \in T_1 \cup T_2$, not necessarily distinct and define $g(n) = \psi(p_1) \cdot \psi(p_2) \cdots \psi(p_k)$. We need to verify that g is well-defined, multiplicative and bijective. First of all, note that there is an even number of elements of T_2 among $p_1, p_2, ..., p_k$. Then there is an even number of elements of P_3 among $\psi(p_i)$ and thus $g(n) \in 4 \cdot \mathbb{N}_0 + 1$. Thus g is well-defined. Clearly g is multiplicative (by the definition itself) and, using the properties of ψ it is immediate to verify that g is also bijective. This proves the existence of a function f with the desired properties.

From a previous observation, we know that the condition that a number is a sum of two squares is quite restrictive. This suggests that the set Q_2 is rather sparse. This conclusion can be translated into the following nice problem.

Example 5. Prove that Q_2 does not have bounded gaps, that is there are arbitrarily long sequences of consecutive integers, none of which can be written as the sum of two perfect squares.

AMM

Solution. The statement of the problem suggests using the Chinese Remainder Theorem, but here the main idea is to use the complete characterization of the set Q_2 we have just discussed: $Q_2 = \{n \in \mathbb{Z} |$ if $p|n$ and $p \in P_3$, then $v_p(n) \in 2\mathbb{Z}\}$. We know what we have to do. We will take long sequences of consecutive integers, each of them having a prime factor that belongs to P_3 and has exponent 1. More precisely, we take different elements of P_3, let them be $p_1, p_2, \ldots, p_n$ (we can take as many as we need, since P_3 is infinite) and then we look for a solution to the system of congruences

$$\begin{cases} x \equiv p_1 - 1 \pmod{p_1^2} \\ x \equiv p_2 - 2 \pmod{p_2^2} \\ \ldots \\ x \equiv p_n - n \pmod{p_n^2} \end{cases}$$

The existence of such a solution follows from the Chinese Remainder Theorem. Thus, the numbers $x+1, x+2, \ldots, x+n$ cannot be written as the sum of two perfect squares, since $p_i | x+i$, but p_i^2 does not divide $x+i$. Because n is as large as we want, the conclusion follows.

The Diophantine equation $x(x+1)(x+2)\cdots(x+n) = y^k$ has been extensively studied by many mathematicians and great results have been obtained by Erdős and Selfridge. But these results are very difficult to prove and we prefer to present a related problem, with a nice flavor of elementary mathematics.

Example 6. For any p in P_3, prove that no set of $p-1$ consecutive positive integers can be partitioned into two subsets, each having the same product of the elements.

Solution. Let us suppose that the positive integers $x+1, x+2, \ldots, x+p-1$ have been partitioned into two classes X, Y, each of them having the same product of the elements. If at least one of the $p-1$ numbers is a multiple of p, then there must be another one divisible by p (since in this case both products of elements from X and Y must be multiples of p), which is clearly impossible. Thus, none of these numbers is a multiple of p, which means that the set of the remainders of these numbers when divided by p is exactly $1, 2, \ldots, p-1$. Also, from the hypothesis it follows that there exists a positive integer n such that

$$(x+1)(x+2)\cdots(x+p-1) = n^2.$$

Hence $n^2 \equiv 1 \cdot 2 \cdots (p-1) \equiv -1 \pmod{p}$, the last congruence following from Wilson's theorem. But from the second example we know that the congruence $n^2 \equiv -1 \pmod{p}$ is impossible for $p \in P_3$ and this is the needed contradiction.

The results in the second example are useful tools in solving nonstandard Diophantine equations. You can see this in the following two examples.

Example 7. Prove that the equation $x^4 = y^2 + z^2 + 4$ does not have integer solutions.

[Reid Barton] Rookie Contest 1999

Solution. Practically, we have to show that $x^4 - 4$ does not belong to Q_2. Hence we need to find an element of P_3 that has an odd exponent in the prime factorization of x^4-4. The first case is when x is odd. Using the factorization $x^4-4=(x^2-2)(x^2+2)$ and the observation that $x^2+2\equiv 3 \pmod 4$, we deduce that there exists $p\in P_3$ such that $v_p(x^2+2)$ is odd. But since p cannot divide x^2-2 (otherwise $p|x^2+2-(x^2-2)$, which is not the case), we conclude that $v_p(x^4-4)$ is odd, and so x^4-4 does not belong to Q_2. We have thus shown that in any solution of the equation, x is even, let us say $x=2k$. Then, we must also have $4k^4-1\in Q_2$, which is clearly impossible since $4k^4-1\equiv 3 \pmod 4$ and thus $4k^4-1$ has a prime factor that belongs to P_3 and has odd exponent. Moreover, it is worth noting that the equation $x^2+y^2=4k+3$ can be solved directly, by working modulo 4.

The following problem is much more difficult, but the basic idea is the same. Yet the details are not so obvious and, most importantly, it is not clear how to begin.

Example 8. Let $p\in P_3$ and suppose that x,y,z,t are integers such that $x^{2p}+y^{2p}+z^{2p}=t^{2p}$. Prove that at least one of the numbers x,y,z,t is a multiple of p.

[Barry Powel] AMM

Solution. Without loss of generality, we may assume that x,y,z,t are relatively prime. Next, we prove that t is odd. Supposing the contrary, we obtain $x^{2p}+y^{2p}+z^{2p}\equiv 0 \pmod 4$. Because $a^2 \pmod 4\in\{0,1\}$, the latter implies that x,y,z are even, contradicting the assumption that $\gcd(x,y,z,t)=1$. Hence t is odd. This implies that at least one of the numbers x,y,z is odd. Suppose that z is odd. We write the equation in the form

$$x^{2p}+y^{2p}=\frac{t^{2p}-z^{2p}}{t^2-z^2}(t^2-z^2)$$

and look for a prime $q\in P_3$ with an odd exponent in the decomposition of a factor that appears in the right-hand side. The best candidate for this factor seems to be

$$\frac{t^{2p}-z^{2p}}{t^2-z^2}=(t^2)^{p-1}+(t^2)^{p-2}z^2+\cdots+(z^2)^{p-1},$$

which is congruent to 3 (mod 4). This follows from the hypothesis $p \in P_3$ and the fact that $a^2 \equiv 1 \pmod 4$ for any odd number a. Hence there is a $q \in P_3$ such that $v_q\left(\frac{t^{2p} - z^{2p}}{t^2 - z^2}\right)$ is odd. Because $x^{2p} + y^{2p} \in Q_2$, it follows that $v_q(x^{2p} + y^{2p})$ is even and so $v_q(t^2 - z^2)$ is odd. In particular, $q|t^2 - z^2$ and, because

$$q|(t^2)^{p-1} + (t^2)^{p-2}z^2 + \cdots + (z^2)^{p-1},$$

we deduce that $q|pt^{2(p-1)}$. If $q \neq p$, then $q|t$, hence $q|z$ and also $q|x^{2p} + y^{2p}$. Because $q \in P_3$, we infer that $q|\gcd(x, y, z, t) = 1$, which is clearly impossible. Therefore $q = p$ and so $p|x^{2p} + y^{2p}$. Because $p \in P_3$, we find that $p|x$ and $p|y$. The conclusion follows.

The previous results are used in the solution of the following problem. Even if the problem is formulated as a functional equation, we will immediately see that it is pure number theory mixed with some simple algebraic manipulations.

Example 9. Find the least nonnegative integer n for which there exists a nonconstant function $f : \mathbb{Z} \to [0, \infty)$ with the following properties:

a) $f(xy) = f(x)f(y)$;

b) $2f(x^2 + y^2) - f(x) - f(y) \in \{0, 1, \ldots, n\}$ for all $x, y \in \mathbb{Z}$.

For this n, find all functions with the above properties.

[Gabriel Dospinescu] Crux Mathematicorum

Solution. First of all, we will prove that for $n = 1$ there are functions which satisfy a) and b). For any $p \in P_3$ define:

$$f_p : \mathbb{Z} \to \mathbb{Z}, \quad f_p(x) = \begin{cases} 0, & \text{if } p|x \\ 1, & \text{otherwise} \end{cases}$$

Using the properties of P_1 and P_3, you can easily verify that f_p satisfies the conditions of the problem. Hence f_p is a solution for all $p \in P_3$.

We will prove now that if f is nonconstant and satisfies the conditions in the problem, then $n > 0$. Suppose not. Then $2f(x^2+y^2) = f(x)+f(y)$ and hence $2f(x)^2 = 2f(x^2+0^2) = f(x)+f(0)$. It is clear that we have $f(0)^2 = f(0)$. Because f is nonconstant, we must have $f(0) = 0$. Consequently, $2f(x)^2 = f(x)$ for every integer x. But if there exists x such that $f(x) = \frac{1}{2}$, then $2f(x^2)^2 \neq f(x^2)$, contradiction. Thus, $f(x) = 0$ for any integer x and f is constant, contradiction. So, $n = 1$ is the least number for which there are nonconstant functions which satisfy a) and b).

We will now prove that any nonconstant function f which satisfies a) and b) must be of the form f_p: or the function sending all nonzero integers to 1 and 0 to 0. We have already seen that $f(0) = 0$. Since $f(1)^2 = f(1)$ and f is nonconstant, we must have $f(1) = 1$. Also,

$$2f(x)^2 - f(x) = 2f(x^2+0^2) - f(x) - f(0) \in \{0,1\}$$

for every integer x. Thus $f(x) \in \{0,1\}$. Because $f(-1)^2 = f(1) = 1$ and $f(-1) \in [0,\infty)$, we must have $f(-1) = 1$ and $f(-x) = f(-1)f(x) = f(x)$ for any integer x. Then, since $f(xy) = f(x)f(y)$, it suffices to find $f(p)$ for any prime p. We prove that there is exactly one prime p for which $f(p) = 0$. Because f is nonconstant and f is not the function sending all nonzero integers to 1, there is a prime number p for which $f(p) = 0$. Suppose there is another prime q for which $f(q) = 0$. Then $2f(p^2+q^2) \in \{0,1\}$, which means $f(p^2+q^2) = 0$. Then for any integers a and b we must have:

$$0 = 2f(a^2+b^2)f(p^2+q^2) = 2f((ap+bq)^2+(aq-bp)^2).$$

Observe that $0 \leq f(x)+f(y) \leq 2f(x^2+y^2)$ for any x and y, so we must have $f(ap+bq) = f(aq-bp) = 0$. But p and q are relatively prime, so there are integers a and b such that $aq-bp = 1$. Then $1 = f(1) = f(aq-bp) = 0$, a contradiction. So, there is exactly one prime p for which $f(p) = 0$. Let us suppose that $p = 2$. Then $f(x) = 0$ for any even x and $2f(x^2+y^2) = 0$ for any odd numbers x and y. This implies that $f(x) = f(y) = 0$ for any odd numbers x and y and thus f is constant, contradiction. Therefore $p \in P_1 \cup P_3$. Suppose $p \in P_1$. According example 1, there are positive integers a and b such that $p = a^2+b^2$. Then we must have $f(a) = f(b) = 0$. But $\max\{a,b\} > 1$ and there is a prime number q such that $q|\max\{a,b\}$ and $f(q)=0$ (otherwise, we would have $f(\max\{a,b\})=1$). But it is clear that $q<p$ and thus we have found two distinct primes p and q such that $f(p)=f(q)=0$, which, as we have already seen, is impossible. Consequently, $p\in P_3$ and we have $f(x)=0$ for any x divisible by p and $f(x)=1$ for any x which is not divisible by p. Hence, f must be f_p and the conclusion follows.

We end this chapter with two beautiful problems concerning properties of prime numbers of the form $4k+1$ or $4k+3$. We saw that Q_2 does not have bounded gaps. In fact, much more is true. We will show that Q_2 has density zero. Define the density of a set of positive integers P_1 as the limit (if it exists) of the sequence $\frac{P_1(x)}{x}$, where $P_1(x)$ is the counting function of the set P_1, that is $P_1(x)=\sum\limits_{a\in P_1, a\leq x} 1$. Before proving that Q_2 has density zero, we want to prove a jewel of mathematics, the first step in analytic number theory:

Example 10. The sets P_1 and P_3 have Dirichlet density $\frac{1}{2}$, that is

$$\lim_{s\to 1}\frac{1}{\ln\frac{1}{s-1}}\cdot\sum_{p\in P_1}\frac{1}{p^s}=\frac{1}{2}$$

and similarly for P_3.

[Dirichlet]

Solution. Let us consider $s>1$ and $L(s)=\sum\limits_{n\geq 1}\frac{\lambda(n)}{n^s}$, where $\lambda(n)=0$ if n is even and $\lambda(n)=(-1)^{\frac{n-1}{2}}$ otherwise. It is clear that $\lambda(n)\cdot\lambda(m)=\lambda(mn)$. Using this, it is not difficult to see that

$$L(s)=\prod_p\left(1+\frac{\lambda(p)}{p^s}+\frac{\lambda(p^2)}{p^{2s}}+\cdots\right)=\prod_p\frac{1}{1-\frac{\lambda(p)}{p^s}}.$$

Indeed, let us define $P(x) = \prod_{p\leq x}\left(1 + \frac{\lambda(p)}{p^s} + \frac{\lambda(p^2)}{p^{2s}} + \cdots\right)$. It is a finite product of absolutely convergent series, so we can write

$$P(x) = \sum_{n\in P_1(x)} \frac{\lambda(n)}{n^s},$$

where $P_1(x)$ is the set of positive integers having all prime divisors not exceeding x. Thus the difference between the sum of the absolutely convergent series $\sum_{n\geq 1} \frac{\lambda(n)}{n^s}$ and $P(x)$ is just the sum of $\frac{\lambda(n)}{n^s}$ taken over the set of all positive integers that have at least one prime divisor greater than x, thus it is certainly bounded in absolute value by $\sum_{n\geq x} \frac{1}{n^s}$. Because this converges to 0 for $x \to \infty$, it follows that $P(x)$ converges to $L(s)$ for $x \to \infty$, so we have

$$L(s) = \prod_p \left(1 + \frac{\lambda(p)}{p^s} + \frac{\lambda(p^2)}{p^{2s}} + \cdots\right) = \prod_p \frac{1}{1 - \frac{\lambda(p)}{p^s}}.$$

Now, observe that

$$L(s) = 1 - \frac{1}{3^s} + \frac{1}{5^s} - \frac{1}{7^s} + \cdots > 0, \qquad (4.1)$$

so we can take logarithms in both sides of (4.1) in order to obtain

$$\ln L(s) = -\sum_p \ln\left(1 - \frac{\lambda(p)}{p^s}\right).$$

Finally, observe that there exists a constant w such that $|-\ln(1-x)-x| \leq Cx^2$ for all $0 \leq x \leq \frac{1}{2}$. Indeed, the function $\frac{-\ln(1-x)-x}{x^2}$ is continuous on $\left[0, \frac{1}{2}\right]$, so it is bounded. Therefore

$$\left|\ln L(s) - \sum_p \frac{\lambda(p)}{p^s}\right| \leq w \cdot \sum_p \frac{1}{p^{2s}} \leq w \sum_p \frac{1}{p^2}.$$

Now, let us prove that $\ln L(s)$ is bounded for $s \to 1$. Indeed, from

$$L(s) = \left(1 - \frac{1}{3^s}\right) + \left(\frac{1}{5^s} - \frac{1}{7^s}\right) + \cdots = 1 - \left(\frac{1}{3^s} - \frac{1}{5^s}\right) - \cdots$$

it follows that for $s > 1$ we have $\ln L(s) \in \left(\ln \frac{2}{3}, 0\right)$. With exactly the same arguments (applied this time for the function $\psi(n) = 1$ for odd n and $\psi(n) = 0$ for even n and $L_1(s) = \sum\limits_{n \geq 1} \frac{\psi(n)}{n^s}$), we can prove that

$$\ln\left(\sum_{n \in 2\mathbb{N}+1} \frac{1}{n^s}\right) - \sum_{p>2} \frac{1}{p^s}$$

is bounded for $s \to 1$. However, it is clear that

$$\sum_{n \in 2\mathbb{N}+1} \frac{1}{n^s} = \left(1 - \frac{1}{2^s}\right) \cdot \zeta(s),$$

where $\zeta(s) = \sum\limits_{n \geq 1} \frac{1}{n^s}$ is the famous Riemann's function. Because $\ln L(s)$ is bounded, it follows from a previous inequality that $\sum\limits_{p>2} \frac{\lambda(p)}{p^s}$ is also bounded near 1. Finally, we deduce from these observations that

$$\sum_{p \in P_1} \frac{1}{p^s} - \sum_{p \in P_3} \frac{1}{p^s} = O(1)$$

and

$$\sum_{p \in P_1} \frac{1}{p^s} + \sum_{p \in P_3} \frac{1}{p^s} = \ln(1 - 2^{-s}) + \ln \zeta(s) + O(1)$$

for $s \to 1$. A simple integral estimation shows that $\ln(1-2^{-s}) + \ln \zeta(s) \approx \ln \frac{1}{s-1}$ for $s \to 1$, which finishes the proof of this beautiful theorem.

Now, let us see why the set Q_2 has zero density. The proof of this result will surely look very complicated. Actually, it is a motivation to give some other very useful results connected to this problem. First of all, let us start with

Theorem 4.2. *Let P be a set of prime numbers. The set of positive integers n divisible by some prime $p \in P$ has density 1 if $\prod_{q\in P}\left(1-\frac{1}{q}\right)=0$.*

Proof. The proof of this result is quite simple, even though in order to make it rigorous we need some technical details. It is clear that P is infinite, so let $p_1 < p_2 < \ldots$ be its elements. Let E be the set of the numbers n divisible by some prime $p \in P$ and let X be the set of positive integers n that are not divisible by any element of P. Also, let $f(x,y)$ be the cardinal of the set of those numbers not exceeding x and which are relatively prime to $\prod_{q\in P, q\le y} q$. Using the Inclusion-Exclusion Principle and the fact that the number of multiples of $p_{i_1}p_{i_2}\ldots p_{i_s}$ not exceeding x differs by at most 1 from $\frac{x}{p_{i_1}p_{i_2}\cdot\ldots\cdot p_{i_s}}$, we deduce that

$$f(x,y) = x \cdot \prod_{q\in P, q\le y}\left(1-\frac{1}{q}\right) + O(2^y)$$

(because in the sum appearing in the Inclusion-Exclusion Principle there are 2^y terms of the form $\frac{x}{p_{i_1}p_{i_2}\cdot\ldots\cdot p_{i_s}} + O(1)$). Now, by choosing $y = \ln x$ we deduce that

$$f(x,\ln x) = x \cdot \prod_{q\in P, q\le \ln x}\left(1-\frac{1}{q}\right) + O(x^{\ln 2}).$$

Because the counting function of X satisfies $R(x) \le f(x,y)$ for all x, y and because $\lim_{x\to\infty} \prod_{q\in P, q\le \ln x}\left(1-\frac{1}{q}\right) = 0$, it follows that $R(x) = o(x)$, that is X has zero density. It is clear then that E has density 1.

□

Now, using the previous theorem due to Dirichlet, we can easily establish that $\sum_{p\in P_3}\frac{1}{p} = \infty$. Because $\ln\left(1-\frac{1}{n}\right)+\frac{1}{n} = O\left(\frac{1}{n^2}\right)$, it easily follows that $\prod_{p\in P_3}\left(1-\frac{1}{p}\right) = 0$. By the previous theorem, it follows that the set of integers divisible by at least an element of P_3 has density 1. Now, let $P_3(x)$ be the counting function of the set of positive integers that are not divisible by 4 or by any element of P_3. They are the only integers that are sums of two coprime squares. Also, we have proved that $P_3(x) = o(x)$. It is also clear that if $Sq(x)$ is the counting function of the set of positive integers that are sums of two squares, then $Sq(x) \leq \sum_{j\geq 1} P_3\left(\frac{x}{j^2}\right)$. Now, for N an arbitrary positive integer, observe that

$$\sum_{\frac{x}{j^2}\leq N} P_3\left(\frac{x}{j^2}\right) \leq \sqrt{x}P_3(N)$$

because $P_3\left(\frac{x}{j^2}\right) \leq P_3(N)$ for these j and the sum has at most $\sqrt{x}$ nonzero terms. On the other hand,

$$\sum_{\frac{x}{j^2}\geq N} P_3\left(\frac{x}{j^2}\right) \leq \sup_{t\geq N} \frac{P_3(t)}{t} \cdot \sum_{j\geq 1} \frac{x}{j^2} \leq 3x \cdot \sup_{t\geq N} \frac{P_4(t)}{t}.$$

Everything should be clear now: for $\epsilon > 0$ choose N such that $\sup_{t\geq N} \frac{P_3(t)}{t} < \frac{\epsilon}{6}$. Then for $x > \frac{4B(N)^2}{\epsilon}$ we have $Sq(x) \leq \epsilon x$, which means that $Sq(x) = x$.

4.2 Practice problems

1. Prove that the number $4mn - m - n$ cannot be a perfect square if m and n are positive integers.

2. Let n be a positive integer. Prove that the equation $x^2 + y^2 = n$ has integer solutions if and only if it has rational solutions.

   Euler

3. Prove that each prime p of the form $4k+1$ can be represented in exactly one way as the sum of the squares of two integers, up to the order and signs of the terms.

   Euler

4. Prove that the equation $3^k = m^2 + n^2 + 1$ has infinitely many solutions in positive integers.

   Saint-Petersburg Olympiad

5. Find all pairs (m, n) of positive integers such that

$$m^2 - 1 \mid 3^m + (n! - 2)^m.$$

   Gabriel Dospinescu

6. Find all pairs (x, y) of positive integers such that the number $\dfrac{x^2 + y^2}{x - y}$ is a divisor of 1995.

   Bulgaria 1995

7. Find all n-tuples $(a_1, a_2, \ldots, a_n)$ of positive integers such that

$$(a_1! - 1)(a_2! - 1) \cdots (a_n! - 1) - 16$$

is a perfect square.

Gabriel Dospinescu

8. Prove that there are infinitely many pairs of consecutive numbers, no two of which have any prime factor of the form $4k + 3$.

9. Prove that the equation $y^2 = x^5 - 4$ has no integer solutions.

Balkan Olympiad 1998

10. Prove that for no integer n is $n^7 + 7$ a perfect square.

Titu Andreescu, USA TST 2008

11. Let $p > 2$ be a prime. Prove that $p \equiv= 1 \pmod 4$ if and only if there are integers x, y such that $x^2 - py^2 = -1$.

12. Find all positive integers n such that the number $2^n - 1$ has a multiple of the form $m^2 + 9$.

IMO 1999 Shortlist

13. It is a long standing conjecture of Erdős that the equation $\frac{4}{n} = \frac{1}{x} + \frac{1}{y} + \frac{1}{z}$ has solutions in positive integers for all positive integers n. Prove that the set of those n for which this statement is true has density 1.

14. Let T the set of the positive integers n for which the equation $n^2 = a^2 + b^2$ has solutions in positive integers. Prove that T has density 1.

Moshe Laub, AMM 6583

15. Let p be a prime number of the form $4k+1$. Prove that

$$\sum_{j=1}^{\frac{p-1}{4}} \left\lfloor \sqrt{jp} \right\rfloor = \frac{p^2-1}{12}.$$

V.Bunyakovski

16. Find all functions $f : \mathbb{Z}^+ \to \mathbb{Z}$ with the properties

    (a) $f(a) \geq f(b)$ whenever a divides b.

    (b) for all positive integers a and b,

$$f(ab) + f(a^2+b^2) = f(a) + f(b).$$

Gabriel Dospinescu, Mathlinks Contest

17. Prove that the equation $x^8 = n! + 1$ has finitely many solutions in non-negative integers.

18. Let $L_0 = 2$, $L_1 = 1$ and $L_{n+2} = L_{n+1} + L_n$ be the Lucas sequence. Then the only $n > 1$ for which L_n is a perfect square is $n = 3$.

Cohn's theorem

Chapter 5

5.1 Theory and examples

$T2$'s lemma is clearly a direct application of the Cauchy-Schwarz inequality. Some will say that it is actually the Cauchy-Schwarz inequality and they are not wrong. Anyway, this particular lemma has become very popular among the American students who attended the training of the USA IMO team. This happened after a lecture delivered by the first author at the Mathematical Olympiad Summer Program (MOSP) held at Georgetown University in June, 2001.

But what exactly does this lemma say? It says that for any real numbers $a_1, a_2, \ldots, a_n$ and any positive real numbers $x_1, x_2, \ldots, x_n$ the inequality

$$\frac{a_1^2}{x_1} + \frac{a_2^2}{x_2} + \cdots + \frac{a_n^2}{x_n} \geq \frac{(a_1 + a_2 + \cdots + a_n)^2}{x_1 + x_2 + \cdots + x_n} \tag{5.1}$$

holds. And now we see why calling it also the Cauchy-Schwarz inequality is natural, since it is practically an equivalent form of this inequality:

$$\left(\frac{a_1^2}{x_1} + \frac{a_2^2}{x_2} + \cdots + \frac{a_n^2}{x_n}\right)(x_1 + x_2 + \cdots + x_n)$$

$$\geq \left(\sqrt{\frac{a_1^2}{x_1}} \cdot \sqrt{x_1} + \sqrt{\frac{a_2^2}{x_2}} \cdot \sqrt{x_2} + \cdots + \sqrt{\frac{a_n^2}{x_n}} \cdot \sqrt{x_n}\right)^2.$$

But there is another nice proof of (5.1), by induction. The inductive step is reduced practically to the case $n = 2$, which is immediate. Indeed, it boils down to $(a_1x_2 - a_2x_1)^2 \geq 0$ and the equality occurs if and only if $\frac{a_1}{x_1} = \frac{a_2}{x_2}$. Applying this result twice it follows that

$$\frac{a_1^2}{x_1} + \frac{a_2^2}{x_2} + \frac{a_3^2}{x_3} \geq \frac{(a_1 + a_2)^2}{x_1 + x_2} + \frac{a_3^2}{x_3} \geq \frac{(a_1 + a_2 + a_3)^2}{x_1 + x_2 + x_3}$$

and we see that a simple inductive argument finishes the proof. With this brief introduction, let us discuss some problems. And there are plenty of them given in mathematical contests or proposed in mathematical magazines!

First, an old problem, that became classical. We will see that with $T2$'s lemma it becomes straightforward and even more, we will obtain a refinement of the inequality.

Example 1. Prove that for any positive real numbers a, b, c

$$\frac{a^3}{a^2+ab+b^2}+\frac{b^3}{b^2+bc+c^2}+\frac{c^3}{c^2+ca+a^2}\geq\frac{a+b+c}{3}.$$

Tournament of the Towns, 1998

Solution. We will change the left-hand side of the inequality so that we could apply $T2$'s lemma. This is not difficult: we just have to write it in the form

$$\frac{a^4}{a(a^2+ab+b^2)}+\frac{b^4}{b(b^2+bc+c^2)}+\frac{c^4}{c(c^2+ca+a^2)}.$$

It follows that the left-hand side is greater than or equal to

$$\frac{(a^2+b^2+c^2)^2}{a^3+b^3+c^3+ab(a+b)+bc(b+c)+ca(c+a)}$$

But we can easily observe that

$$a^3+b^3+c^3+ab(a+b)+bc(b+c)+ca(c+a)=(a+b+c)(a^2+b^2+c^2),$$

so we have proved an even stronger inequality, that is

$$\frac{a^3}{a^2+ab+b^2}+\frac{b^3}{b^2+bc+c^2}+\frac{c^3}{c^2+ca+a^2}\geq\frac{a^2+b^2+c^2}{a+b+c}.$$

The second example also became representative for a whole class of problems. There are countless examples of this type in numerous contests and mathematical magazines, so we find it necessary to discuss it at this point.

Example 2. For arbitrary positive real numbers a, b, c, d prove the inequality

$$\frac{a}{b+2c+3d}+\frac{b}{c+2d+3a}+\frac{c}{d+2a+3b}+\frac{d}{a+2b+3c}\geq\frac{2}{3}.$$

[Titu Andreescu] IMO 1993 Shortlist

Solution. If we write the left-hand side in the form

$$\frac{a^2}{a(b+2c+3d)}+\frac{b^2}{b(c+2d+3a)}+\frac{c^2}{c(d+2a+3b)}+\frac{d^2}{d(a+2b+3c)},$$

then the way to continue is clear, since from the lemma we obtain

$$\frac{a}{b+2c+3d}+\frac{b}{c+2d+3a}+\frac{c}{d+2a+3b}+\frac{d}{a+2b+3c}$$
$$\geq\frac{(a+b+c+d)^2}{4(ab+bc+cd+da+ac+bd)}.$$

Hence it suffices to prove the inequality

$$3(a+b+c+d)^2\geq 8(ab+bc+cd+da+ac+bd).$$

But it is not difficult to see that

$$(a+b+c+d)^2=a^2+b^2+c^2+d^2+2(ab+bc+cd+da+ac+bd),$$

implies

$$8(ab+bc+cd+da+ac+bd)=4(a+b+c+d)^2-4(a^2+b^2+c^2+d^2).$$

Consequently, we are left with the inequality

$$4(a^2+b^2+c^2+d^2)\geq(a+b+c+d)^2,$$

which is just the Cauchy-Schwarz inequality for four variables.
The problem below, given at the IMO 1995, was discussed extensively in many publications. It could be also solved by using the above lemma.

Example 3. Let a, b, c be positive real numbers such that $abc = 1$. Prove that

$$\frac{1}{a^3(b+c)} + \frac{1}{b^3(c+a)} + \frac{1}{c^3(a+b)} \geq \frac{3}{2}.$$

Solution. We have:

$$\frac{1}{a^3(b+c)} + \frac{1}{b^3(c+a)} + \frac{1}{c^3(a+b)} = \frac{\frac{1}{a^2}}{a(b+c)} + \frac{\frac{1}{b^2}}{b(c+a)} + \frac{\frac{1}{c^2}}{c(c+a)}$$

$$\geq \frac{\left(\frac{1}{a}+\frac{1}{b}+\frac{1}{c}\right)^2}{2(ab+bc+ca)} = \frac{(ab+bc+ca)^2}{2(ab+bc+ca)} = \frac{ab+bc+ca}{2} \geq \frac{3}{2},$$

the last inequality following from the AM-GM inequality.

The following problem is also not difficult, but it uses a nice combination of this lemma and the Power-Mean inequality. It is another example in which proving the intermediate inequality (that is, the inequality that remains to be proved after using the lemma) is not difficult.

Example 4. Let $n \geq 2$. Find the minimal value of the expression

$$\frac{x_1^5}{x_2+x_3+\cdots+x_n} + \frac{x_2^5}{x_1+x_3+\cdots+x_n} + \cdots + \frac{x_n^5}{x_1+x_2+\cdots+x_{n-1}},$$

where $x_1, x_2, \ldots, x_n$ are positive real numbers satisfying

$$x_1^2 + x_2^2 + \cdots + x_n^2 = 1.$$

Turkey, 1997

Solution. Usually, in such problems the minimal value is attained when the variables are equal. So, we conjecture that the minimal value is $\frac{1}{n(n-1)}$ attained when $x_1 = x_2 = \cdots = x_n = \frac{1}{\sqrt{n}}$. Indeed, by using the lemma, it follows that the left-hand side is greater than or equal to

$$\frac{\left(\sum_{i=1}^{n} x_i^3\right)^2}{\sum_{i=1}^{n} x_i(x_1 + \cdots + x_{i-1} + x_{i+1} + \cdots + x_n)}.$$

But it is not difficult to observe that

$$\sum_{i=1}^{n} x_i(x_1 + \cdots + x_{i-1} + x_{i+1} + \cdots + x_n) = \left(\sum_{i=1}^{n} x_i\right)^2 - 1.$$

So, proving that

$$\frac{x_1^5}{x_2 + x_3 + \cdots + x_n} + \frac{x_2^5}{x_1 + x_3 + \cdots + x_n} + \cdots + \frac{x_n^5}{x_1 + x_2 + \cdots + x_{n-1}} \geq \frac{1}{n(n-1)}$$

reduces to proving the inequality

$$\left(\sum_{i=1}^{n} x_i^3\right)^2 \geq \frac{\left(\sum_{i=1}^{n} x_i\right)^2 - 1}{n(n-1)}.$$

But this is a simple consequence of the Power-Mean inequality. Indeed, we have

$$\left(\sum_{i=1}^{n} \frac{x_i^3}{n}\right)^{\frac{1}{3}} \geq \left(\sum_{i=1}^{n} \frac{x_i^2}{n}\right)^{\frac{1}{2}} \geq \sum_{i=1}^{n} \frac{x_i}{n},$$

implying

$$\sum_{i=1}^{n} x_i^3 \geq \frac{1}{\sqrt{n}} \text{ and } \sum_{i=1}^{n} x_i \leq \sqrt{n}.$$

The conclusion follows.

In 1954, H.S.Shapiro asked whether the following inequality is true for any positive real numbers $a_1, a_2, \ldots, a_n$:

$$\frac{a_1}{a_2 + a_3} + \frac{a_2}{a_3 + a_4} + \cdots + \frac{a_n}{a_1 + a_2} \geq \frac{n}{2}.$$

The question turned out to be extremely difficult. The answer is really unexpected: the inequality holds for all odd integers smaller than or equal to 23 and all even integers smaller than or equal to 12, but fails for all the others. Let us examine the case $n = 5$, a problem proposed for MOSP 2001.

Example 5. Prove that for any positive real numbers a_1, a_2, a_3, a_4, a_5,

$$\frac{a_1}{a_2 + a_3} + \frac{a_2}{a_3 + a_4} + \frac{a_3}{a_4 + a_5} + \frac{a_4}{a_5 + a_1} + \frac{a_5}{a_1 + a_2} \geq \frac{5}{2}.$$

Solution. Again, we apply the lemma and we conclude that it suffices to prove the inequality

$$(a_1 + a_2 + a_3 + a_4 + a_5)^2$$

$$\geq \frac{5}{2}[a_1(a_2 + a_3) + a_2(a_3 + a_4) + a_3(a_4 + a_5) + a_4(a_5 + a_1) + a_5(a_1 + a_2)]$$

Let us denote $a_1 + a_2 + a_3 + a_4 + a_5 = S$. Then we observe that

$$a_1(a_2 + a_3) + a_2(a_3 + a_4) + a_3(a_4 + a_5) + a_4(a_5 + a_1) + a_5(a_1 + a_2)$$

$$= \frac{a_1(S - a_1) + a_2(S - a_2) + a_3(S - a_3) + a_4(S - a_4) + a_5(S - a_5)}{2}$$

$$= \frac{S^2 - a_1^2 - a_2^2 - a_3^2 - a_4^2 - a_5^2}{2}.$$

With this identity, we infer that the intermediate inequality is in fact

$$(a_1 + a_2 + a_3 + a_4 + a_5)^2 \geq \frac{5}{4}(S^2 - a_1^2 - a_2^2 - a_3^2 - a_4^2 - a_5^2),$$

equivalent to $5(a_1^2 + a_2^2 + a_3^2 + a_4^2 + a_5^2) \geq S^2$, which is nothing else then the Cauchy-Schwarz inequality.

Another question arises: is there a positive real number such that for any positive real numbers $a_1, a_2, \ldots, a_n$ and any $n \geq 3$ the following inequality holds:

$$\frac{a_1}{a_2 + a_3} + \frac{a_2}{a_3 + a_4} + \cdots + \frac{a_n}{a_1 + a_2} \geq cn.$$

This time, the answer is positive, but finding the best such constant is an extremely difficult task. It was first solved by Drinfield (who, by the way, is a Fields' medalist). The answer is quite complicated and we will not discuss it here (for a detailed presentation of Drinfield's method the interested reader can consult the written examination given at ENS in 1997). The following problem, given at the Moldavian TST in 2005, shows that $c = \sqrt{2} - 1$ is such a constant (not optimal). The optimal constant is quite complicated, but an approximation is 0.49456682.

For any $a_1, a_2, \ldots, a_n$ and any $n \geq 3$ the following inequality holds:

$$\frac{a_1}{a_2 + a_3} + \frac{a_2}{a_3 + a_4} + \cdots + \frac{a_n}{a_1 + a_2} \geq (\sqrt{2} - 1)n.$$

The proof is completely elementary, yet very difficult to find. An ingenious argument using the arithmetic-geometric means inequality does the job: let us write the inequality in the form

$$\frac{a_1 + a_2 + a_3}{a_2 + a_3} + \frac{a_2 + a_3 + a_4}{a_3 + a_4} + \cdots + \frac{a_n + a_1 + a_2}{a_1 + a_2} \geq \sqrt{2} \cdot n.$$

Now, using the AM-GM inequality, we see that it suffices to prove the stronger inequality:

$$\frac{a_1 + a_2 + a_3}{a_2 + a_3} \cdot \frac{a_2 + a_3 + a_4}{a_3 + a_4} \cdots \frac{a_n + a_1 + a_2}{a_1 + a_2} \geq (\sqrt{2})^n.$$

Observe that

$$(a_i + a_{i+1} + a_{i+2})^2 = \left(a_i + \frac{a_{i+1}}{2} + \frac{a_{i+1}}{2} + a_{i+2}\right)^2$$

$$\geq 4\left(a_i + \frac{a_{i+1}}{2}\right)\left(\frac{a_{i+1}}{2} + a_{i+2}\right)$$

(the last inequality being another consequence of the AM-GM inequality). Thus,

$$\prod_{i=1}^{n}(a_i + a_{i+1} + a_{i+2})^2 \geq \prod_{i=1}^{n}(2a_i + a_{i+1})\prod_{i=1}^{n}(2a_{i+2} + a_{i+1}).$$

Now, the real trick is to rewrite the last products appropriately. Let us observe that

$$\prod_{i=1}^{n}(2a_{i+2} + a_{i+1}) = \prod_{i=1}^{n}(2a_{i+1} + a_i),$$

so

$$\prod_{i=1}^{n}(2a_i + a_{i+1})\prod_{i=1}^{n}(2a_{i+2} + a_{i+1}) = \prod_{i=1}^{n}[(2a_i + a_{i+1})(a_i + 2a_{i+1})]$$

$$\geq \prod_{i=1}^{n}(2(a_i + a_{i+1})^2) = 2^n\left(\prod_{i=1}^{n}(a_i + a_{i+1})\right)^2.$$

The conclusion now follows.

This lemma came handy even at the IMO 2005 (problem 3). In order to prove that for any positive real numbers x, y, z such that $xyz \geq 1$ the following inequality holds

$$\sum \frac{x^2 + y^2 + z^2}{x^5 + y^2 + z^2} \leq 3,$$

a few students successfully used the above mentioned lemma. For example, a student from Ireland applied this result and called it "SQ Lemma". During the coordination, the Irish deputy leader explained what "SQ" stood for: "...escu". A typical solution using this lemma is as follows:

$$x^5 + y^2 + z^2 = \frac{x^4}{\frac{1}{x}} + \frac{y^4}{y^2} + \frac{z^4}{z^2} \geq \frac{(x^2+y^2+z^2)^2}{\frac{1}{x} + y^2 + z^2},$$

hence

$$\sum \frac{x^2+y^2+z^2}{x^5+y^2+z^2} \leq \sum \frac{\frac{1}{x} + y^2 + z^2}{x^2+y^2+z^2} = 2 + \frac{xy+yz+zx}{xyz(x^2+y^2+z^2)} \leq 3.$$

It is now time for the champions. We begin with a difficult geometric inequality for which we have found a direct solution using $T2$'s lemma. Here it is.

Example 6. Let $m_a, m_b, m_c, r_a, r_b, r_c$ be the lengths of the medians and the radii of the circumscribed circles in a triangle ABC. Prove that the following inequality holds

$$\frac{r_a r_b}{m_a m_b} + \frac{r_b r_c}{m_b m_c} + \frac{r_c r_a}{m_c m_a} \geq 3.$$

[Ji Chen] Crux Mathematicorum

Solution. Of course, we start by translating the inequality into an algebraic one. Fortunately, this is not difficult, since using Heron's relation and the formulas

$$r_a = \frac{K}{s-a}, \quad m_a = \frac{\sqrt{2b^2+2c^2-a^2}}{2}$$

and the like, the desired inequality takes the equivalent form

$$\frac{(a+b+c)(b+c-a)}{\sqrt{2a^2+2b^2-c^2}\cdot\sqrt{2a^2+2c^2-b^2}} + \frac{(a+b+c)(c+a-b)}{\sqrt{2b^2+2a^2-c^2}\cdot\sqrt{2b^2+2c^2-a^2}}$$

$$+\frac{(a+b+c)(a+b-c)}{\sqrt{2c^2+2b^2-a^2}\cdot\sqrt{2c^2+2a^2-b^2}} \geq 3.$$

In this form, the inequality is more than monstrous, so we try to see if a simpler form holds, by applying the AM-GM inequality to each denominator. So, let us try to prove the stronger inequality

$$\frac{2(a+b+c)(c+b-a)}{4a^2+b^2+c^2}+\frac{2(a+b+c)(c+a-b)}{4b^2+c^2+a^2}$$

$$+\frac{2(a+b+c)(a+b-c)}{4c^2+a^2+b^2}\geq 3.$$

Written in the more appropriate form

$$\frac{c+b-a}{4a^2+b^2+c^2}+\frac{c+a-b}{4b^2+c^2+a^2}+\frac{a+b-c}{4c^2+a^2+b^2}\geq\frac{3}{2(a+b+c)}$$

we see that by $T2$'s lemma the left-hand side is at least

$$\frac{(a+b+c)^2}{(b+c-a)(4a^2+b^2+c^2)+(c+a-b)(4b^2+a^2+c^2)+(a+b-c)(4c^2+a^2+b^2)}.$$

Basic computations show that the denominator of the last expression is equal to

$$4a^2(b+c)+4b^2(c+a)+4c^2(a+b)-2(a^3+b^3+c^3)$$

and consequently the intermediate inequality reduces to the simpler form

$$3(a^3+b^3+c^3)+(a+b+c)^3\geq 6[a^2(b+c)+b^2(c+a)+c^2(a+b)].$$

Again, we expand $(a+b+c)^3$ and obtain the equivalent inequality

$$4(a^3+b^3+c^3)+6abc\geq 3[a^2(b+c)+b^2(c+a)+c^2(a+b)],$$

which is not difficult at all. Indeed, it follows from the inequalities

$$4(a^3+b^3+c^3)\geq 4[a^2(b+c)+b^2(c+a)+c^2(a+b)]-12abc$$

and

$$a^2(b+c)+b^2(c+a)+c^2(a+b)\geq 6abc.$$

The first one is just an equivalent form of Schur's inequality, while the second follows immediately from the identity

$$a^2(b+c)+b^2(c+a)+c^2(a+b)-6abc = a(b-c)^2+b(c-a)^2+c(a-b)^2.$$

Finally, we have managed to prove the intermediate inequality, and hence the problem is solved.

The journey continues with a very difficult problem, given at the Japanese Mathematical Olympiad in 1997, and which became infamous due to its difficulty. We will present two solutions for this inequality. The first one uses a nice combination between the $T2$ lemma and the substitution discussed in the unit "Two useful substitutions".

Example 7. Prove that for any positive real numbers a, b, c the following inequality holds

$$\frac{(b+c-a)^2}{a^2+(b+c)^2}+\frac{(c+a-b)^2}{b^2+(c+a)^2}+\frac{(a+b-c)^2}{c^2+(a+b)^2} \geq \frac{3}{5}.$$

Japan 1997

Solution. Of course, from the introduction to this problem, the reader has already noticed that it is useless to try a direct application of the lemma, since any such approach is doomed. But with the substitution

$$x=\frac{b+c}{a}, \quad y=\frac{c+a}{b}, \quad z=\frac{a+b}{c},$$

we have to prove that for any positive real numbers x, y, z satisfying $xyz = x+y+z+2$, the inequality

$$\frac{(x-1)^2}{x^2+1}+\frac{(y-1)^2}{y^2+1}+\frac{(z-1)^2}{z^2+1} \geq \frac{3}{5}$$

holds. It is now time to use $T2$'s lemma in the form

$$\frac{(x-1)^2}{x^2+1}+\frac{(y-1)^2}{y^2+1}+\frac{(z-1)^2}{z^2+1} \geq \frac{(x+y+z-3)^2}{x^2+y^2+z^2+3}.$$

Hence it is enough to prove the inequality

$$\frac{(x+y+z-3)^2}{x^2+y^2+z^2+3} \geq \frac{3}{5}.$$

But this is equivalent to

$$(x+y+z)^2 - 15(x+y+z) + 3(xy+yz+zx) + 18 \geq 0.$$

This is not an easy inequality. We will use the proposed problem 6 from the chapter **Two Useful Substitutions** to reduce the above inequality to the form

$$(x+y+z)^2 - 9(x+y+z) + 18 \geq 0,$$

which follows from the inequality $x+y+z \geq 6$. And the problem is solved. But here is another original solution.

Alternative solution. Let us apply $T2$'s lemma in the following form:

$$\frac{(b+c-a)^2}{a^2+(b+c)^2} + \frac{(c+a-b)^2}{b^2+(c+a)^2} + \frac{(a+b-c)^2}{c^2+(a+b)^2}$$

$$= \frac{((b+c)^2-a(b+c))^2}{a^2(b+c)^2+(b+c)^4} + \frac{((c+a)^2-b(c+a))^2}{b^2(c+a)^2+(c+a)^4} + \frac{((a+b)^2-c(a+b))^2}{c^2(a+b)^2+(a+b)^4}$$

$$\geq \frac{4(a^2+b^2+c^2)^2}{a^2(b+c)^2+b^2(c+a)^2+c^2(a+b)^2+(a+b)^4+(b+c)^4+(c+a)^4}.$$

Consequently, it suffices to prove that the last quantity is greater than or equal to $\frac{3}{5}$. This can be done by expanding everything, but here is an elegant proof using the observation that

$$\begin{aligned} a^2(b+c)^2 + b^2(c+a)^2 + c^2(a+b)^2 + (a+b)^4 + (b+c)^4 + (c+a)^4 \\ = [(a+b)^2+(b+c)^2+(c+a)^2](a^2+b^2+c^2) \\ +2ab(a+b)^2 + 2bc(b+c)^2 + 2ca(c+a)^2. \end{aligned}$$

Because

$$(a+b)^2 + (b+c)^2 + (c+a)^2 \leq 4(a^2+b^2+c^2),$$

we observe that the desired inequality reduces to

$$2ab(a+b)^2 + 2bc(b+c)^2 + 2ca(c+a)^2 \le \frac{8}{3}(a^2+b^2+c^2)^2.$$

But this inequality is not so difficult. Indeed, first we observe that

$$2ab(a+b)^2 + 2bc(b+c)^2 + 2ca(c+a)^2$$

$$\le 4ab(a^2+b^2) + 4bc(b^2+c^2) + 4ca(c^2+a^2).$$

Then, we also find that

$$4ab(a^2+b^2) \le a^4 + b^4 + 6a^2b^2,$$

since $(a-b)^4 \ge 0$. Hence

$$4ab(a^2+b^2) + 4bc(b^2+c^2) + 4ca(c^2+a^2) \le 2(a^2+b^2+c^2)^2$$

$$+2(a^2b^2 + b^2c^2 + c^2a^2) \le \frac{8}{3}(a^2+b^2+c^2)^2$$

and so the problem is solved. With minor changes, we can readily see that this solution even works without the assumption that a, b, c are positive.

We end this discussion (which remains probably permanently open) with a series of more difficult problems, based on less obvious applications of $T2$'s lemma.

Example 8. Let $a_1, a_2, \ldots, a_n > 0$ such that $a_1 + a_2 + \cdots + a_n = 1$. Prove that:

$$(a_1a_2 + \cdots + a_na_1)\left(\frac{a_1}{a_2^2 + a_2} + \cdots + \frac{a_n}{a_1^2 + a_1}\right) \ge \frac{n}{n+1}.$$

[Gabriel Dospinescu]

Solution. How can we get to $a_1a_2 + a_2a_3 + \cdots + a_na_1$? Probably from

$$\frac{a_1^2}{a_1a_2} + \frac{a_2^2}{a_2a_3} + \cdots + \frac{a_n^2}{a_na_1}$$

after we use the lemma. So, let us try the following estimation:

$$\frac{a_1}{a_2} + \frac{a_2}{a_3} + \cdots + \frac{a_n}{a_1} = \frac{a_1^2}{a_1a_2} + \frac{a_2^2}{a_2a_3} + \cdots + \frac{a_n^2}{a_na_1} \geq \frac{1}{a_1a_2 + a_2a_3 + \cdots + a_na_1}.$$

The new problem, proving that

$$\frac{a_1}{a_2^2 + a_2} + \frac{a_2}{a_3^2 + a_3} + \cdots + \frac{a_n}{a_1^2 + a_1} \geq \frac{n}{n+1}\left(\frac{a_1}{a_2} + \frac{a_2}{a_3} + \cdots + \frac{a_n}{a_1}\right)$$

seems even more difficult, but we will see that we have to make one more step in order to solve it. Again, we look at the right-hand side and we write $\frac{a_1}{a_2} + \frac{a_2}{a_3} + \cdots + \frac{a_n}{a_1}$ as

$$\frac{\left(\frac{a_1}{a_2} + \frac{a_2}{a_3} + \cdots + \frac{a_n}{a_1}\right)^2}{\frac{a_1}{a_2} + \frac{a_2}{a_3} + \cdots + \frac{a_n}{a_1}}.$$

After applying $T2$'s lemma, we find that

$$\frac{a_1}{a_2^2 + a_2} + \frac{a_2}{a_3^2 + a_3} + \cdots + \frac{a_n}{a_1^2 + a_1} = \frac{\left(\frac{a_1}{a_2}\right)^2}{a_1 + \frac{a_1}{a_2}} + \frac{\left(\frac{a_2}{a_3}\right)^2}{a_2 + \frac{a_2}{a_3}} + \cdots + \frac{\left(\frac{a_n}{a_1}\right)^2}{a_n + \frac{a_n}{a_1}}$$

$$\geq \frac{\left(\frac{a_1}{a_2} + \frac{a_2}{a_3} + \cdots + \frac{a_n}{a_1}\right)^2}{1 + \frac{a_1}{a_2} + \frac{a_2}{a_3} + \cdots + \frac{a_n}{a_1}}.$$

We are left with an easy problem: if $t = \frac{a_1}{a_2} + \cdots + \frac{a_n}{a_1}$, then $\frac{t^2}{1+t} \geq \frac{nt}{n+1}$, or $t \geq n$. But this follows immediately from the AM-GM inequality.

Example 9. Prove that for any positive real numbers a, b, c the following inequality holds

$$\frac{(a+b)^2}{c^2+ab}+\frac{(b+c)^2}{a^2+bc}+\frac{(c+a)^2}{b^2+ca}\geq 6.$$

[Darij Grinberg, Peter Scholze]

Solution. We do not hide from you that things become really complicated here. However, let us try to use $T2$'s lemma again, but of course not in a direct form, since that one is doomed. Trying to make the numerators as strong as possible, we may first try the choice $(a+b)^4$. And so, we know that the left hand side is at least

$$\frac{(\sum(a+b)^2)^2}{\sum(a+b)^2(c^2+ab)}.$$

So, we should see whether the inequality

$$\left(\sum(a+b)^2\right)^2\geq 6\sum(a+b)^2(c^2+ab)$$

holds. However, this is not easy, at least not without computations. With some courage, we can develop everything and reach the equivalent inequality

$$2(a^4+b^4+c^4)+ab(a^2+b^2)+bc(b^2+c^2)+ca(c^2+a^2)+$$

$$2abc(a+b+c)\geq 6(a^2b^2+b^2c^2+c^2a^2).$$

Fortunately, this can be broken into pieces: because $bc(b^2+c^2)\geq 2b^2c^2$, it is enough to prove that

$$a^4+b^4+c^4+abc(a+b+c)\geq 2(a^2b^2+b^2c^2+c^2a^2).$$

Now, if you know Heron's formula for the area of a triangle,

$$2(a^2b^2+b^2c^2+c^2a^2)-(a^4+b^4+c^4)$$

should ring a bell! It is actually equal to $(a+b+c)(a+b-c)(b+c-a)(c+a-b)$. So, we are left with the classical inequality

$$(a+b-c)(b+c-a)(c+a-b) \leq abc.$$

If one of $a+b-c$, $b+c-a$, $c+a-b$ is negative, we are done. Otherwise, observe that

$$a = (a+b-c) + (c+a-b) \geq 2\sqrt{(a+b-c)(c+a-b)}.$$

Multiplying this and two similar inequalities easily yields the conclusion.

Do you like inequalities that can be solved with identities? Here is one which combines $T2$'s lemma with a very strange identity. Do not worry, things like that do not appear too often. Fortunately...

Example 10. Prove that if $a, b, c, d > 0$ satisfy

$$abc + bcd + cda + dab = a + b + c + d,$$

then

$$\sqrt{\frac{a^2+1}{2}} + \sqrt{\frac{b^2+1}{2}} + \sqrt{\frac{c^2+1}{2}} + \sqrt{\frac{d^2+1}{2}} \leq a + b + c + d.$$

[Gabriel Dospinescu]

Solution. The following solution is very difficult to find, but it is the only one that the authors have. The idea is to apply $T2$'s lemma to an identity which is almost impossible to find. We will prove that

$$\frac{a^2+1}{a+b} + \frac{b^2+1}{b+c} + \frac{c^2+1}{c+d} + \frac{d^2+1}{d+a} = a + b + c + d$$

and after that $T2$'s lemma will do the rest.

To prove the identity, just observe that

$$(a+b)(a+c)(a+d) = a^2(a+b+c+d) + abc + bcd + cda + dab = (a^2+1)(a+b+c+d).$$

Use similar identities and add them up.

5.2 Practice problems

1. Let $x_1, x_2, \ldots, x_n, y_1, y_2, \ldots, y_n$ be positive real numbers such that

$$x_1 + x_2 + \cdots + x_n \geq x_1y_1 + x_2y_2 + \cdots + x_ny_n.$$

Prove that

$$x_1 + x_2 + \cdots + x_n \leq \frac{x_1}{y_1} + \frac{x_2}{y_2} + \cdots + \frac{x_n}{y_n}.$$

Romeo Ilie, Romania 1999

2. Let a, b, c be positive real numbers. Prove that

$$\frac{a^3}{b^2 + c^2} + \frac{b^3}{c^2 + a^2} + \frac{c^3}{a^2 + b^2} \geq \frac{a + b + c}{2}.$$

Mircea Becheanu, Mathematical Reflections

3. Let a, b, c be nonzero real numbers such that $ab + bc + ca \geq 0$. Prove that

$$\frac{ab}{a^2 + b^2} + \frac{bc}{b^2 + c^2} + \frac{ca}{c^2 + a^2} \geq -\frac{1}{2}.$$

Titu Andreescu

4. If a, b, c, d are positive real numbers such that $ab + bc + cd + da = 1$, then

$$\frac{a^3}{b + c + d} + \frac{b^3}{c + d + a} + \frac{c^3}{d + a + b} + \frac{d^3}{a + b + c} \geq \frac{1}{3}.$$

IMO 1990 Shortlist

5. Let a, b, c be real numbers such that

$$\frac{1}{a^2 + 1} + \frac{1}{b^2 + 1} + \frac{1}{c^2 + 1} \leq 2.$$

Prove that $ab + bc + ca \leq \frac{3}{2}$.

6. Prove that if the positive real numbers a, b, c satisfy $abc = 1$, then

$$\frac{a}{b+c+1} + \frac{b}{c+a+1} + \frac{c}{a+b+1} \geq 1.$$

Vasile Cîrtoaje, Gazeta Matematică

7. Prove that for any positive real numbers a, b, c,

$$\left(\frac{a}{b+c}\right)^2 + \left(\frac{b}{c+a}\right)^2 + \left(\frac{c}{a+b}\right)^2 \geq \frac{3}{4} \cdot \frac{a^2+b^2+c^2}{ab+bc+ca}.$$

Gabriel Dospinescu

8. Prove that for all positive real numbers a, b, c satisfying $a + b + c = 1$,

$$\frac{a}{1+bc} + \frac{b}{1+ca} + \frac{c}{1+ab} \geq \frac{9}{10}.$$

9. Let a, b, c be positive real numbers such that $abc = 1$. Prove that

$$\frac{a+b+1}{a+b^2+c^3} + \frac{b+c+1}{b+c^2+a^3} + \frac{c+a+1}{c+a^2+b^3} \leq \frac{(a+1)(b+1)(c+1)+1}{a+b+c}.$$

Titu Andreescu, Mathematical Reflections

10. Prove that for any positive real numbers a, b, c,

$$\frac{1}{3a+b} + \frac{1}{3b+c} + \frac{1}{3c+a} \geq \frac{1}{2a+b+c} + \frac{1}{2b+c+a} + \frac{1}{2c+a+b}.$$

11. Prove that for any $n \geq 4$ and any nonnegative real numbers $x_1, x_2, \ldots, x_n$,

$$\frac{x_1}{x_n+x_2} + \frac{x_2}{x_1+x_3} + \cdots + \frac{x_n}{x_{n-1}+x_1} \geq 2.$$

Tournament of the Towns 1982

12. Let $n \geq 4$ an integer and let $a_1, a_2, \ldots, a_n$ be positive real numbers such that $a_1^2 + a_2^2 + \cdots + a_n^2 = 1$. Prove that

$$\frac{a_1}{a_2^2+1} + \frac{a_2}{a_3^2+1} + \cdots + \frac{a_n}{a_1^2+1} \geq \frac{4}{5}(a_1\sqrt{a_1} + a_2\sqrt{a_2} + \cdots + a_n\sqrt{a_n})^2.$$

Mircea Becheanu and Bogdan Enescu, Romanian TST 2002

13. Prove that for any positive real numbers a, b, c the following inequality holds

$$\frac{a}{\sqrt{a^2+8bc}} + \frac{b}{\sqrt{b^2+8ca}} + \frac{c}{\sqrt{c^2+8ab}} \geq 1.$$

Hojoo Lee, IMO 2001

14. Let a, b, c be positive real numbers such that $ab + bc + ca = 3$. Prove that

$$\frac{a}{2a+b^2} + \frac{b}{2b+c^2} + \frac{c}{2c+a^2} \leq 1.$$

T.Q.Anh

15. Determine the best constant k_n such that for all positive real numbers $a_1, a_2, \ldots, a_n$ satisfying $a_1 a_2 \cdots a_n = 1$,

$$\frac{a_1a_2}{(a_1^2+a_2)(a_2^2+a_1)} + \frac{a_2a_3}{(a_2^2+a_3)(a_3^2+a_2)} + \cdots + \frac{a_na_1}{(a_n^2+a_1)(a_1^2+a_n)} \leq k_n.$$

Gabriel Dospinescu and Mircea Lascu

16. Prove that for any positive real numbers a, b, c,

$$\frac{(2a+b+c)^2}{2a^2+(b+c)^2} + \frac{(2b+c+a)^2}{2b^2+(c+a)^2} + \frac{(2c+a+b)^2}{2c^2+(a+b)^2} \leq 8.$$

Titu Andreescu and Zuming Feng, USAMO 2003

17. Let a, b, c, d be positive real numbers such that $abcd = 1$. Prove that

$$\frac{1}{(1+a)^2} + \frac{1}{(1+b)^2} + \frac{1}{(1+c)^2} + \frac{1}{(1+d)^2} \geq 1.$$

Vasile Cartoaje

18. Let $n \geq 13$ be a positive integer and let $a_1, a_2, \ldots, a_n$ be positive real numbers such that $a_1 + a_2 + \cdots + a_n = 1$ and $a_1 + 2a_2 + \cdots + na_n = 2$. Prove that

$$(a_2 - a_1)\sqrt{2} + (a_3 - a_2)\sqrt{3} + \cdots + (a_n - a_{n-1})\sqrt{n} < 0.$$

Gabriel Dospinescu

19. Prove that for all positive numbers a, b, c,

$$\sqrt{\frac{a}{8b+c}} + \sqrt{\frac{b}{8c+a}} + \sqrt{\frac{c}{8a+b}} \geq 1.$$

Vo Quoc Ba Can

20. Let $a_n = \frac{1}{\sqrt{2\cos\frac{2\pi}{n} - 1}}$. Prove that for all $x_1, x_2, \ldots, x_n \in \left[\frac{1}{a_n}, a_n\right]$, Shapiro's inequality holds

$$\frac{x_1}{x_2+x_3} + \frac{x_2}{x_3+x_4} + \cdots + \frac{x_n}{x_1+x_2} \geq \frac{n}{2}.$$

Vasile Cartoaje, Gabriel Dospinescu

Chapter

6

6.1 Theory and examples

You have already seen quite a few strategies and ideas, and you might say: "Enough with these tricks! When will we go to serious facts?" We will try to convince you that the following results are more than simple tools or tricks. They help to create a good base, which is absolutely indispensable for someone who enjoys mathematics, and moreover, they are the first steps to some really beautiful and difficult theorems or problems. And you must admit that the last problems discussed in the previous units are quite serious facts. It is worth mentioning that these strategies are not a panacea. This assertion is proved by the fact that every year problems that are based on well-known tricks prove to be very difficult in contests.

We will "disappoint" you again in this unit by focusing on a very familiar theme: graphs without complete subgraphs. Why do we say familiar? Because there are hundreds of problems proposed in different mathematics competitions around the world and in professional journals that deal with this subject. And each such problem seems to add something. Before passing to the first problem, we will assume that the basic knowledge about graphs is known and we will denote by $d(V)$ and $C(V)$ the number, and the set of vertices adjacent to V, respectively. Also, we will say that a graph has a complete $k-$subgraph if there are k vertices any two of which are connected. For simplicity, we will say that G is k-free if it does not contain a complete $k-$subgraph. First we will discuss one famous classical result about k-free graphs, namely Turan's theorem. Before that, though, we prove a useful lemma, also known as Zarankiewicz's lemma, which is the main step in the proof of Turan's theorem.

Example 1. If G is a k-free graph, then there exists a vertex having degree at most $\left\lfloor \frac{k-2}{k-1} n \right\rfloor$.

[Zarankiewicz]

Solution. Suppose not and take an arbitrary vertex V_1. Then

$$|C(V_1)| > \left\lfloor \frac{k-2}{k-1} n \right\rfloor,$$

so there exists $V_2 \in C(V_1)$. Moreover,

$$\begin{aligned}|C(V_1) \cap C(V_2)| &= d(V_1) + d(V_2) - |C(V_1) \cup C(V_2))| \\ &\geq 2\left(1 + \left\lfloor \frac{k-2}{k-1} n \right\rfloor\right) - n > 0.\end{aligned}$$

Pick a vertex $V_3 \in C(V_1) \cap C(V_2)$. A similar argument shows that

$$|C(V_1) \cap C(V_2) \cap C(V_3)| \geq 3\left(1 + \left\lfloor \frac{k-2}{k-1} n \right\rfloor\right) - 2n.$$

Repeating this argument, we find

$$\begin{aligned}&V_4 \in C(V_1) \cap C(V_2) \cap C(V_3) \\ &\vdots \\ &V_{k-1} \in \bigcap_{i=1}^{k-2} C(V_i).\end{aligned}$$

Also,

$$\left|\bigcap_{i=1}^{j} C(V_i)\right| \geq j\left(1 + \left\lfloor \frac{k-2}{k-1} n \right\rfloor\right) - (j-1)n.$$

This can be proved easily by induction. Thus

$$\left|\bigcap_{i=1}^{k-1} C(V_i)\right| \geq (k-1)\left(1 + \left\lfloor \frac{k-2}{k-1} n \right\rfloor\right) - (k-2)n > 0,$$

and, consequently, we can choose

$$V_k \in \bigcap_{i=1}^{k-1} C(V_i).$$

But it is clear that $V_1, V_2, \ldots, V_k$ form a complete k graph, which contradicts the assumption that G is k-free.

We are now ready to prove Turan's theorem.

Example 2. The greatest number of edges of a k-free graph with n vertices is

$$\frac{k-2}{k-1}\cdot\frac{n^2-r^2}{2}+\binom{r}{2},$$

where r is the remainder left by n when divided to $k-1$.

[Turan]

Solution. We will use induction on n. The first case is trivial, so let us assume the result true for all k-free graphs having $n-1$ vertices. Let G be a k-free graph with n vertices. Using Zarankiewicz's lemma, we can find a vertex V such that

$$d(V)\leq\left\lfloor\frac{k-2}{k-1}n\right\rfloor.$$

Because the subgraph determined by the other $n-1$ vertices is clearly k-free, using the inductive hypothesis we find that G has at most

$$\left\lfloor\frac{k-2}{k-1}n\right\rfloor+\frac{k-2}{k-1}\cdot\frac{(n-1)^2-r_1^2}{2}+\binom{r_1}{2}$$

edges, where $r_1=n-1 \pmod{k-1}$.
Let $n=q(k-1)+r=q_1(k-1)+r_1+1$. Then $r_1\in\{r-1,r+k-2\}$ (this is because $r-r_1\equiv 1 \pmod{k-1}$) and it is easy to check that

$$\left\lfloor\frac{k-2}{k-1}n\right\rfloor+\frac{k-2}{k-1}\cdot\frac{(n-1)^2-r_1^2}{2}+\binom{r_1}{2}=\frac{k-2}{k-1}\cdot\frac{n^2-r^2}{2}+\binom{r}{2}$$

The inductive step is proved. Now, it remains to construct a k-free graph with n vertices and $\frac{k-2}{2}\cdot\frac{n^2-r^2}{k-1}+\binom{r}{2}$ edges. This is not difficult. Just consider $k-1$ classes of vertices, r of them having $q+1$ elements and the rest q elements, where $q(k-1)+r=n$ and join the vertices situated in different groups. It is immediate that this graph is k-free, has $\frac{k-2}{k-1}\cdot\frac{n^2-r^2}{2}+\binom{r}{2}$

edges and also the minimal degree of the vertices is $\left\lfloor \frac{k-2}{k-1} n \right\rfloor$. This graph is called Turan's graph and is denoted by $T(n, k)$.
These two theorems generate numerous beautiful and difficult problems. For example, using these results yields a straightforward solution for the following Bulgarian problem.

Example 3. There are 2001 towns in a country, each of which is connected with at least 1600 towns by a direct bus line. Find the largest n for which it must be possible to find n towns, any two of which are connected by a direct bus line.

Spring Mathematics Tournament 2001

Solution. Practically, the problem asks to find the greatest n such that any graph G with 2001 vertices and minimum degree at least 1600 is not n-free. But Zarankiewicz's lemma implies that if G is n-free, then at least one vertex has degree at most $\left\lfloor \frac{n-2}{n-1} 2001 \right\rfloor$. So, we need the greatest n for which $\left\lfloor \frac{n-2}{n-1} 2001 \right\rfloor < 1600$. It is immediate to see that $n = 5$. Thus for $n = 5$ any such graph G is not n-free. It suffices to construct a graph with all degrees of the vertices at least 1600, which is 6-free. We will take of course $T(2001, 6)$, whose minimal degree is $\left\lfloor \frac{4}{5} 2001 \right\rfloor = 1600$ and which is (as shown before) 6-free. Thus, the answer is $n = 5$.

Here is a beautiful application of Turan's theorem in combinatorial geometry.

Example 4. Consider 21 points on a circle. Show that at least 100 pairs of points subtend an angle less than or equal to $120°$ at the center.

Tournament of the Towns 1986

Solution. In such problems, it is more important to choose the right graph than to apply the theorem, because as soon as the graph is appropriately chosen, the solution is more or less straightforward. Here we will consider the graph with vertices at the given points and we will connect two points if they subtend an angle less than or equal to $120°$ at the center. Therefore we need to prove that this graph has at least 100 edges. It seems that this is a reversed form of Turan's theorem, which maximizes the number of edges in a k-free graph. Yet, the reversed form of the reversed form is the natural one. Applying this principle, let us look at the "reversed" graph, the complementary one. We must show that it has at most $\binom{21}{2} - 100 = 110$ edges. But this is immediate, since it is clear that this new graph does not have triangles and so, by Turan's theorem, it has at most $\dfrac{21^2 - 1}{4} = 110$ edges, and the problem is solved.

At first glance, the following problem seems to have no connection with the previous examples, but, as we will immediately see, it is a simple consequence of Zarankiewicz's lemma. It is an adaptation of an USAMO 1978 problem. Anyway, this is trickier than the actual contest problem.

Example 5. There are n delegates at a conference, each of them knowing at most k languages. Among any three delegates, at least two speak a common language. Find the least number n such that for any distribution of the languages satisfying the above properties, it is possible to find a language spoken by at least three delegates.

Solution. We will prove that $n = 2k+3$. First, we prove that if there are $2k+3$ delegates, then the conclusion of the problem holds. The condition "among any three of them there are at least two who can speak the same language" suggests taking the 3-free graph with vertices the persons and whose edges join persons that do not speak a common language. From Zarankiewicz's lemma, there exists a vertex whose degree is at most $\left\lfloor \dfrac{n}{2} \right\rfloor = k + 1$. Thus, it is not connected with at least $k+1$ other vertices. Hence there exists a person A and $k+1$ persons $A_1, A_2, \ldots, A_{k+1}$ that can communicate with A. Because A speaks at most k languages, there are two persons among $A_1, A_2, \ldots, A_{k+1}$ that speak with A in the same language. But that language is spoken by at least three delegates and we are done. It remains to prove now that we can create a situation in which there are $2k+2$ delegates, but no language is spoken by more than two delegates. We use again Turan's graph, by creating two groups of $k+1$ delegates. Assign to each pair of persons in the first group a common language, so that the language associated is different for any two pairs in that group. Do the same for the second group, taking care that no language associated with a pair in the second group is identical to a language associated with a pair in the first group. Persons in different groups do not communicate. Then it is clear that among three persons, two will be in the same group and therefore will have a common language. Of course, any language is spoken by at most two delegates.

The following problem turned out to be an upset at one of the Romanian Team Selection Tests for 2004 IMO, being solved by only four contestants. The idea is even easier than in the previous problems, but this time we need a little observation that is not so obvious.

Example 6. Let $A_1, A_2, \ldots, A_{101}$ be different subsets of the set $\{1, 2, \ldots, n\}$. Suppose that the union of any 50 subsets has more than $\frac{50}{51}n$ elements. Prove that among them there are three any two of which having common elements.

[Gabriel Dospinescu] Romanian TST 2004

Solution. As the conclusion suggests, we should take a graph with vertices the subsets, connecting two subsets if they have common elements. Let us assume that this graph is 3-free. The main idea is not to use Zarankiewicz's lemma, but to find many vertices with small degrees. In fact, we will prove that there are at least 51 vertices all of them having degree at most 50. Suppose this is not the case, so there are at least 51 vertices whose degrees are greater than

51. Let us pick such a vertex A. It is connected with at least 51 vertices, so it must be adjacent to a vertex B whose degree is at least 51. Because A and B are each connected with at least 51 vertices, there is a vertex adjacent to both, so we have a triangle, contradicting our assumption. Therefore, we can find $A_{i_1}, \ldots, A_{i_{51}}$, all of them having degrees at most 50. Consequently, A_{i_1} is disjoint from at least 50 subsets. Because the union of these fifty subsets has more than $\frac{50}{51}n$ elements, we infer that $|A_{i_1}| < n - \frac{50}{51}n = \frac{n}{51}$. In a similar way, we obtain $|A_{i_j}| \le \frac{n}{51}$ for all $j \in \{1, 2, \ldots, 51\}$ and so

$$|A_{i_1} \cup A_{i_2} \cup \cdots \cup A_{i_{50}}| \le |A_{i_1}| + \cdots + |A_{i_{50}}| < \frac{50}{51}n,$$

which contradicts the hypothesis.

We continue with an adaptation of a very nice and quite challenging problem from the American Mathematical Monthly.

Example 7. Prove that the complement of any 3-free graph with n vertices and m edges has at least

$$\frac{n(n-1)(n-5)}{24} + \frac{2}{n}\left(m - \frac{n^2 - n}{4}\right)^2$$

triangles.

[A.W Goodman] AMM

Solution. Believe it or not, the number of triangles from the complementary graph can be expressed in terms of the degrees of the vertices of the graph only. More precisely, if G is a 3−free graph, then the number of triangles from the complementary graph is

$$\binom{n}{3} - \frac{1}{2}\sum_{x \in X} d(x)(n-1-d(x)),$$

where X is the set of vertices of G. Indeed, consider all triples (x, y, z) of vertices of G. We will count the triples that do not form a triangle in the complementary graph $\overline{G}$. Indeed, consider the sum $\sum\limits_{x\in X} d(x)(n-1-d(x))$. It counts twice every triple (x, y, z) in which x and y are connected, while z is not adjacent to any of x and y: once for x and once for y. But it also counts twice every triple (x, y, z) in which y is connected with both x and z: once for x and once for z. Therefore, $\frac{1}{2}\sum\limits_{x\in X} d(x)(n-1-d(x))$ is exactly the number of triples (x, y, z) that do not form a triangle in the complementary graph. (Here we have used the fact that G is 3-free.) Now, it is enough to prove that

$$\binom{n}{3} - \frac{1}{2}\sum_{x\in X} d(x)(n-1-d(x)) \geq \frac{n(n-1)(n-5)}{24} + \frac{2}{n}\left(m - \frac{n^2-n}{4}\right)^2.$$

Because $\sum\limits_{x\in X} d(x) = 2m$, after a few computations the inequality reduces to

$$\sum_{x\in X} d^2(x) \geq \frac{4m^2}{n} \qquad (6.1)$$

But this is the Cauchy-Schwarz inequality combined with $\sum_{x\in X} d(x) = 2m$. Finally, two chestnuts. The following problem is not directly related to our topic at first glance, but it gives a very beautiful proof of Turan's theorem:

Example 8. Let G be a simple graph. To every vertex of G one assigns a nonnegative real number such that the sum of the numbers assigned to all vertices is 1. For any two vertices connected by an edge, compute the product of the numbers associated to these vertices. What is the maximal value of the sum of these products?

Solution. The answer is not obvious at all, so let us start by making a few remarks. If the graph is complete of order n then the problem reduces to finding the maximum of $\sum\limits_{1\le i<j\le n} x_i x_j$ knowing that $x_1 + x_2 + \cdots + x_n = 1$. This is easy, since

$$\sum_{1\le i<j\le n} x_i x_j = \frac{1}{2}(1 - \sum_{i=1}^{n} x_i^2) \le \frac{1}{2}\left(1 - \frac{1}{n}\right).$$

The last inequality is just the Cauchy-Schwarz inequality and we have equality when all variables are $\frac{1}{n}$. Unfortunately, the problem is much more difficult in other cases, but at least we have an idea of a possible answer: indeed, it is easy now to find a lower bound for the maximum: if H is the complete subgraph with maximal number of vertices k, then by assigning these vertices $\frac{1}{k}$, and to all other vertices 0, we find that the desired maximum is at least $\frac{1}{2}(1 - \frac{1}{k})$. We still have to solve the difficult part: showing that the desired maximum is at most $\frac{1}{2}(1 - \frac{1}{k})$. Let us proceed by induction on the number n of vertices of G. If $n = 1$ everything is clear, so assume the result true for all graphs with at most $n - 1$ vertices and take a graph G with n vertices, numbered $1, 2, ..., n$. Let A be a set of vectors with nonnegative coordinates and whose components add up to 1 and E the set of edges of G. Because the function $f(x_1, x_2, ..., x_n) = \sum\limits_{(i,j)\in E} x_i x_j$ is continuous on the compact set A, it attains its maximum in a point $(x_1, x_2, ..., x_n)$. Denote by $f(G)$ the maximum value of this function on A. If at least one of the x_i is zero, then $f(G) = f(G_1)$ where G_1 is the graph obtained by erasing vertex i and all edges that are incident to this vertex. It suffices to apply the induction hypothesis to G_1 (clearly, the maximal complete subgraph of G_1 has at most as many vertices as the maximal complete subgraph of G). So, suppose that all x_i are positive. We may assume that G is not complete, since this case has already been discussed. So, let us assume for example that vertices 1 and 2 are not connected. Choose any number $0 < a \le x_1$ and assign to vertices $1, 2, ..., n$ of G the numbers $x_1 - a, x_2 + a, x_3, ..., x_n$. By maximality of $f(G)$, we must have

$$\sum_{i\in C_1} x_i \le \sum_{i\in C_2} x_i,$$

where C_1 is the set of vertices that are adjacent to vertex 2 and not adjacent to vertex 1 (the definition of C_2 being clear). By symmetry, we deduce that we must actually have

$$\sum_{i\in C_1} x_i = \sum_{i\in C_2} x_i,$$

which shows that $f(x_1, x_2, ..., x_n) = f(0, x_1 + x_2, x_3, ..., x_n)$. Hence we can apply the previous case and the problem is solved. Observe that the inequality in Turan's theorem follows by taking all x_i to be $\frac{1}{n}$.

The final problem is a very beautiful result on the number of complete subgraphs of a graph:

Example 9. What is the maximal number of complete maximal subgraphs that a graph on n vertices can have?

[Leo Moser, J. W. Moon]

Solution. Let us suppose that $n \geq 5$, the other cases being easy to check. Let $f(n)$ be the desired number and G a graph for which this maximum is attained. Clearly, this graph is not complete, so there are two vertices x and y not connected by an edge. In order to simplify the solution, we need several notations. Let $V(x)$ be the set of vertices that are adjacent to x, $G(x)$ the subgraph obtained by erasing vertex x and $G(x,y)$ the graph obtained by erasing all edges incident to x and replacing them with edges from x to any vertex in $V(y)$. Finally, let $a(x)$ be the number of complete subgraphs with vertices in $V(x)$, maximal with respect to $G(x)$ and let $c(x)$ be the number of complete maximal subgraphs of G that contain x.

Now, we pass to serious things: by erasing edges incident to x, exactly $c(x) - a(x)$ complete maximal subgraphs vanish, and by joining x with all vertices of $V(y)$, exactly $c(y)$ complete maximal subgraphs appear. So, if $c(G)$ is the number of complete maximal subgraphs in the graph G, then we have the relation

$$c(G(x,y)) = c(G) + c(y) - c(x) + a(x).$$

By symmetry, we can assume that $c(y) \geq c(x)$. By maximality of $c(G)$, we must have $c(G(x,y)) \leq c(G)$, which is the same as $c(y) = c(x)$ and $a(x) = 0$. Therefore $G(x,y)$ also has $f(n)$ complete maximal subgraphs. In the same way, we deduce that $c(G(x,y)) = c(G(y,x)) = c(G)$. Now take a vertex x and let $x_1, x_2, ..., x_k$ be the vertices not adjacent to x. By performing the previous operations, we change G into $G_1 = G(x_1, x)$, then into $G_2 = G_1(x_2, x)$ and so on until $G_k = G_{k-1}(x_k, x)$, by conserving the number $f(n)$ of maximal complete subgraphs. Observe now that G_k has the property that $x, x_1, ..., x_k$ are not joined by edges, yet $V(x_1) = V(x_2) = \cdots = V(x_k) = V(x)$. Now, we know what to do: if $V(x)$ is void, we stop the process. Otherwise, consider a vertex of $V(x)$ and apply the previous transformation. In the end, we obtain a complete multipartite graph G' whose vertices can be partitioned into r classes with $n_1, n_2, ..., n_r$ vertices, two vertices being connected by an edge if and only if they do not belong to the same class. Because G' has $f(n)$ maximal complete subgraphs, we deduce that

$$f(n) = \max_{r} \max_{n_1+n_2+\cdots+n_r=n} n_1 n_2 ... n_r. \tag{6.2}$$

(6.2) can be easily computed. Indeed, let $(n_1, n_2, ..., n_r)$ the r-tuple for which the maximum is attained. If one of these numbers is at least equal to 4, let us say n_1, we consider $(2, n_1 - 2, n_3, ..., n_r)$ for which the product of the components is at least the desired maximum. So none of the n_i exceed 3. Even more, since $2 \cdot 2 \cdot 2 < 3 \cdot 3$, there are at most two numbers equal to 2 among $n_1, n_2, ..., n_r$. This shows that $f(n) = 3^{\frac{n}{3}}$ if n is a multiple of 3, $f(n) = 4 \cdot 3^{\frac{n-4}{3}}$ if $n - 1$ is a multiple of 3 and $f(n) = 2 \cdot 3^{\frac{n-2}{3}}$ otherwise.

6.2 Practice problems

1. Let $x_1, x_2, \ldots, x_n$ be real numbers. Prove that there are at most $\dfrac{n^2}{4}$ pairs $(i, j) \in \{1, 2, \ldots, n\}^2$ such that $i < j$ and $1 < |x_i - x_j| < 2$.

   MOSP 2001

2. Prove that if n points lie on a unit circle, then at most $\dfrac{n^2}{3}$ segments connecting them have length greater than $\sqrt{2}$.

   Poland 1997

3. There are 1999 people participating in an exhibition. Out of any 50 people, at least two do not know each other. Prove that we can find at least 41 people who each know at most 1958 other people.

   Taiwan 1999

4. We are given $5n$ points in a plane and we connect some of them so that $10n^2 + 1$ segments are drawn. We color these segments in 2 colors. Prove that we can find a monochromatic triangle.

5. A group of people is called n-balanced if the following two conditions are satisfied

   (a) among any three people, there are two who know each other;

   (b) among any n people, there are at least two not knowing each other.

   Prove that there are always at most $\frac{(n-1)(n+2)}{2}$ people in an n-balanced group.

   Dorel Mihet, Romanian TST 2008

6. Let A be a subset of the set $S = \{1, 2, \ldots, 1000000\}$ having exactly 101 elements. Prove that there exist $t_1, t_2, \ldots, t_{100} \in S$ such that the sets $A_j = \{x + t_j \mid x \in A\}$ are pairwise disjoint.

IMO 2003

7. Prove that a graph with n vertices and k edges has at least $\dfrac{k}{3n}(4k - n^2)$ triangles.

APMO 1989

8. A graph G has n vertices and contains no complete subgraph with four vertices. Prove that G contains at most $\frac{n^3}{27}$ triangles.

Ivan Borsenco, Mathematical Reflections

9. (a) Let p be a prime. Consider the graph whose vertices are the ordered pairs (x, y) with $x, y \in \{0, 1, \ldots, p-1\}$ and whose edges join vertices (x, y) and (x', y') if and only if $xx' + yy' \equiv 1 \pmod{p}$. Prove that this graph does not contain a 4-cycle .

   (b) Prove that for infinitely many values n there is a graph G_n with at least $\frac{n\sqrt{n}}{2} - n$ edges that does not contain a 4-cycle.

Hungary-Israel Competition 2001

10. A graph with n vertices and k edges has no triangles. Prove that we can choose a vertex such that the subgraph obtained by deleting this vertex and all its neighbors has at most $k\left(1 - \dfrac{4k}{n^2}\right)$ edges.

USAMO 1995

11. A graph has $2n$ vertices and $n^2 + 1$ edges. Prove that it contains at least n triangles.

12. A graph with n^2+1 edges and $2n$ vertices is given. Prove that it contains two triangles sharing a common edge.

Chinese TST 1987

13. There are n inhabitants on an island. Any two of them are either friends or enemies. One day they receive an order saying that all citizens should make and wear a necklace with zero or more stones so that

    (a) for any pair of friends there exists a color such that each of the two persons has a stone of that color;

    (b) for any pair of enemies there does not exist such a color. What is the least number of colors of stones required (considering all possible relationships between the inhabitants of the island)?

Belarus 2001

14. What is the least number of edges in a connected n-vertex graph such that any edge belongs to a triangle?

Paul Erdős, AMM E 3255

15. Prove that for every n one can construct a graph with no triangles and whose chromatic number is at least n.

Mycielski's theorem

16. For a finite graph G let $f(G)$ (respectively $g(G)$) be the number of triangles (respectively tetrahedra) formed by the edges of G. Find the least constant c such that $g^3(G) \leq c \cdot f(G)^4$ for any finite graph G.

IMO Shortlist 2004

17. For a pair $A = (x_1, y_1)$ and $B = (x_2, y_2)$ of points on the coordinate plane, let

$$d(A, B) = |x_1 - x_2| + |y_1 - y_2|.$$

Among all configurations of 100 points in the plane, determine the maximum number of pairs (A, B) of (unordered) points such that $1 < d(A, B) \leq 2$.

USA TST 2006

18. Let k be a positive integer. A graph whose vertex set is the set of positive integers does not contain any complete $k \times k$ bipartite subgraph. Prove that there exist arbitrarily long arithmetic progressions of positive integers such that no two elements of the same progression are joined by an edge in this graph.

Komal

Chapter

7

7.1 Theory and examples

When reading the title, you will perhaps expect a difficult unit, reflecting the complexity of combinatorics. But, this was not our intention. We just wanted to discuss some combinatorial problems that can be solved elegantly by using complex numbers. At this moment, the reader will probably say that we are crazy, but we will support our idea and prove that complex numbers can play a significant role in solving counting problems, and also in problems related to tilings. They also have numerous applications in combinatorial number theory, so our purpose is to illustrate a little bit from each of these situations. After that, you will surely have the pleasure of solving the proposed problems using this technique. To avoid repetition, we will present in the beginning of the discussion a useful result

Lemma 7.1. *If p is a prime number and $a_0, a_1, \ldots, a_{p-1}$ are rational numbers satisfying*

$$a_0 + a_1\varepsilon + a_2\varepsilon^2 + \cdots + a_{p-1}\varepsilon^{p-1} = 0,$$

where

$$\varepsilon = \cos\frac{2\pi}{p} + i\sin\frac{2\pi}{p} = e^{\frac{2\pi i}{p}},$$

then $a_0 = a_1 = \cdots = a_{p-1}$.

Proof. We will just sketch the proof, which is not difficult. It is enough to observe that the polynomials $a_0+a_1X+a_2X^2+\cdots+a_{p-1}X^{p-1}$ and $1+X+X^2+\cdots+X^{p-1}$ are not relatively prime–because they share a common root–and since $1+X+X^2+\cdots+X^{p-1}$ is irreducible over $\mathbb{Q}$ (you can find a proof in the chapter concerning the irreducibility of polynomials), $1+X+X^2+\cdots+X^{p-1}$ must divide $a_0 + a_1X + a_2X^2 + \cdots + a_{p-1}X^{p-1}$, which can only happen if $a_0 = a_1 = \cdots = a_{p-1}$. Therefore, the lemma is proved and it is time to solve some nice problems. □

Note, in the following examples, $m(A)$ will denote the sum of the elements of the set A. By convention $m(\emptyset) = 0$.

The first example is an adaptation from a problem given in the Romanian Contest "Traian Lalescu". Of course, there is a solution using recursive sequences, but it is by far less elegant than the following one.

Example 1. How many n-digit numbers, all of whose digits are 1, 3, 4, 6, 7, or 9 have the digit sum a multiple of 7?

Solution. Let $a_n^{(k)}$ be the number of n-digit numbers, all of whose digits are 1, 3, 4, 6, 7, 9 and whose digit sum is congruent to k modulo 7. It is clear that

$$\sum_{k=0}^{6} a_n^{(k)} \varepsilon^k = \sum_{x_1, x_2, \ldots, x_n \in \{1,3,4,6,7,9\}} \varepsilon^{x_1+x_2+\cdots+x_n}$$
$$= (\varepsilon + \varepsilon^3 + \varepsilon^4 + \varepsilon^6 + \varepsilon^7 + \varepsilon^9)^n,$$

where $\varepsilon = \cos\frac{2\pi}{7} + i\sin\frac{2\pi}{7}$. Observing that $1 + \varepsilon + \varepsilon^2 + \cdots + \varepsilon^6 = 0$ and $\varepsilon^9 = \varepsilon^2$ helps us bring $(\varepsilon + \varepsilon^3 + \varepsilon^4 + \varepsilon^6 + \varepsilon^7 + \varepsilon^9)^n$ to the simpler form $(-\varepsilon^5)^n$. Let us assume, for example, that n is divisible by 7 (the other cases can be discussed similarly). Then

$$\sum_{k=0}^{6} a_n^{(k)} \varepsilon^k = (-1)^n$$

and from the lemma we infer that $a_n^{(0)} - (-1)^n = a_n^{(1)} = \cdots = a_n^{(6)}$. Let q be the common value. Then $7q = \sum_{k=0}^{6} a_n^{(k)} - (-1)^n = 6^n - (-1)^n$ - this is because exactly 6^n numbers have n digits, all equal to 1, 3, 4, 6, 7, 9. In this case we have $a_n^{(0)} = (-1)^n + \frac{6^n - (-1)^n}{7}$. We leave you with the other cases: $n \equiv 1, 2, 3, 4, 5, 6 \pmod 7$.

Following this trick, here is a slightly more difficult problem, which appeared on the Balkan Olympiad Shortlist in 2005, and which was used for the selection of the Romanian IMO 2005 team:

Example 2. Let $(a_n)_{n\geq 1}$ be a sequence of distinct positive integers such that $a_n \leq 4.999n$ for all n. Prove that there are infinitely many n for which the sum of digits of a_n is not a multiple of 5. Does the result remain true if the condition is relaxed to $a_n \leq 5n$ for all n?

[Gabriel Dospinescu]

Solution. Let $s(x)$ be the sum of digits of x, and suppose that for all $n > M$ we have $5|s(a_n)$. Let n be such that $\left\lfloor \frac{10^n-1}{4.999} \right\rfloor > M + 3$ and let A be the set of the first 10^n nonnegative integers. The numbers a_k with $1 \leq k \leq \left\lfloor \frac{10^n-1}{4.999} \right\rfloor$ are in A because $1 \leq a_k \leq 4.999k \leq 10^n - 1$ for these numbers k. It follows that A contains at least $\left\lfloor \frac{10^n-1}{4.999} \right\rfloor - M$ numbers with digit sum divisible by 5. Now fix a number $2 \leq i \leq n$ and observe that if x_j is the number of elements of A with i digits and having digit sum congruent to j mod 5, then

$$x_0 + x_1\varepsilon + x_2\varepsilon^2 + x_3\varepsilon^3 + x_4\varepsilon^4 = \sum_{\substack{0\leq a_2,\ldots,a_i\leq 9\\ 1\leq a_1\leq 9}} \varepsilon^{a_1+a_2+\cdots+a_i} =$$

$$= (\varepsilon + \varepsilon^2 + \cdots + \varepsilon^9)(1 + \varepsilon + \cdots + \varepsilon^9)^{i-1} = 0.$$

Using the lemma, and taking into account that $x_0 + x_1 + \cdots + x_4 = 9 \cdot 10^{i-1}$, we deduce that there are at most $1 + \sum_{i=2}^{n} \frac{9\cdot 10^{i-1}}{5} = 2 \cdot 10^{n-1} - 1$ elements of A with the digit sum a multiple of 5. Thus $\left\lfloor \frac{10^n-1}{4.999} \right\rfloor - M \leq 2 \cdot 10^{n-1} - 1$ for all sufficiently large n, which is certainly impossible.

For the second part of the problem, the answer is negative. Indeed, consider the sequence starting with 1 and containing the positive integers (in increasing order) whose digit sum is divisible by 5. Let us prove that $a_n < 5n$ for all n. Indeed, this is clear for $n = 1, 2, 3$ because $a_1 = 1, a_2 = 5, a_3 = 14$. The crucial observation is that clearly among any 10 consecutive positive integers, exactly two are terms of the sequence. Thus $a_{2n} < 10n$ and $a_{2n-1} = a_{2n} - 5 < 5(2n-1)$. This proves that for $a_n \leq 5n$ the statement is no longer true.

The same simple, but tricky, idea can offer probably the most beautiful solution for the difficult IMO 1995 problem 6. It is worth mentioning that Nikolai Nikolov won a special prize for the following magnificent solution.

Example 3. Let $p > 2$ be a prime number and let $A = \{1, 2, \ldots, 2p\}$. Find the number of subsets of A each having p elements and whose sum is divisible by p.

IMO 1995

Solution. Consider $\varepsilon = \cos\dfrac{2\pi}{p} + i\sin\dfrac{2\pi}{p}$ and let x_j be the number of subsets X of A such that $|X| = p$ and $m(X) \equiv j \pmod{p}$. Then it is not difficult to see that

$$\sum_{j=0}^{p-1} x_j \varepsilon^j = \sum_{B \subset A, |B|=p} \varepsilon^{m(B)} = \sum_{1 \le c_1 < c_2 < \cdots < c_p \le 2p} \varepsilon^{c_1 + c_2 + \cdots + c_p}.$$

But $\sum\limits_{1 \le c_1 < c_2 < \cdots < c_p \le 2p} \varepsilon^{c_1 + c_2 + \cdots + c_p}$ is precisely the coefficient of X^p in the polynomial $(X + \varepsilon)(X + \varepsilon^2) \ldots (X + \varepsilon^{2p})$. Because

$$X^p - 1 = (X - 1)(X - \varepsilon) \ldots (X - \varepsilon^{p-1}),$$

we easily find that

$$(X + \varepsilon)(X + \varepsilon^2) \ldots (X + \varepsilon^{2p}) = (X^p + 1)^2.$$

Thus $\sum\limits_{j=0}^{p-1} x_j \varepsilon^j = 2$, and the lemma implies $x_0 - 2 = x_1 = \cdots = x_{p-1}$. Since there are $\dbinom{2p}{p}$ subsets with p elements, it follows that

$$x_0 + x_1 + \cdots + x_{p-1} = \binom{2p}{p}.$$

Therefore

$$x_0 = 2 + \frac{1}{p}\left(\binom{2p}{p} - 2\right),$$

and we are done.

The following problem deals with a little more general case, even though the restriction imposed on the cardinality is no longer maintained.

Example 4. Let $f(n)$ be the number of subsets of $1, 2, 3, ..., n$ whose elements sum to $0 \pmod{n}$. The empty set is included, having the element sum equal to zero. Prove that

$$f(n) = \frac{1}{n} \cdot \sum_{\substack{d|n \\ d \text{ odd}}} \varphi(d) 2^{\frac{n}{d}}.$$

Solution. Let

$$g(X) = \prod_{i=1}^{n} (1 + X^i) = \sum_{k \geq 0} a_k X^k$$

and let $\varepsilon = e^{\frac{2i\pi}{n}}$. It is clear that $f(n) = \sum_{j \geq 0} a_{jn}$. On the other hand, the last sum can easily be computed in terms of $g(\varepsilon^j)$. Indeed, one can verify the identity

$$\frac{1}{n} \cdot \sum_{j=1}^{n} g(\varepsilon^j) = \sum_{j \geq 0} a_{jn}.$$

Now, let us compute $g(\varepsilon^j)$. If $d = \frac{n}{\gcd(j,n)}$ (that is, ε^j is a primitive d-th root of unity), then

$$X^d - 1 = (X - \varepsilon^j)(X - \varepsilon^{2j}) \cdots (X - \varepsilon^{dj})$$

and so

$$(1 + \varepsilon^j)(1 + \varepsilon^{2j}) \cdots (1 + \varepsilon^{dj}) = 2$$

if d is odd and 0 otherwise. This shows that $g(\varepsilon^j) = 2^{\frac{n}{d}}$ if d is odd and 0 otherwise. But there are exactly $\varphi(d)$ values of j for which ε^j is a primitive d–th root of unity, so

$$\frac{1}{n} \cdot \sum_{j=1}^{n} g(\varepsilon^j) = \frac{1}{n} \cdot \sum_{\substack{d|n \\ d \text{ odd}}} \varphi(d) 2^{\frac{n}{d}}.$$

With a somewhat different but closely related idea we can solve the following nice problem.

Example 5. Let $n > 1$ be an integer and let $a_1, a_2, \ldots, a_m$ be positive integers. Denote by $f(k)$ the number of m-tuples $(c_1, c_2, \ldots, c_m)$ such that $1 \le c_i \le a_i$ for all i and $c_1 + c_2 + \cdots + c_m \equiv k \pmod{n}$. Prove that $f(0) = f(1) = \cdots = f(n-1)$ if and only if there exists an index $i \in \{1, 2, \ldots, m\}$ such that $n | a_i$.

[Reid Barton] Rookie Contest 1999

Solution. Observe that

$$\sum_{k=0}^{n-1} f(k)\varepsilon^k = \sum_{1 \le c_i \le a_i} \varepsilon^{c_1 + c_2 + \cdots + c_m} = \prod_{i=1}^{m} (\varepsilon + \varepsilon^2 + \cdots + \varepsilon^{a_i})$$

for any complex number ε such that $\varepsilon^{n-1} + \varepsilon^{n-2} + \cdots + \varepsilon + 1 = 0$. Hence one implication of the problem is already verified, since if $f(0) = f(1) = \cdots = f(n-1)$ then we can find $i \in \{1, 2, \ldots, m\}$ such that $\varepsilon + \varepsilon^2 + \cdots + \varepsilon^{a_i} = 0$ (we have chosen here a primitive root ε of unity). We infer that $\varepsilon^{a_i} = 1$ and so $n | a_i$. Now, suppose there exists an index $i \in \{1, 2, \ldots, m\}$ such that $n | a_i$. Then for any zero ε of the polynomial $\sum_{k=0}^{n-1} X^k$ we have $\sum_{k=0}^{n-1} f(k)\varepsilon^k = 0$ and so the polynomial $\sum_{k=0}^{n-1} X^k$ divides $\sum_{k=0}^{n-1} f(k)X^k$. This is because $\sum_{k=0}^{n-1} X^k$ has only simple roots. By a simple degree consideration, this is possible only if $f(0) = f(1) = \cdots = f(n-1)$.

The enthusiasm generated by the above solutions might be inhibited by the following problem, where we additionally need several tricky manipulations.

Example 6. Let $p > 2$ be a prime number and let m and n be multiples of p, with n odd. For any function $f : \{1, 2, \ldots, m\} \to \{1, 2, \ldots, n\}$ satisfying $\sum\limits_{k=1}^{m} f(k) \equiv 0 \pmod{p}$, consider the product $\prod\limits_{k=1}^{m} f(k)$. Prove that the sum of these products is divisible by $\left(\dfrac{n}{p}\right)^m$.

[Gabriel Dospinescu]

Solution. Let $\varepsilon = \cos\dfrac{2\pi}{p} + i\sin\dfrac{2\pi}{p}$ and let x_k be the sum of $\prod\limits_{k=1}^{m} f(k)$ over all functions $f : \{1, 2, \ldots, m\} \to \{1, 2, \ldots, n\}$ such that $\sum\limits_{i=1}^{m} f(i) \equiv k \pmod{p}$. It is clear that

$$\sum_{k=0}^{p-1} x_k \varepsilon^k = \sum_{c_1, c_2, \ldots, c_m \in \{1, 2, \ldots, n\}} c_1 c_2 \ldots c_m \varepsilon^{c_1 + c_2 + \cdots + c_m}$$

$$= (\varepsilon + 2\varepsilon^2 + \cdots + n\varepsilon^n)^m.$$

Recall the identity

$$1 + 2x + 3x^2 + \cdots + nx^{n-1} = \frac{nx^{n+1} - (n+1)x^n + 1}{(x-1)^2}.$$

Plugging ε in the previous identity, we find that

$$\varepsilon + 2\varepsilon^2 + \cdots + n\varepsilon^n = \frac{n\varepsilon^{n+2} - (n+1)\varepsilon^{n+1} + \varepsilon}{(\varepsilon - 1)^2} = \frac{n\varepsilon}{\varepsilon - 1}.$$

Consequently,

$$\sum_{k=0}^{p-1} x_k\varepsilon^k = \frac{n^m}{(\varepsilon-1)^m}.$$

On the other hand, it is not difficult to justify that

$$\varepsilon^{p-1}+\varepsilon^{p-2}+\cdots+\varepsilon+1=0 \Leftrightarrow$$

$$\frac{1}{\varepsilon-1} = -\frac{1}{p}(\varepsilon^{p-2}+2\varepsilon^{p-3}+\cdots+(p-2)\varepsilon+p-1).$$

Considering

$$(X^{p-2}+2X^{p-3}+\cdots+(p-2)X+p-1)^m = b_0+b_1X+\cdots+b_{m(p-2)}X^{m(p-2)},$$

we have

$$\frac{n^m}{(\varepsilon-1)^m} = \left(-\frac{n}{p}\right)^m (c_0+c_1\varepsilon+\cdots+c_{p-1}\varepsilon^{p-1}),$$

where

$$c_k = \sum_{j\equiv k \pmod p} b_j.$$

Setting $r=\left(-\dfrac{n}{p}\right)^m$, we deduce that

$$x_0-rc_0+(x_1-rc_1)\varepsilon+\cdots+(x_{p-1}-rc_{p-1})\varepsilon^{p-1}=0.$$

From the lemma, it follows that $x_0-rc_0=x_1-rc_1=\cdots=x_{p-1}-rc_{p-1}=k$. Because clearly $c_0,c_1,\ldots,c_{p-1}$ are integers, it remains to prove that $r|k$. Because

$$\begin{aligned} pk &= x_0+x_1+\cdots+x_{p-1}-r(c_0+c_1+\cdots+c_{p-1}) \\ &= (1+2+\cdots+n)^m - r(b_0+b_1+\cdots+b_{m(p-2)}) \\ &= \left(\frac{n(n+1)}{2}\right)^m - r\left(\frac{p(p-1)}{2}\right)^m, \end{aligned}$$

it is clear that $r|k$. Here we have used the conditions in the hypothesis. The problem is solved.

It is now time to leave these kinds of problems and to talk a little bit about some nice applications of complex numbers in tilings. Before presenting some examples, let us make some conventions: consider a rectangular table with edges parallel to two fixed (orthogonal) lines Ox and Oy. An $a \times b$ rectangle is a figure consisting of ab unit squares, with edges parallel to Ox and Oy and such that the edge parallel to Ox has length a and the one parallel to Oy has length b. For instance, the rectangle with vertices $(0,0), (2,0), (2,1), (0,1)$ is a 2×1 rectangle, while the rectangle with vertices $(0,0), (1,0), (1,2), (0,2)$ is a 1×2 rectangle. Now, the idea is to put a complex number in each square of a table and then to reformulate the hypothesis and the conclusion of a particular tiling problem in terms of complex numbers. We will see how this technique works better by solving a few actual problems. First, some easy examples.

Example 7. Consider a rectangle that can be tiled by a finite combination of $1 \times m$ and $n \times 1$ rectangles, where m, n are positive integers. Prove that it is possible to tile this rectangle using only $1 \times m$ rectangles or only $n \times 1$ rectangles.

[Gabriel Carrol] BMC Contest 2000

Solution. Let the dimensions of the initial rectangle be $a \times b$, for the positive integers a and b. Now let us partition the rectangle into 1×1 squares and denote these squares by

$$(1,1), (1,2), \ldots, (1,b), \ldots, (a,1), (a,2), \ldots, (a,b).$$

Next, put the number $\varepsilon_1^x \varepsilon_2^y$ in the square labeled (x,y), where

$$\varepsilon_1 = \cos\frac{2\pi}{n} + i\sin\frac{2\pi}{n}, \ \varepsilon_2 = \cos\frac{2\pi}{m} + i\sin\frac{2\pi}{m}.$$

The main observation is that the sum of the numbers in any $1 \times m$ or $n \times 1$ rectangle is 0. This is immediate, but the consequence of this simple observation is really surprising. Indeed, it follows that the sum of the numbers in all the squares is 0, and so

$$0 = \sum_{\substack{1\le x\le a\\ 1\le y\le b}} \varepsilon_1^x \varepsilon_2^y = \sum_{i=1}^{a} \varepsilon_1^i \cdot \sum_{j=1}^{b} \varepsilon_2^j.$$

Hence at least one of the numbers $\sum_{i=1}^{a} \varepsilon_1^i$ and $\sum_{j-1}^{b} \varepsilon_2^j$ is 0. But this means that $n|a$ or $m|b$. In either case, the conclusion of the problem follows.

The idea in the previous problem is quite useful, helping many tiling problems become straightforward. Here is one more example:

Example 8. Can we tile a 13×13 table from which we remove the central unit square using only 1×4 or 4×1 rectangles?

Baltic Contest 1998

Solution. Suppose such a tiling is possible, and label the squares of the table as in the previous problem. Next, associate to square (k, j) the number i^{k+2j}. Clearly, the sum of the numbers from each 1×4 or 4×1 rectangle is 0. Therefore the sum of all labels is equal to the number corresponding to the central unit square. Hence

$$i^{21} = (i + i^2 + \cdots + i^{13})(i^2 + i^4 + \cdots + i^{26}) = i \cdot \frac{i^{13} - 1}{i - 1} \cdot i^2 \cdot \frac{i^{26} - 1}{i^2 - 1} = i^3,$$

which clearly cannot hold. Thus the assumption we made is wrong, and such a tiling is not possible.

The example we are going to discuss now is based on the same idea, and here complex numbers are even more involved.

Example 9. On an 8×9 table we place 3×1 rectangles and "broken" 1×3 rectangles, obtained by removing their central unit square. The rectangles and the "broken" rectangles do not overlap and cannot be rotated. Prove that there exists a set S consisting of 18 squares of the table such that if 70 unit squares of the table are covered, then the remaining two belong to S.

[Gabriel Dospinescu]

Solution. Again, we label the squares of the table $(1,1)$, $(1,2), \ldots,$ $(8,9)$ by starting from the upper left corner. In the square labeled (k,j) we will place the number $i^j \cdot \varepsilon^k$, where $i^2 = -1$ and $\varepsilon^2 + \varepsilon + 1 = 0$. The sum of the numbers from any rectangle or "broken" rectangle is 0. The sum of all numbers is

$$\left(\sum_{k=1}^{8} \varepsilon^k\right)\left(\sum_{j=1}^{9} i^j\right) = -i.$$

Let us suppose that (a_1, b_1) and (a_2, b_2) are the only uncovered squares. Then $i^{b_1}\varepsilon^{a_1} + i^{b_2}\varepsilon^{a_2} = -i$. Let $z_1 = i^{b_1-1}\varepsilon^{a_1}$ and $z_2 = i^{b_2-1}\varepsilon^{a_2}$. We have $|z_1| = |z_2| = 1$ and $z_1 + z_2 = -1$. It follows that $\dfrac{1}{z_1} + \dfrac{1}{z_2} = -1$ and so $z_1^3 = z_2^3 = 1$. This in turn implies the equalities $i^{3(b_1-1)} = i^{3(b_2-1)} = 1$, from which we conclude that $b_1 \equiv b_2 \equiv 1 \pmod 4$. Therefore the relation $z_1 + z_2 = -1$ becomes $\varepsilon^{a_1} + \varepsilon^{a_2} = -1$, which is possible if and only if the remainders of a_1, a_2 when divided by 3 are 1 and 2. Thus we can choose S to be the set of squares that lie at the intersection of the lines 1, 2, 4, 5, 7, 8 with the columns 1, 5, 9. From the above argument, if two squares remain uncovered, then they belong to S. The conclusion is immediate.

Example 10. Let m and n be integers greater than 1 and let $a_1, a_2, \ldots, a_n$ be integers, none of which is divisible by m^{n-1}. Prove that we can find integers $e_1, e_2, \ldots, e_n$, not all zero, such that $|e_i| < m$ for all i and $m^n | e_1a_1 + e_2a_2 + \cdots + e_na_n$.

IMO 2002 Shortlist

Solution. Look at the numbers $\sum_{i=1}^{n} e_i a_i$, where $0 \le e_i \le m-1$ for all i. Observe that we have a collection of m^n numbers (denote this collection by A). We can assume that this is a complete system of residues modulo m^n (otherwise, the conclusion is immediate). Now, consider $f(x) = \sum_{a \in A} x^a$. Then

$$f(x) = \prod_{i=1}^{n} \left(\sum_{j=0}^{m-1} x^{ja_i} \right) = \prod_{i=1}^{n} \frac{1 - x^{ma_i}}{1 - x^{a_i}}.$$

Now take $\varepsilon = e^{\frac{2i\pi}{m^n}}$. Since the m^n numbers we previously considered form a complete system of residues modulo m^n, we must have $f(\varepsilon) = 0$. Therefore (the hypothesis ensures that $\varepsilon^{a_i} \neq 1$) $\prod_{i=1}^{n}(1 - \varepsilon^{ma_i}) = 0$. But this clearly contradicts the fact that none of the numbers $a_1, a_2, \ldots, a_n$ is a multiple of m^{n-1}.

Example 11. Let p be a prime number and let $f_k(x_1, x_2, ..., x_n) = a_{k1}x_1 + a_{k2}x_2 + \cdots + a_{kn}x_n$ be linear forms with integer coefficients for $k = 1, 2, ..., p^n$. Suppose that for all systems of integers $(x_1, x_2, ..., x_n)$, not all divisible by p,

$$f_1(x_1, x_2, ..., x_n), f_2(x_1, x_2, ..., x_n), ..., f_{p^n}(x_1, x_2, ..., x_n)$$

represent every remainder mod p exactly p^{n-1} times. Prove that $\{(a_{k1}, a_{k2}, ..., a_{kn}) | k = 1, 2, ..., p^n\}$ is equal to

$$\{(i_1, i_2, ..., i_n) | i_1, i_2, ..., i_n = 0, 1, ..., p-1\}.$$

Miklos Schweitzer Competition

Solution. Let $\varepsilon = e^{\frac{2i\pi}{p}}$ and observe that the hypothesis implies the identities

$$\sum_{k=1}^{p^n} \varepsilon^{f_k(x_1, \ldots, x_n)} = 0$$

for all $x_1, ..., x_n$, not all multiples of p. Now fix $i_1, i_2, ..., i_n$. By multiplying both sides of the equality by $\varepsilon^{-i_1x_1-i_2x_2-\cdots-i_nx_n}$ we deduce that

$$\sum_{k=1}^{p^n} \varepsilon^{(a_{k1}-i_1)x_1+\cdots+(a_{kn}-i_n)x_n} = 0.$$

By making the sum of all these equalities corresponding to all $(x_1, x_2, ..., x_n) \in \{0, 1, ..., p-1\}^n$ and by taking into account that for $(x_1, x_2, ..., x_n) = (0, 0, ..., 0)$ the left-hand side equals p^n, we deduce that

$$p^n = \sum_{k=1}^{p^n} \prod_{j=1}^{n} (\sum_{x_j=0}^{p-1} \varepsilon^{x_j(a_{kj}-i_j)}). \tag{7.1}$$

Because the sum in the right-hand side of (7.1) is not zero, at least one term is not zero. Observe however that every term of the sum equals 0 or p^n. Therefore there exists an unique k such that $a_{kj} = i_j \pmod p$ for all j. This is just another way of saying that

$$\{(a_{k1}, \ldots, a_{kn}) | k = 1, \ldots, p^n\} = \{(i_1, \ldots, i_n) | i_1, \ldots, i_n = 0, \ldots, p-1\}.$$

The following problem, communicated by Vesselin Dimitrov, is a very special one. It concerns a concept introduced by Erdős in a paper dating back to 1952: the covering systems of congruences. More precisely, the family of ordered pairs $(a_1, d_1), (a_2, d_2), \ldots, (a_k, d_k)$, where $1 < d_1 < d_2 < \cdots < d_k$ is called a covering set of congruences if $x \equiv a_i \pmod{d_i}$ is solvable for any integer x. Erdős immediately realized that this new concept can be a source of difficult questions, and that became source of intensive research. Erdős conjectured that there exists no covering set of congruences in which all the moduli are odd. This remains open. On the other hand, Erdős used covering sets (more precisely, the set $(0, 2), (0, 3), (1, 4), (3, 8), (7, 12), (23, 24)$) to prove the existence of an infinite arithmetic progression of odd positive integers, none of which is of the form $2^n + p$. Schinzel also studied these systems in his researches concerning the irreducibility of polynomials. Conjectured by Erdős, the example that comes next was proved by Sun, and what is really strange is that the solution is absolutely elementary. We thank Vesselin Dimitrov for pointing out this jewel of number theory.

Example 12. Let F be a family of k infinite arithmetic progressions $a_i + d_i\mathbb{Z}$, where $1 < d_1 < \cdots < d_k$. Assume that F covers 2^k consecutive integers (that is, there exists an integer x such that every number in the sequence $x, x+1, ..., x+2^k-1$ belongs to at least one member of the family F). Then F is a covering system of congruences.

[Erdős-Sun]

Solution. The magical idea is to rewrite the condition that a number belongs to a union of arithmetic progressions in a more algebraic way. For instance, the fact that $x+t$ belongs to the union of members of F can be written in the form

$$\prod_{1\leq j\leq k}\left(1-e^{\frac{2i\pi}{d_j}(x+t-a_j)}\right)=0.$$

Now, all we have to do is to develop this product and observe that the same relation can be expressed in the form

$$\sum_{I\subseteq S}\alpha_I\cdot e^{(x+t)\beta_I}=0, \tag{7.2}$$

where $S=\{1,2,\ldots,k\}$ (including the void set, in case of which the term is 0), α_I and β_I depending only on the cover itself. Indeed, this is clear, because

$$\prod_{1\leq j\leq k}(1-e^{\frac{2i\pi}{d_j}(x+t-a_j)})=\sum_{I\subset\{1,2,\ldots,k\}}(-1)^{|I|}\cdot e^{-2i\pi\cdot\sum_{j\in I}\frac{a_j}{d_j}}\cdot e^{2i\pi\cdot\sum_{j\in I}\frac{1}{d_j}(x+t)}.$$

If we manage to prove that the same relation (7.2) holds for any integer y instead of x, we are done, since it would follow that any integer belongs to at least one member of F. If we consider $z_I = e^{\beta_I}$, then we know that $\sum_I \alpha_I z_I^{t+x} = 0$ for all $0 \leq t \leq 2^k - 1$. Define $u_n = \sum_I \alpha_I z_I^n$ and observe that u_n satisfies a linearly recurrent relation of order 2^k, the coefficient of u_n being nonzero. Indeed, consider the polynomial $\prod_I (X - z_I)$, which has degree 2^k and nonzero free term (because all z_I are nonzero), and write it in the form $X^{2^k} + A_{2^k-1}X^{2^k-1} + \cdots + A_1X + A_0$. Then we know that

$$z_I^{2^k} + A_{2^k-1}z_I^{2^k-1} + \cdots + A_0 = 0.$$

By multiplying this relation by $\alpha_I \cdot z_I^n$ (we allow here negative exponents as well) and by adding up these relations, we obtain a recurrence relation

$$u_{n+2^k} + A_{2^k-1}u_{n+2^k-1} + \cdots + A_0 = 0.$$

And now... we are done: from the hypothesis, 2^k consecutive terms of this sequence vanish. Since the sequence satisfies a recurrence relation of order 2^k with nonzero free term, it follows by a trivial induction that all terms are zero. This finishes the proof.

7.2 Practice Problems

1. Three persons A, B, C play the following game: a subset with k elements of the set $\{1, 2, \dots, 1986\}$ is selected randomly, all selections having the same probability. The winner is A, B, or C, according to whether the sum of the elements of the selected subset is congruent to 0, 1, or 2 modulo 3. Find all values of k for which A, B, C have equal chances of winning.

   IMO 1987 Shortlist

2. We roll a regular die n times. What is the probability that the sum of the numbers shown is a multiple of 5?

   IMC 1999

3. Let a_k, b_k, c_k be integers, $k = 1, 2, \dots, n$ and let $f(x)$ be the number of ordered triples (A, B, C) of subsets (not necessarily nonempty) of the set $S = \{1, 2, \dots, n\}$ whose union is S and for which

$$\sum_{i \in S \setminus A} a_i + \sum_{i \in S \setminus B} b_i + \sum_{i \in S \setminus C} c_i \equiv x \pmod{3}.$$

   Suppose that $f(0) = f(1) = f(2)$. Prove that there exists $i \in S$ such that $3 \mid a_i + b_i + c_i$.

   Gabriel Dospinescu

4. Let k be an integer greater than 2. For which odd positive integers n can we tile a $n \times n$ table by $1 \times k$ or $k \times 1$ rectangles such that only the central unit square is uncovered?

   Gabriel Dospinescu

5. Let $n \geq 2$ be an integer. At each point (i, j) having integer coordinates we write the number $i + j \pmod{n}$. Find all pairs (a, b) of positive integers such that any residue modulo n appears the same number of times on the sides of the rectangle with vertices $(0,0)$, $(a,0)$, (a,b), $(0,b)$ and also any residue modulo n appears the same number of times in the interior of this rectangle.

Bulgaria 2001

6. Let $p > 2$ be a prime. How many subsets of $\{1, 2, \ldots, p-1\}$ have the sum of their elements divisible by p?

Ivan Landjev, Bulgaria TST 2006

7. Prove that the number of subsets with n elements of $\{1, 2, \ldots, 2n\}$ whose sum is a multiple of n is

$$\frac{(-1)^n}{n} \cdot \sum_{d|n} (-1)^d \varphi\left(\frac{n}{d}\right) \binom{2d}{d}.$$

Adapted after IMO 1995

8. Let p be an odd prime. Find the number of 6-tuples (a, b, c, d, e, f) of integers between 0 and $p - 1$ such that

$$a^2 + b^2 + c^2 \equiv d^2 + e^2 + f^2 \pmod{p}.$$

MOSP 1997

9. Let d and n be positive integers such that $d \mid n$. Consider all sequences $(x_1, x_2, \ldots, x_n)$ such that $0 \leq x_1 \leq x_2 \leq \cdots \leq x_n \leq n$ and $d \mid x_1 + x_2 + \cdots + x_n$. Prove that among these sequences, exactly half satisfy $x_n = n$.

Chinese IMO training program

10. Let p be an odd prime and $n \geq 2$. For a permutation σ of the set $\{1, 2, \ldots, n\}$ define

$$S(\sigma) = \sigma(1) + 2\sigma(2) + \cdots + n\sigma(n).$$

Let A_j be the set of even permutations σ such that $S(\sigma) \equiv j \pmod{p}$ and B_j be the set of odd permutations σ for which $S(\sigma) \equiv j \pmod{p}$. Prove that $n > p$ if and only if A_j and B_j have the same number of elements for all j.

Gabriel Dospinescu

11. Let p be an odd prime. Prove that the $2^{\frac{p-1}{2}}$ numbers $\pm 1 \pm 2 \pm \cdots \pm \frac{p-1}{2}$ represent each nonzero residue class mod p the same number of times. Compute this number.

R. L. McFarland, AMM 6457

12. Each element of the set $M = \{1, 2, \ldots, n\}$ is colored in one of three colors. Let A be the set of triples (x, y, z) of elements of M such that n divides $x + y + z$ and x, y, z have the same color. Define B similarly, by asking that x, y, z have pairwise distinct colors. Prove that $2|A| \geq |B|$.

Chinese TST 2010

13. Color the numbers $1, 2, \ldots, N$ using 3 colors such that there are at most $\frac{N}{2}$ numbers of each color. Let A be the set of 4-tuples $(a, b, c, d) \in \{1, 2, \ldots, n\}^4$ such that $a + b + c + d = 0 \pmod{N}$ and a, b, c, d have the same color. Let B be the set of 4-tuples $(a, b, c, d) \in \{1, 2, \ldots, n\}^4$ such that $a + b + c + d = 0 \pmod{N}$, a, b and c, d have the same color, but these colors are distinct. Prove that $|A| \leq |B|$.

Komal

14. (a) Let n be an odd integer. Find the number of sequences $(a_0, a_1, \ldots, a_n)$ such that $a_i \in \{1, 2, \ldots, n\}$ for all i, $a_n = a_0$ and $a_i - a_{i-1} \not\equiv i \pmod{n}$ for all $i = 1, 2, \ldots, n$.

    (b) Let n be an odd prime. Find the number of sequences $(a_0, a_1, ..., a_n)$ such that $a_i \in \{1, 2, \ldots, n\}$ for all i, $a_n = a_0$ and $a_i - a_{i-1} \not\equiv i, 2i \pmod{n}$ for all $i = 1, 2, \ldots, n$.

USA TST 2004

15. Let $p > 3$ be a prime number and let $f(X)$ be the number of sequences $(a_1, a_2, \ldots, a_{p-1})$ such that $a_j \in X$ for all j and p divides $\sum_{j=1}^{p-1} ja_j$. Here X is a nonempty subset of $\{0, 1, \ldots, p-1\}$. Prove that $f(\{0, 1, 3\}) \geq f(\{0, 1, 2\})$, with equality if and only if $p = 5$.

IMO 1999 Shortlist

16. Is there a positive integer k such that $p = 6k + 1$ is a prime and

$$\binom{3k}{k} \equiv 1 \pmod{p}?$$

USA TST 2010

17. Let p be an odd prime and let a, b, c, d be integers not divisible by p such that

$$\left\{\frac{ra}{p}\right\} + \left\{\frac{rb}{p}\right\} + \left\{\frac{rc}{p}\right\} + \left\{\frac{rd}{p}\right\} = 2$$

for all integers r not divisible by p (here $\{\}$ is the fractional part). Prove that at least two of the numbers $a+b, a+c, a+d, b+c, b+d, c+d$ are divisible by p.

Kiran Kedlaya, USAMO 1999

18. Let p be a prime and let $S \subset (\mathbb{Z}/p\mathbb{Z})^d$ be a subset containing no line of the affine space $(\mathbb{Z}/p\mathbb{Z})^d$. Prove that S has at most $\frac{2\cdot 3^d}{d}$ elements. However, prove that we can find such a set with at least $3^{\frac{2d}{3}-1}$ elements.

Meshulam-Roth theorem

Chapter

8

8.1 Theory and examples

We start with a riddle and a challenge: what is the connection between the following problems?
1. The set of nonnegative integers is partitioned into $n \geq 1$ infinite arithmetical sequences with common differences $r_1, r_2, \ldots, r_n$ and first terms $a_1, a_2, \ldots, a_n$. Then

$$\frac{a_1}{r_1} + \frac{a_2}{r_2} + \cdots + \frac{a_n}{r_n} = \frac{n-1}{2}.$$

2. The vertices of a regular polygon are colored such that each set of vertices having the same color is the set of vertices of a regular polygon. Prove that there are two congruent polygons among them.

The first problem was discussed during the preparation of the USA IMO team, but it seems to be a classical result. As for the second one, well, it is a famous problem given at a Russian Olympiad, proposed by N. Vasiliev. If you have no clue, then we will give you a small hint: the methods used to solve both problems are very similar and can be included into a larger field, that of formal series. What are those? Well, given a commutative ring A, we can define another ring, called the ring of formal series with coefficients in A and denoted $A[[X]]$. An element of $A[[X]]$ is of the form $\sum\limits_{n\geq 0} a_n X^n$, where $a_n \in A$, and it is also called the generating function of the sequence $(a_n)_{n\geq 0}$. The addition and multiplication are the natural ones, defined as the similar operations with polynomials:

$$\left(\sum_{n\geq 0} a_n X^n\right) + \left(\sum_{n\geq 0} b_n X^n\right) = \sum_{n\geq 0} (a_n + b_n) X^n$$

and

$$\left(\sum_{n\geq 0} a_n X^n\right) \cdot \left(\sum_{n\geq 0} b_n X^n\right) = \sum_{n\geq 0} c_n X^n,$$

where $c_n = \sum\limits_{p+q=n} a_p b_q$. Yet, for an entire function $g(z) = \sum\limits_{n\geq 0} g_n z^n$ and a formal series $f(X) = \sum\limits_{n\geq 0} a_n X^n$ we can define the formal series $g(f(X)) = \sum\limits_{n\geq 0} b_n X^n$

obtained "formally" from the formula $g(f(X)) = \sum_{n\geq 0} g_n f^n(X)$ by developing $f^n(X)$ and grouping terms according to successive powers of X. You can (and you should, if it is the first time you encounter this object) easily prove that all formulae of the type $e^f \cdot e^g = e^{f+g}$, $\sin(f+g) = \sin(f)\cos(g) + \sin(g)\cos(f)$ and so on are valid in the ring of formal series. Also, one can define a derivative on this ring, similarly defined as the usual derivative of polynomials, by $f'(X) = \sum_{n\geq 1} na_n X^{n-1}$ and check that all the properties that the derivative has on the space of polynomials are preserved. Actually, all operations that are allowed on polynomials can be transferred formally to the ring of power series, and preserve their properties, as long as they are expressed purely in terms of the coefficients (this excludes of course speaking about zeros of a formal series). As we will see in what follows, formal series have some very nice applications in different fields: algebra, combinatorics, and number theory. But let's start working now, assuming familiarity with some basic analysis tools. We warn the reader that from time to time we will insist on some questions of convergence or continuity, but at other times we will work only in this ring of formal series, therefore adopting only the operations of this ring, with no further reference to questions of convergence.

Example 1. Let $a_1, \dots, a_n$ be complex numbers such that $a_1^k + \cdots + a_n^k = 0$ for all $1 \leq k \leq n$. Then all numbers are equal to 0.

Solution. The experienced reader has already noticed that this problem is an immediate consequence of Newton's relations. But what can we do if we are not familiar with these relations? Here is a nice way to solve the problem (and a way to prove Newton's relations, too). First of all, observe that the given condition implies

$$a_1^k + a_2^k + \cdots + a_n^k = 0$$

for all positive integers k. Indeed, let

$$f(X) = X^n + b_{n-1}X^{n-1} + \cdots + b_1 X + b_0 = \prod_{i=1}^{n}(X - a_i).$$

Then

$$a_i^k + b_{n-1}a_i^{k-1} + \cdots + b_0 a_i^{k-n} = 0$$

for all $k \geq n+1$. It suffices to add these relations and to prove the statement by strong induction. Now, let us consider the function

$$f(z) = \sum_{i=1}^{n} \frac{1}{1 - za_i}.$$

Developing it by using

$$\frac{1}{1-x} = 1 + x + x^2 + \cdots \quad (\text{for } |x| < 1),$$

we obtain that $f(z) = n$ for all sufficiently small z (meaning for such z that satisfy $|z| \max_{1 \leq i \leq n}\{|a_i|\} < 1$). Assume that not all numbers are zero and take $a_1, \ldots, a_s$ $(1 \leq s \leq n)$ to be the collection of numbers of maximal absolute value among the n numbers and let this maximal absolute value be r. By taking a sequence $z_p \to \dfrac{1}{r}$ such that $|z_p \cdot r| < 1$, we obtain a contradiction with the relation $\displaystyle\sum_{i=1}^{n} \frac{1}{1 - z_p a_i} = n$ (indeed, it suffices to observe that the left-hand side is unbounded, while the right one is bounded). This shows that all numbers are equal to 0.

We are going to discuss a nice number theory problem whose solution is practically based on the same idea. This result is an important step in proving that the order of any finite subgroup of $GL_n(\mathbb{Z})$ divides $(2n)!$. Indeed, it is not difficult to prove that if G is a finite subgroup of $GL_n(\mathbb{Z})$ then $|G|$ divides $\sum_{g \in G} \operatorname{Tr}(g)$ (all you need is to note that $\frac{1}{|G|} \sum_{g \in G} g$ is idempotent, which is an immediate consequence of the fact that in a finite group the translations are actually permutations; or, the trace of an idempotent matrix is just its rank, and thus an integer). Working with the tensorial product matrices $A \otimes A$ where $A \in G$ and repeating the above argument yields $|G| \mid \sum_{g \in G} (\operatorname{Tr}(g))^k$ for all $k \geq 0$. Now, all we need is to apply the result below in order to conclude that

$$|G| \mid (n - \operatorname{Tr}(g_1))(n - \operatorname{Tr}(g_2)) \cdots (n - \operatorname{Tr}(g_s)),$$

where $\mathrm{Tr}(g_1), \mathrm{Tr}(g_2), ..., \mathrm{Tr}(g_s)$ are the distinct traces that appear in the list $(\mathrm{Tr}(g))_{g\in G, g\neq I_n}$. Because $n - \mathrm{Tr}(g_i)$ are distinct integers between 1 and $2n$, it follows that $|G|$ divides $(2n)!$.

Example 2. Let $a_1, a_2, \ldots, a_q, x_1, x_2, \ldots, x_q$ and m be integers such that $m|a_1x_1^k + a_2x_2^k + \cdots + a_qx_q^k$ for all $k \geq 0$. Then

$$m|a_1\prod_{i=2}^{q}(x_1 - x_i).$$

Solution. Consider this time the formal series

$$f(z) = \sum_{i=1}^{q}\frac{a_i}{1 - zx_i}.$$

By using the same formula as in the first problem, we obtain

$$f(z) = \sum_{i=1}^{q} a_i + \left(\sum_{i=1}^{q} a_ix_i\right) z + \cdots,$$

which shows that all coefficients of this formal series are integers divisible by m. It follows that the formal series

$$\sum_{i=1}^{q} a_i\prod_{j\neq i}(1 - x_jz)$$

also has all of its coefficients divisible by m. Now consider $S_t^{(i)}$, the t-th fundamental symmetric sum in x_j $(j \neq i)$. Because all coefficients of $\sum_{i=1}^{q} a_i\prod_{j\neq i}(1-x_jz)$ are multiples of m, a simple computation shows that we have the divisibility relation

$$m|x_1^{q-1}\sum_{i=1}^{q} a_i - x_1^{q-2}\sum_{i=1}^{q} a_iS_1^{(i)} + \cdots + (-1)^{q-1}\sum_{i=1}^{q} a_iS_{q-1}^{(i)}.$$

This can also be rewritten as

$$m \mid \sum_{i=1}^{q} a_i(x_1^{q-1} - x_1^{q-2}S_1^{(i)} + \cdots + (-1)^{q-1}S_{q-1}^{(i)}).$$

Now, the trivial identity

$$(x_1 - x_1)\dots(x_1 - x_{i-1})(x_1 - x_{i+1})\dots(x_1 - x_n) = 0$$

gives us the not-so obvious relation

$$x_1^{q-1} - x_1^{q-2}S_1^{(i)} + \cdots + (-1)^{q-1}S_{q-1}^{(i)} = 0$$

for $i \geq 2$. Therefore

$$x_1^{q-1} - x_1^{q-2}S_1^{(1)} + \cdots + (-1)^{q-1}S_{q-1}^{(1)} = (x_1 - x_2)\dots(x_1 - x_n)$$

and we are done.

In order to solve the problem announced at the very beginning of the presentation, we need a lemma, which is interesting itself, and which we prefer to present as a separate problem.

Example 3. Suppose that the set of nonnegative integers is partitioned into a finite number of infinite arithmetical progressions with common differences $r_1, r_2, \dots, r_n$ and first terms $a_1, a_2, \dots, a_n$. Then

$$\frac{1}{r_1} + \frac{1}{r_2} + \cdots + \frac{1}{r_n} = 1.$$

Solution. Let us observe that for any $|x| < 1$ we have the identity:

$$\sum_{k\geq 0} x^{a_1+kr_1} + \sum_{k\geq 0} x^{a_2+kr_2} + \cdots + \sum_{k\geq 0} x^{a_n+kr_n} = \sum_{k\geq 0} x^k.$$

Indeed, all we did was to write the fact that each nonnegative integer is in exactly one of the arithmetical sequences. The above relation becomes:

$$\frac{x^{a_1}}{1-x^{r_1}} + \frac{x^{a_2}}{1-x^{r_2}} + \cdots + \frac{x^{a_n}}{1-x^{r_n}} = \frac{1}{1-x} \tag{8.1}$$

Let us multiply (8.1) by $1-x$ and use the fact that $\lim_{x\to 1} \frac{1-x^a}{1-x} = a$. We find the desired relation

$$\frac{1}{r_1} + \frac{1}{r_2} + \cdots + \frac{1}{r_n} = 1.$$

It is now time to solve the first problem. We will just take a small, but far from obvious, step and we'll be done. The fundamental relation is again (8.1).

Example 4. The set of nonnegative integers is partitioned into $n \geq 1$ infinite arithmetical progressions with common differences $r_1, \ldots, r_n$ and first terms $a_1, a_2, \ldots, a_n$. Then

$$\frac{a_1}{r_1} + \frac{a_2}{r_2} + \cdots + \frac{a_n}{r_n} = \frac{n-1}{2}.$$

MOSP

Solution. Let us write the relation (8.1) in the more appropriate form:

$$\frac{x^{a_1}}{1+x+\cdots+x^{r_1-1}} + \cdots + \frac{x^{a_n}}{1+x+\cdots+x^{r_n-1}} = 1 \qquad (8.2)$$

Now, let us differentiate (8.2) and then make $x \to 1$ in the resulting expression. An easy computation, which is left to the reader, shows that

$$\sum_{i=1}^{n} \frac{a_i r_i - \frac{r_i(r_i-1)}{2}}{r_i^2} = 0.$$

It suffices now to use the result proved in example 3 in order to conclude that

$$\frac{a_1}{r_1} + \frac{a_2}{r_2} + \cdots + \frac{a_n}{r_n} = \frac{n-1}{2}.$$

Some comments about these two relations are necessary. First of all, using a beautiful and difficult result due to Erdős, we can say that the relation

$$\frac{1}{r_1}+\frac{1}{r_2}+\cdots+\frac{1}{r_n}=1$$

implies that $\max(r_1, r_2, \ldots, r_n) < 2^{2^{n-1}}$. Indeed, this remarkable theorem due to Erdős asserts that if $x_1, x_2, \ldots, x_k$ are positive integers whose sum of reciprocals is less than 1, then

$$\frac{1}{x_1}+\frac{1}{x_2}+\cdots+\frac{1}{x_k}\leq\frac{1}{u_1}+\frac{1}{u_2}+\cdots+\frac{1}{u_k},$$

where $u_1 = 2$, $u_{n+1} = u_n^2 - u_n + 1$. But the reader can verify immediately by induction that

$$\frac{1}{u_1}+\frac{1}{u_2}+\cdots+\frac{1}{u_k}=1-\frac{1}{u_1u_2\ldots u_k}.$$

Thus we can write

$$1-\frac{1}{r_n}\leq 1-\frac{1}{u_1u_2\ldots u_{n-1}},$$

or, even better, $r_n \leq u_1u_2\cdots u_{n-1} = u_n - 1$ (the last relation following again by a simple induction). Another inductive argument proves that $u_n \leq 2^{2^{n-1}}$. Hence $\max(r_1, r_2, \ldots, r_n) < 2^{2^{n-1}}$. Using the relation proved in example 4, we also deduce that

$$\max(a_1, a_2, \ldots, a_n) < (n-1)\cdot 2^{2^{n-1}-1}.$$

This shows that for fixed n not only is there a finite number of ways to partition the set of positive integers into n arithmetical progressions, but we also have some explicit (even though huge) bound on the common differences and first terms.

It is now time to solve the remarkable problem discussed at the beginning of this chapter. We will see that using the previous results proved here, the solution becomes natural. However, the problem is still really difficult.

Example 5. The vertices of a regular polygon are colored in such a way that each set of vertices having the same color is the set of vertices of a regular polygon. Prove that there are two congruent polygons among them.

[N. Vasiliev] Russian Olympiad

Solution. Let us assume that the initial polygon (which we will call big from now on) has n edges, and that it is inscribed in the unit circle, the vertices having as coordinates the numbers $1, \varepsilon, \varepsilon^2, \dots, \varepsilon^{n-1}$, where $\varepsilon = e^{\frac{2i\pi}{n}}$ (of course, we will not lose generality with all these restrictions). Let $n_1, n_2, \dots, n_k$ be the number of edges of the monochromatic polygons, and assume that all these numbers are distinct. Let $\varepsilon_j = e^{\frac{2i\pi}{n_j}}$ and observe that the coordinates of the vertices of each monochromatic polygon are $z_j, z_j\varepsilon_j, \dots, z_j\varepsilon_j^{n_j-1}$, for some complex numbers z_j on the unit circle. First, a technical result.

Lemma 8.1. *For any complex number z and $\zeta = e^{\frac{2i\pi}{p}}$ we have the identity*

$$\frac{1}{1-z} + \frac{1}{1-z\zeta} + \cdots + \frac{1}{1-z\zeta^{p-1}} = \frac{p}{1-z^p}.$$

Proof. Proving this lemma is a simple task. Indeed, it suffices to observe that $z, z\zeta, \dots, z\zeta^{p-1}$ are exactly the zeros of $P(X) = X^p - z^p$. Or, observe that

$$\frac{P'(X)}{P(X)} = \frac{1}{X-z} + \cdots + \frac{1}{X-z\zeta^{p-1}},$$

thus by taking $X = 1$ we obtain the desired result.

Now, the hypothesis of the problem and lemma allow us to write

$$\frac{n_1}{1-(zz_1)^{n_1}} + \frac{n_2}{1-(zz_2)^{n_2}} \cdots + \frac{n_k}{1-(zz_k)^{n_k}} = \frac{n}{1-z^n}.$$

Also, the simple observation $n_1 + n_2 + \cdots + n_k = n$ yields the new identity

$$\frac{n_1 z_1^{n_1}}{1-(zz_1)^{n_1}} z^{n_1} + \frac{n_2 z_2^{n_2}}{1-(zz_2)^{n_2}} z^{n_2} + \cdots + \frac{n_k z_k^{n_k}}{1-(zz_k)^{n_k}} z^{n_k} = \frac{nz^n}{1-z^n}. \quad (1)$$

Let us assume now that $n_1 < \min(n_2, \ldots, n_k)$ and divide (1) by z^{n_1}. It follows that for any nonzero z we have

$$\frac{n_1 z_1^{n_1}}{1-(zz_1)^{n_1}} + \frac{n_2 z_2^{n_2}}{1-(zz_2)^{n_2}} z^{n_2-n_1} + \cdots + \frac{n_k z_k^{n_k}}{1-(zz_k)^{n_k}} z^{n_k-n_1} = \frac{nz^{n-n_1}}{1-z^n}. \quad (2)$$

We are done: it suffices to observe that if we make $z \to 0$ (by nonzero values) in (2), we obtain $z_1^{n_1} = 0$, which is clearly impossible, since $|z_1| = 1$. The proof ends here. □

The problem that we are going to discuss now has appeared in various contests in different forms. It is a very nice identity that can be proved in quite messy but elementary ways. Here is a magical proof using formal series.

Example 6. For any complex numbers $a_1, a_2, \ldots, a_n$ the following identity holds:

$$\left(\sum_{i=1}^n a_i\right)^n - \sum_{i=1}^n \left(\sum_{j\neq i} a_j\right)^n$$

$$+ \sum_{1\le i<j\le n} \left(\sum_{k\neq i,j} a_k\right)^n - \cdots + (-1)^{n-1}\sum_{i=1}^n a_i^n = n!\prod_{i=1}^n a_i.$$

Solution. Consider the formal series

$$f(z) = \prod_{i=1}^n (e^{za_i} - 1).$$

We are going to compute it in two different ways. First of all, it is clear that

$$f(z) = \prod_{i=1}^n \left(za_i + \frac{z^2a_i^2}{2!} + \ldots\right),$$

hence the coefficient of z^n is $\prod_{i=1}^{n} a_i$. On the other hand, we can write

$$f(z) = e^{z\sum_{i=1}^{n} a_i} - \sum_{i=1}^{n} e^{z\sum_{j\neq i} a_j} + \cdots + (-1)^{n-1}\sum_{i=1}^{n} e^{za_i} + (-1)^n.$$

Indeed, you are right: everything is now clear, since the coefficient of z^n in e^{kz} is $\frac{k^n}{n!}$. The conclusion follows.

Here are two applications of this formula. The first one is a recent Putnam problem (2004), which asked competitors to prove that for any n there exists N and some rational numbers $c_1, c_2, ..., c_N$ such that

$$x_1x_2\cdots x_n = \sum_{i=1}^{N} c_i(a_{i1}x_1 + a_{i2}x_2 + \cdots + a_{in}x_n)^n$$

holds identically in complex variables $x_1, x_2, ..., x_n$, and a_{ij} are equal to $-1, 0$ or 1. It is clear that the above identity furnishes an answer $N = 2^n$ where we have even more, $a_{ij} \in \{0, 1\}$. Some twenty years before the Putnam Competition, the following problem was proposed at the Saint Petersburg Olympiad: A calculator can perform the following: add or subtract two numbers, divide any number by any nonzero integer and raise any number to the tenth power. Prove that using this calculator one can compute the product of any ten numbers. As you can immediately see, the solution follows by the above identity. Without using it, it is really difficult to solve this problem.

Not only algebra problems can be solved in an elegant manner using formal series, but also some beautiful number theory and combinatorics problems. We shall focus a little more on each type of problem in the sequel.

Example 7. Let $0 = a_0 < a_1 < a_2 < \ldots$ be a sequence of nonnegative integers such that for all n the equation $a_i + 2a_j + 4a_k = n$ has a unique solution (i, j, k). Find a_{1998}.

IMO 1998 Shortlist

Solution. Here is the very nice answer: 9817030729. Let $A = \{a_0, a_1, \dots\}$ and let $b_n = 1$ if $n \in A$ and 0 otherwise. Next, consider the formal series $f(x) = \sum_{n\geq 0} b_n x^n$, the generating function of the set A (we can write it in a more intuitive way $f(x) = \sum_{n\geq 0} x^{a_n}$). The hypothesis imposed on the set A translates into

$$f(x)f(x^2)f(x^4) = \frac{1}{1-x}.$$

Replace x by x^{2^k}. We obtain the recursive relation

$$f(x^{2^k})f(x^{2^{k+1}})f(x^{2^{k+2}}) = \frac{1}{1-x^{2^k}}.$$

Now, observe that

$$\prod_{k\geq 0} f(x^{2^k}) = \prod_{k\geq 0}\left(f(x^{2^{3k}})f(x^{2^{3k+1}})f(x^{2^{3k+2}})\right) = \prod_{k\geq 0}\frac{1}{1-x^{2^{3k}}}$$

and

$$\prod_{k\geq 1} f(x^{2^k}) = \prod_{k\geq 0}\left(f(x^{2^{3k+1}})f(x^{2^{3k+2}})f(x^{2^{3k+3}})\right) = \prod_{k\geq 0}\frac{1}{1-x^{2^{3k+1}}}.$$

Therefore (you have observed that rigor was not the strong point in establishing these relations),

$$f(x)=\prod_{k\geq 0}\frac{1-x^{2^{3k+1}}}{1-x^{2^{3k}}} = \prod_{k\geq 0}(1+x^{8^k})$$

This shows that the set A is exactly the set of nonnegative integers that use only the digits 0 and 1 when written in base 8. A quick computation based on this observation shows that the magical term asked for by the problem is 9817030729.

The following problem is an absolute classic. It has appeared under different forms in Olympiads from all over the world. We will present the latest one, given at the 2003 Putnam Competition:

Example 8. Find all partitions with two classes A, B of the set of nonnegative integers having the property that for all nonnegative integers n the equation $x + y = n$ with $x < y$ has as many solutions $(x, y) \in A \times A$ as in $B \times B$.

Solution. Let f and g be the generating functions of A and B respectively. Then

$$f(x) = \sum_{n \geq 0} a_n x^n, \quad g(x) = \sum_{n \geq 0} b_n x^n$$

where, as in the previous problem, a_n equals 1 if $n \in A$ and 0 otherwise. The fact that A and B form a partition of the set of nonnegative integers can be also rewritten as

$$f(x) + g(x) = \sum_{n \geq 0} x^n = \frac{1}{1-x}.$$

Also, the hypothesis on the number of solutions of the equation $x + y = n$ implies that

$$f^2(x) - f(x^2) = g^2(x) - g(x^2).$$

Hence

$$f(x^2) - g(x^2) = \frac{f(x) - g(x)}{1-x},$$

which can be rewritten as

$$\frac{f(x) - g(x)}{f(x^2) - g(x^2)} = 1 - x.$$

Now, the idea is the same as in the previous problems: replace x by x^{2^k} and iterate. After multiplication, we deduce that

$$f(x) - g(x) = \prod_{k \geq 0} (1 - x^{2^k}) \lim_{n \to \infty} (f(x^{2^n}) - g(x^{2^n})).$$

Let us assume without loss of generality that $0 \in A$. You can easily verify that

$$\lim_{n \to \infty} f(x^{2^n}) = 1 \text{ and } \lim_{n \to \infty} g(x^{2^n}) = 0,$$

which follows from the observation that $1 \le f(x) \le 1 + \frac{x}{1-x}$ and $0 \le g(x) \le \frac{x}{1-x}$ if $0 < x < 1$.
This shows that actually

$$f(x) - g(x) = \prod_{k\ge 0}(1 - x^{2^k}) = \sum_{k\ge 0}(-1)^{s_2(k)}x^k,$$

where $s_2(x)$ is the sum of the digits in the binary representation of x. Taking into account the relation

$$f(x) + g(x) = \frac{1}{1-x},$$

we finally deduce that A and B are respectively the set of nonnegative integers having even (respectively odd) sum of digits when written in base 2.

We will discuss a nice problem in which formal series and complex numbers appear in a quite spectacular way:

Example 9. Let n and k be positive integers such that $n \ge 2^{k-1}$ and let $S = \{1, 2, \dots, n\}$. Prove that the number of subsets A of S for which $\sum_{x\in A} x \equiv m \pmod{2^k}$ does not depend on $m \in \{0, 1, \dots, 2^k - 1\}$.

Solution. Let us consider the function (call it formal series, if you want):

$$f(x) = \prod_{i=1}^{n}(1 + x^i).$$

If we prove that $1 + x + \cdots + x^{2^k-1}$ divides $f(x)$, then we have certainly done the job. In order to prove this, it suffices to prove that any 2^kth root of unity, except for 1, is a root of f. But it suffices to observe that for any $l \in \{1, 2, \dots, 2^{k-1} - 1\}$ we have

$$\left(\cos\frac{2l\pi}{2^k} + i\sin\frac{2l\pi}{2^k}\right)^{2^{k-1-v_2(l)}} = -1$$

and so

$$f\left(\cos\frac{2l\pi}{2^k} + i\sin\frac{2l\pi}{2^k}\right) = 0,$$

which settles our claim.

Finally, it is time for a tough problem, solved by Constantin Tănăsescu.

Example 10. Let S be the set of all words which can be formed using $m \geq 1$ given letters. For any $w \in S$, let $l(w)$ be its length. Also, let $W \subseteq S$ be a set of words. We know that any word in S can be obtained in at most one way by concatenating words from W. Prove that

$$\sum_{w\in W} \frac{1}{m^{l(w)}} \leq 1.$$

[Adrian Zahariuc]

Solution. Let A be the set of all words which can be obtained by concatenating words from W. Let

$$f(x) = \sum_{w\in W} x^{l(w)}, \quad g(x) = \sum_{w\in A} x^{l(w)}.$$

By the definition of A,

$$g(x) = 1 + f(x) + f^2(x) + \cdots = \frac{1}{1 - f(x)}.$$

Hence

$$f(x)g(x) = g(x) - 1. \tag{8.3}$$

Now, A (and W) has at most m^k elements of length k, thus $g(x) < \infty$ and $f(x) < \infty$ for $x < \dfrac{1}{m}$. Thus for all $x \in \left(0, \dfrac{1}{m}\right)$ the expression in (8.3) is less than $g(x)$ and so $f(x) < 1$ for all $x \in \left(0, \frac{1}{m}\right)$. All we need now is to make x tend to $\frac{1}{m}$ and we will obtain $f\left(\frac{1}{m}\right) \leq 1$, which is precisely the desired inequality. Indeed, observe that f can be written as $f(x) = \sum_{n\geq 0} a_n x^n$ for some nonnegative real numbers a_n. Fix a positive integer N. Because

$$\sum_{k=0}^{N} a_k x^k \leq f(x) \leq 1,$$

for all $0 < x < \frac{1}{m}$, by continuity of the polynomials it follows that $\sum_{k=0}^{N} \frac{a_k}{m^k} \leq 1$, and because N is arbitrary, we have $\sum_{k\geq 0} \frac{a_k}{m^k} \leq 1$, that is $f\left(\frac{1}{m}\right) \leq 1$.

There is a very short solution for the following result using group theory. However, this is not the natural approach. The following solution may seem very involved and technical, but it was written in order to convince the reader that from time to time we need to work with composition of formal series, not merely with their sum and product.

Example 11. Let $c(\sigma)$ be the number of cycles (including those of length 1) in the decomposition of σ into disjoint cycles. Prove that

$$\frac{1}{m!} \cdot \sum_{\sigma \in S_m} n^{c(\sigma)} = \binom{m+n-1}{m},$$

where S_m is the set of permutations of the set $\{1, 2, ..., m\}$.

[Marvin Marcus] AMM 5751

Solution. Let us start with a Lemma:

Lemma 8.2. *For given nonnegative integers $k_1, k_2, \ldots, k_n$ such that $k_1 + 2k_2 + \cdots + nk_n = n$, the number of permutations of $\{1, 2, \ldots, n\}$ which have k_i cycles of length i for all i is*

$$\frac{n!}{k_1! k_2! \cdots k_n! 1^{k_1} 2^{k_2} \cdots n^{k_n}}.$$

Proof. Indeed, there are $n!$ ways to fill in the elements of all cycles, but observe that every cycle of length j can be rotated around j ways and be the same cycle (so we must divide $n!$ by j^{k_j}) and also there are $k_j!$ ways to permute the cycles of length j in order to obtain the same permutation. All these operations being independent, the statement of the lemma follows. $\square$

Thus the sum we need to evaluate is

$$\sum_{k_1+2k_2+\cdots+mk_m=m} \frac{m!}{1^{k_1}2^{k_2}\cdots m^{k_m}k_1!k_2!\cdots k_m!} n^{k_1+k_2+\cdots+k_m}.$$

You will probably say that this is much more difficult than the initial problem, but you are not right, because the latter sum can also be written as

$$m! \cdot \sum_p \frac{1}{p!} \cdot \sum_{\substack{k_1+2k_2+\cdots+mk_m=m \\ k_1+k_2+\cdots+k_m=p}} \frac{p!}{k_1!k_2!\cdots k_m!} \cdot \left(\frac{n}{1}\right)^{k_1} \cdot \left(\frac{n}{2}\right)^{k_2} \cdots \left(\frac{n}{m}\right)^{k_m}.$$

Now, observe that the multinomial formula implies that

$$\sum_{\substack{k_1+2k_2+\cdots+mk_m=m \\ k_1+k_2+\cdots+k_m=p}} \frac{p!}{k_1!k_2!\cdots k_m!} \cdot \left(\frac{n}{1}\right)^{k_1} \cdot \left(\frac{n}{2}\right)^{k_2} \cdots \left(\frac{n}{m}\right)^{k_m}$$

is the coefficient of X^m in the formal series $\left(\frac{nX}{1} + \frac{nX^2}{2} + \cdots + \frac{nX^m}{m} + \cdots\right)^p$. Therefore the sum to be evaluated is the coefficient of X^m in the formal series

$$m! \cdot \sum_p \frac{1}{p!}\left(\frac{nX}{1} + \frac{nX^2}{2} + \cdots + \frac{nX^m}{m} + \cdots\right)^p = m! \cdot e^{\frac{nX}{1}+\frac{nX^2}{2}+\cdots+\frac{nX^m}{m}+\cdots}.$$

Finally, observe that $\frac{nX}{1} + \frac{nX^2}{2} + \cdots + \frac{nX^m}{m} + \cdots = -n\ln(1-X)$, so

$$e^{\frac{nX}{1}+\frac{nX^2}{2}+\cdots+\frac{nX^m}{m}+\cdots} = \frac{1}{(1-X)^n}.$$

But using the binomial formula for $(1-x)^{-n}$ we easily find the coefficient of X^m in $\frac{1}{(1-x)^n}$ to be $\binom{n+m-1}{m}$. This finishes the solution.

We should also mention the beautiful solution using group theory. Remember that when a group G is acting on a set Y (that is, we can define for all $g \in G$ and $x \in Y$ an element $g \cdot x \in Y$ such that for all g, h, x we have $g \cdot (h \cdot x) = (gh) \cdot x$ and $1 \cdot x = x$), the number of orbits for the action of G on Y, that is the number of distinct sets of the form $\{g \cdot x | g \in G\}$, is equal to

$$\frac{1}{|G|} \cdot \sum_{g \in G} |\mathrm{Fix}(g)|,$$

where $\mathrm{Fix}(g)$ is the set of $x \in Y$ such that $g \cdot x = x$. This is called Burnside's lemma and it is very useful, even though its proof is really simple: all you need to do is to count in two ways the pairs (g, x) such that $g \cdot x = x$. Now, consider Y the set of the first m positive integers, and G the set of permutations of its elements. G acts obviously on the set of colorings of Y with n colors $C_1, C_2, ..., C_n$ (that is, on the set Y of functions from Y to $\{1, 2, ..., n\}$). The number of orbits is just the number of pairwise inequivalent classes of colorings, where two colorings are equivalent if they can be obtained by a permutation of G. Clearly, there are $\binom{n+m-1}{m}$ such classes of equivalence (because they are determined by the nonnegative integers $(k_1, k_2, ..., k_n)$ which add up to m, where k_i is the number of objects colored with the color C_i; there are $\binom{n+m-1}{m}$ solutions of the equation $k_1 + k_2 + \cdots + k_n = m$ in nonnegative integers). On the other hand, we can use Burnside's lemma to count these pairwise inequivalent colorings. Observe that a permutation g fixes a coloring if and only if the numbers belonging to the cycles of g have the same color. Therefore, $\mathrm{Fix}(g)$ is the set of colorings which are constant on each cycle of g. There are $n^{c(g)}$ such colorings. Thus, there are

$$\frac{1}{m!} \sum_{g \in G} n^{c(g)}$$

classes of colorings, and this finishes the proof of the identity.

In order to see whether you understood this type of argument, try to show (using this technique) that n divides $\sum_{k=1}^{n} N^{\gcd(k,n)}$ for all integers N. (Hint: count the number of classes of colorings of the vertices of a regular n-gon, two colorings being equivalent if they are obtained by a rotation keeping the polygon invariant.)

8.2 Practice problems

1. Let $a_1, a_2, \ldots, a_n$ be relatively prime positive integers. Find an equivalent as $k \to \infty$ for the number of positive integral solutions of the equation $a_1x_1 + a_2x_2 + \cdots + a_nx_n = k$.

2. Prove that if we partition the set of nonnegative integers into a finite number of infinite arithmetical sequences, then there will be two of them having the same common difference.

3. For $n \geq 3$ and $A \subset \{1, 2, \ldots, n\}$, say A is even if the sum of the elements of A is an even number. Otherwise, say that A is odd. By convention, the empty set is even.

   (a) Find the number of even, respectively odd subsets of $\{1, 2, \ldots, n\}$.

   (b) Find the sum of the elements of the even, respectively odd subsets of $\{1, 2, \ldots, n\}$.

Romanian TST 1994

4. Prove that for each positive integer n

$$\sum_{k=1}^{n} \binom{n+k-1}{2k-1} = F_{2n},$$

where F_n is the Fibonacci sequence (with $F_1 = F_2 = 1$).

Iran 2008

5. For positive integers m and n, let $f(m, n)$ denote the number of n-tuples $(x_1, x_2, \ldots, x_n)$ of integers such that $|x_1| + |x_2| + \cdots + |x_n| \leq m$. Show that $f(m, n) = f(n, m)$.

Putnam 2005

6. Let A be a finite set of nonnegative integers. Define a sequence of sets by $A_0 = A$ and for all $n \geq 0$, an integer a is in A_{n+1} if and only if exactly one of the integers $a - 1$ and a is in A_n. Prove that for infinitely many positive integers k, A_k is the union of A with the set of numbers of the form $k + a$ with $a \in A$.

Putnam 2000

7. How many polynomials P with coefficients 0, 1, 2, or 3 satisfy $P(2) = n$, where n is a given positive integer?

Romanian TST 1994

8. Let n and k be positive integers. For any sequence of nonnegative integers $(a_1, a_2, \ldots, a_k)$ which adds up to n, compute the product $a_1 a_2 \cdots a_k$. Prove that the sum of all these products is

$$\frac{n(n^2 - 1^2)(n^2 - 2^2) \cdots (n^2 - (k-1)^2)}{(2k-1)!}.$$

9. In how many different ways can we parenthesize a non-associative product $a_1 a_2 \ldots a_n$?

Catalan

10. Let $F(n)$ be the number of functions $f : \{1, 2, \ldots, n\} \to \{1, 2, \ldots, n\}$ with the property that if i is in the range of f, then so is j for all $j \leq i$. Prove that

$$F(n) = \sum_{k \geq 0} \frac{k^n}{2^{k+1}}.$$

L. Lovasz, Miklos Schweitzer Competition

11. Let p be a prime and let $d \in \{0, 1, \ldots, p\}$. Prove that

$$\sum_{k=0}^{p-1} \binom{2k}{k+d} \equiv r \pmod{p},$$

where $r \equiv p - d \pmod{3}$, $r \in \{-1, 0, 1\}$.

Mathlinks Contest

12. For which positive integers n can we find real numbers $a_1, a_2, \ldots, a_n$ such that

$$\{|a_i - a_j| \mid 1 \le i < j \le n\} = \left\{1, 2, \ldots, \binom{n}{2}\right\}?$$

Chinese TST 2002

13. Find all positive integers n with the following property: for any real numbers $a_1, a_2, \ldots, a_n$, knowing the numbers $a_i + a_j$, $i < j$, determines the values $a_1, a_2, \ldots, a_n$ uniquely.

Erdős and Selfridge

14. Let $A_1 = \emptyset$, $B_1 = \{0\}$ and

$$A_{n+1} = \{1 + x \mid x \in B_n\}, B_{n+1} = (A_n \setminus B_n) \cup (B_n \setminus A_n).$$

Find all positive integers n such that $B_n = \{0\}$.

Chinese Olympiad

15. Suppose that $a_0 = a_1 = 1$ and $(n+3)a_{n+1} = (2n+3)a_n + 3na_{n-1}$ for $n \ge 1$. Prove that all terms of this sequence are integers.

Kömal

16. Let m, n be positive integers with $m \geq n$, and let S be the set of all n-term sequences of positive integers $(a_1, a_2, \ldots a_n)$ such that $a_1 + a_2 + \cdots + a_n = m$. Show that

$$\sum_{(a_1,\ldots,a_n)\in S} 1^{a_1}2^{a_2}\cdots n^{a_n} = \sum_{i=1}^{n}(-1)^{n-i}\binom{n}{i}i^m.$$

Palmer Mebane, USA TST 2010

17. Is it possible to partition the set of all 12-digit numbers into groups of four numbers such that the numbers in each group have the same digits in 11 places and four consecutive digits in the remaining place?

St. Petersburg Olympiad

18. Consider $(b_n)_{n\geq 1}$ a sequence of integers such that $b_1 = 0$ and define $a_1 = 0$ and $a_n = nb_n + a_1b_{n-1} + \cdots + a_{n-1}b_1$ for all $n \geq 2$. Prove that $p|a_p$ for any prime number p.

Komal

19. Let p be a prime and let $n \geq p$ and $a_1, a_2, \ldots, a_n$ be integers. Define $f_0 = 1$ and f_k the number of subsets $B \subset \{1, 2, \ldots, n\}$ having k elements and such that p divides $\sum_{i\in B} a_i$. Show that $f_0 - f_1 + f_2 - \cdots + (-1)^n f_n$ is a multiple of p.

Saint Petersburg 2003

20. Let A be an infinite set of positive integers. Let $f(n)$ be the number of pairs $(a, b) \in A \times A$ such that $a < b$ and $a + b = n$. Prove that the sequence $(f(n))_n$ is not eventually constant.

Donald J. Newman

21. Let n be a positive integer. Prove the equivalence of the following statements:

    (a) there exists $S \subset \{1, 2, \ldots, n\}$ such that each of the numbers $0, 1, 2, \ldots, n-1$ has an odd number of representations as $x - y$ with $x, y \in S$;

    (b) $2n - 1$ has a multiple of the form $2^{2k+1} - 1$.

Miklos Schweitzer Competition

22. Let $p > 3$ be a prime. Prove that

$$\sum_{k=1}^{p^2-1} \binom{2k}{k} \equiv 0 \pmod{p^2}.$$

David Callan, AMM 11292

23. Let x and y be noncommutative variables. Express in terms of n the constant term of the expression $(x + y + x^{-1} + y^{-1})^n$.

M. Haiman, D. Richman, AMM 6458

Chapter

9

9.1 Theory and examples

We have already seen some topics where algebra, number theory and combinatorics were mixed in order to obtain some beautiful results. We are aware that such topics are not so easy to digest by the unexperienced reader, but we also think that it is fundamental to have a unified vision of elementary mathematics. This is why we have decided to combine algebra and number theory in this chapter. Your effort and patience will be tested again. The purpose of this chapter is to survey some classical results concerning algebraic numbers and their applications, as well as some connections between number theory and linear algebra.

First, we recall some basic facts about matrices, determinants, and systems of linear equations. For example, the fact that any homogeneous linear system

$$\begin{cases} a_{11}x_1 + a_{12}x_2 + \cdots + a_{1n}x_n = 0 \\ a_{21}x_1 + a_{22}x_2 + \cdots + a_{2n}x_n = 0 \\ \dots\dots\dots\dots\dots\dots\dots\dots\dots\dots\dots \\ a_{n1}x_1 + a_{n2}x_2 + \cdots + a_{nn}x_n = 0 \end{cases}$$

in which

$$\begin{vmatrix} a_{11} & a_{12} & \dots & a_{1n} \\ a_{21} & a_{22} & \dots & a_{2n} \\ \dots & \dots & \dots & \dots \\ a_{n1} & a_{n2} & \dots & a_{nn} \end{vmatrix} \neq 0$$

has only the trivial solution. Second, we need Vandermonde's identity

$$\begin{vmatrix} 1 & x_1 & x_1^2 & \dots & x_1^{n-1} \\ 1 & x_2 & x_2^2 & \dots & x_2^{n-1} \\ \dots & \dots & \dots & \dots & \dots \\ 1 & x_n & x_n^2 & \dots & x_n^{n-1} \end{vmatrix} = \prod_{1 \leq i < j \leq n} (x_j - x_i). \tag{9.1}$$

Finally, when studying the algebraic numbers, we will need two more specific results. The first one is due to Hamilton and Cayley, while the second one is known as the fundamental theorem of symmetric polynomials

Theorem 9.1. *For any field F and any matrix $A \in M_n(F)$, if p_A is the characteristic polynomial of A: $p_A(X) = det(XI_n - A)$, then $p_A(A) = O_n$.*

Theorem 9.2. *Let A be a ring and let $f \in A[X_1, X_2, ..., X_n]$ be a symmetric polynomial with coefficients in A, that is for any permutation $\sigma \in S_n$ we have $f(X_1, X_2, ..., X_n) = f(X_{\sigma(1)}, X_{\sigma(2)}, ..., X_{\sigma(n)})$. Then we can find a polynomial $g \in A[X_1, X_2, ..., X_n]$ such that $f(X_1, X_2, ..., X_n) = g(X_1 + X_2 + \cdots + X_n, X_1X_2 + X_1X_3 + \cdots + X_{n-1}X_n, ..., X_1X_2 \cdots X_n)$.*

This means that any symmetric polynomial with coefficients in a ring is a polynomial (with coefficients in the same ring) in the symmetric fundamental sums:

$$S_k(X_1, ..., X_n) = \sum_{1 \le i_1 < i_2 < ... < i_k \le n} X_{i_1} \cdots X_{i_k}.$$

As usual, we start with some easy examples. Here is a nice (and direct) application of theorem 2:

Example 1. Given a polynomial with complex coefficients, can one decide if it has a double zero only by performing additions, multiplications, and divisions on its coefficients?

Solution. Yes, one can, even though at first glance this does not seem natural. Let $f(x) = a_0 + a_1x + \cdots + a_nx^n$. Then this polynomial has a double zero if and only if

$$F(x_1, x_2, \ldots, x_n) = 0,$$

where $F(x_1, x_2, \ldots, x_n) = \prod_{1 \le i < j \le n} (x_i - x_j)^2$ and $x_1, x_2, \ldots, x_n$ are the zeros of the polynomial. At the same time

$$F(x_1, x_2, \ldots, x_n)$$

is symmetric with respect to $x_1, x_2, ..., x_n$, so by theorem 2 it is a polynomial in the fundamental symmetric sums in $x_1, x_2, ..., x_n$. By Vieta's formulas, these fundamental sums are just the coefficients of f (up to a sign), so

$$F(x_1, x_2, \ldots, x_n)$$

is a polynomial on the coefficients of f. Consequently, we can decide whether

$$F(x_1, x_2, \ldots, x_n) = 0$$

only by using the operations on the coefficients of the polynomial mentioned in the hypothesis. This shows that the answer to the problem is positive.

You may know the following classical problem: if $a, b, c \in \mathbb{Q}$ satisfy $a + b\sqrt[3]{2} + c\sqrt[3]{4} = 0$, then $a = b = c = 0$. Have you ever thought about the general case? This cannot be done with only simple tricks. We need much more. Of course, there is a direct solution using Eisenstein's criterion applied to the polynomial $f(X) = X^n - 2$, but here is a beautiful proof using linear algebra. This time we need to be careful and work in the most appropriate field.

Example 2. Prove that if $a_0, a_1, \ldots, a_{n-1} \in \mathbb{Q}$ satisfy

$$a_0 + a_1 \sqrt[n]{2} + \cdots + a_{n-1} \sqrt[n]{2^{n-1}} = 0,$$

then $a_0 = a_1 = \cdots = a_{n-1} = 0$.

Solution. If $a_0 + a_1 \sqrt[n]{2} + \cdots + a_{n-1} \sqrt[n]{2^{n-1}} = 0$, then

$$ka_0 + ka_1 \sqrt[n]{2} + \cdots + ka_{n-1} \sqrt[n]{2^{n-1}} = 0$$

for any real number k. Hence we may assume that $a_0, a_1, \ldots, a_{n-1} \in \mathbb{Z}$. The idea is to choose n values for k to obtain a system of linear equations having nontrivial solutions. Then the determinant of the system must be zero, and this will imply $a_0 = a_1 = \cdots = a_{n-1} = 0$. Now, let us fill in the blanks. What are good values for k? This can be seen by noticing that $\sqrt[n]{2^{n-1}} \cdot \sqrt[n]{2} = 2 \in \mathbb{Z}$. So, the values $(k_1, k_2, \ldots, k_n) = (1, \sqrt[n]{2}, \ldots, \sqrt[n]{2^{n-1}})$ are good, and the system becomes

$$\begin{cases} a_0 + a_1 \cdot \sqrt[n]{2} + \cdots + a_{n-1} \cdot \sqrt[n]{2^{n-1}} = 0 \\ a_0 \cdot \sqrt[n]{2} + a_1 \cdot \sqrt[n]{2^2} + \cdots + 2a_{n-1} = 0 \\ \cdots\cdots\cdots\cdots\cdots\cdots\cdots\cdots\cdots\cdots \\ a_0 \cdot \sqrt[n]{2^{n-1}} + 2a_1 + \cdots + 2a_{n-1} \cdot \sqrt[n]{2^{n-2}} = 0. \end{cases}$$

Viewing $(1, \sqrt[n]{2}, \ldots, \sqrt[n]{2^{n-1}})$ as a nontrivial solution to the system, we conclude that

$$\begin{vmatrix} a_0 & a_1 & \ldots & a_{n-1} \\ 2a_{n-1} & a_0 & \ldots & a_{n-2} \\ \ldots & \ldots & \ldots & \ldots \\ 2a_1 & 2a_2 & \ldots & a_0 \end{vmatrix} = 0.$$

But what can we do now? Expanding the determinant leads nowhere. As we said before passing to the solution, we should always work in the most appropriate field. This time the field is $\mathbb{Z}/2\mathbb{Z}$, since in this case the determinant can be easily computed; it equals $\overline{a}_0^n = \overline{0}$, where $\overline{x}$ means the residue class of the integer x modulo 2. Hence a_0 must be even, that is $a_0 = 2b_0$ and we have

$$\begin{vmatrix} b_0 & a_1 & \ldots & a_{n-1} \\ a_{n-1} & a_0 & \ldots & a_{n-2} \\ \ldots & \ldots & \ldots & \ldots \\ a_1 & 2a_2 & \ldots & a_0 \end{vmatrix} = 0.$$

Now, we interchange the first two lines of the determinant. Its value remains 0, but when we expand it in $\mathbb{Z}_2$, it yields $\overline{a}_1^n = \overline{0}$. Similarly, we find that all a_i are even. Let us write $a_i = 2b_i$. Then we also have $b_0 + b_1 \cdot \sqrt[n]{2} + \cdots + b_{n-1} \cdot \sqrt[n-1]{2^{n-1}} = 0$ and with the same reasoning we conclude that all b_i are even. But of course, we can repeat this as long as we want. By the method of infinite descent, we find that $a_0 = a_1 = \cdots = a_{n-1} = 0$.

The above solution might seem exaggeratedly difficult compared with the one using Eisenstein's criterion, but the idea was too nice not to be presented here. The following problem can become a nightmare despite its apparent simplicity.

Example 3. Let $A = \{a^3+b^3+c^3-3abc|\ a, b, c \in \mathbb{Z}\}$. Prove that if $x, y \in A$, then $xy \in A$.

Solution. The observation that

$$a^3 + b^3 + c^3 - 3abc = \begin{vmatrix} a & c & b \\ b & a & c \\ c & b & a \end{vmatrix}$$

leads to a quick solution. Indeed, it suffices to note that

$$\begin{pmatrix} a & c & b \\ b & a & c \\ c & b & a \end{pmatrix} \begin{pmatrix} x & z & y \\ y & x & z \\ z & y & x \end{pmatrix} =$$

$$= \begin{pmatrix} ax + cy + bz & az + by + cx & ay + bx + cz \\ ay + bx + cz & ax + cy + bz & az + by + cx \\ az + by + cx & ay + bx + cz & ax + cy + bz \end{pmatrix}$$

and thus

$$(a^3 + b^3 + c^3 - 3abc)(x^2 + y^3 + z^3 - 3xyz) = A^3 + B^3 + C^3 - 3ABC,$$

where $A = ax+bz+cy$, $B = ay+bx+cz$, $C = az+by+cx$. You see, identities are not so hard to find...

We all know the famous Bezout's theorem, stating that if $a_1, a_2, \ldots, a_n$ are relatively prime, then one can find integers $k_1, k_2, \ldots, k_n$ such that $k_1a_1 + k_2a_2 + \cdots + k_na_n = 1$. The following problem claims more, at least for $n = 3$.

Example 4. Prove that if a, b, c are relatively prime integers, then there are integers x, y, z, u, v, w such that

$$a(yw - zv) + b(zu - xw) + c(xv - yu) = 1.$$

Solution. The given condition can be written in the form $\det A = 1$, where

$$A = \begin{pmatrix} a & x & u \\ b & y & v \\ c & z & w \end{pmatrix}.$$

So, let us prove a much more general result.

Theorem 9.3. *Any vector v whose integer components are relatively prime is the first column of an integral matrix with determinant equal to 1.*

Proof. We induct on the dimension n of the vector v. Indeed, for $n = 2$ it is exactly Bezout's theorem. Now, assume that it is true for vectors in $\mathbb{Z}^{n-1}$ and take $v = (v_1, v_2, \dots, v_n)$ such that v_i are relatively prime. Consider the numbers $\frac{v_1}{g}, \frac{v_2}{g}, \dots, \frac{v_{n-1}}{g}$, where g is the greatest common divisor of $v_1, v_2, \dots, v_{n-1}$. They are relatively prime and the matrix

$$\begin{pmatrix} \frac{v_1}{g} & a_{12} & \dots & a_{1,n-1} \\ \dots & \dots & \dots & \dots \\ \frac{v_{n-1}}{g} & a_{n-1,2} & \dots & a_{n-1,n-1} \end{pmatrix}$$

has determinant equal to 1. We can find α, β such that $\alpha g + \beta v_n = 1$ and verify that the following matrix has integral entries and determinant 1:

$$\begin{pmatrix} v_1 & a_{12} & \dots & a_{1,n-1} & (-1)^{n-1}\beta\frac{v_1}{g} \\ \dots & \dots & \dots & \dots & \dots \\ v_{n-1} & a_{n-1,2} & \dots & a_{n-1,n-1} & (-1)^{n-1}\beta\frac{v_{n-1}}{g} \\ v_n & 0 & \dots & 0 & (-1)^{n-1}\alpha \end{pmatrix}.$$

□

In the chapter **Look at the Exponent** we have seen a rather complicated solution for the following problem. This one is much easier, but difficult to find:

Example 5. Prove that for any integers $a_1, a_2, \ldots, a_n$, the number

$$\prod_{1 \le i < j \le n} \frac{a_j - a_i}{j - i}$$

is an integer.

[Armond Spencer] AMM E 2637

Solution. With this introduction, the way to proceed is clear. What does the expression $\prod_{1 \le i < j \le n} (a_j - a_i)$ suggest? It is the Vandermonde's identity (9.1), associated with $a_1, a_2, \ldots, a_n$. But we have a hurdle here. We might want to use the same formula for the expression $\prod_{1 \le i < j \le n} (j - i)$. This is a dead end. But it is easy to prove that $\prod_{1 \le i < j \le n} (j - i)$ equals $(n-1)!(n-2)! \cdots 1!$. Now, we can write

$$\prod_{1 \le i < j \le n} \frac{a_j - a_i}{j - i} = \frac{1}{1! \cdot 2! \cdots (n-1)!} \begin{vmatrix} 1 & 1 & 1 & \ldots & 1 \\ a_1 & a_2 & a_3 & \ldots & a_n \\ \ldots & \ldots & \ldots & \ldots & \ldots \\ a_1^{n-1} & a_2^{n-1} & a_3^{n-1} & \ldots & a_n^{n-1} \end{vmatrix}.$$

As usual, the last step is the most important. The above formula can be rewritten as

$$\prod_{1 \le i < j \le n} \frac{a_j - a_i}{j - i} = \begin{vmatrix} 1 & 1 & 1 & \ldots & 1 \\ \dfrac{a_1}{1!} & \dfrac{a_2}{1!} & \dfrac{a_3}{1!} & \ldots & \dfrac{a_n}{1!} \\ \ldots & \ldots & \ldots & \ldots & \ldots \\ \dfrac{a_1^{n-1}}{(n-1)!} & \dfrac{a_2^{n-1}}{(n-1)!} & \dfrac{a_3^{n-1}}{(n-1)!} & \ldots & \dfrac{a_n^{n-1}}{(n-1)!} \end{vmatrix}.$$

And now we recognize the form

$$\prod_{1\leq i<j\leq n} \frac{a_j - a_i}{j - i} = \begin{vmatrix} 1 & 1 & \dots & 1 \\ \binom{a_1}{1} & \binom{a_2}{1} & \dots & \binom{a_n}{1} \\ \binom{a_1}{2} & \binom{a_2}{2} & \dots & \binom{a_n}{2} \\ \dots & \dots & \dots & \dots \\ \binom{a_1}{n-1} & \binom{a_2}{n-1} & \dots & \binom{a_n}{n-1} \end{vmatrix},$$

which can be proved easily by subtracting lines. Because each number $\binom{a_i}{j}$ is an integer, the determinant itself is an integer and the conclusion follows.

At this point, you might be disappointed because we did not keep our promise: no trace of algebraic numbers appeared until now! Yet, we considered that a small introduction featuring easy problems and applications of linear algebra in number theory was absolutely necessary. Now, we can pass to the real purpose of this chapter, a small study of algebraic numbers. But what are they? Let us start with some definitions: we say that a complex number x is algebraic if it is a zero of a polynomial with rational coefficients. The monic polynomial of least degree, with rational coefficients and having x as a zero is called the minimal polynomial of x. Its other complex zeros are called the conjugates of x. Using the division algorithm, it is not difficult to prove that any polynomial with rational coefficients which has x as zero is a multiple of the minimal polynomial of x. Also, it is clear that the minimal polynomial of an algebraic number is irreducible in $\mathbb{Q}[X]$. We say that the complex number x is an algebraic integer if it is zero of a monic polynomial with integer coefficients. You can prove, using Gauss's lemma, that an algebraic number is an algebraic integer if and only if its minimal polynomial has integer coefficients. In order to avoid confusion, we will call the usual integers "rational" integers in this chapter. There are two very important results concerning algebraic integers that you should know:

Theorem 9.4. *The sum or product of two algebraic numbers is algebraic. The sum or product of two algebraic integers is an algebraic integer.*

Proof. This result is extremely important, because it shows that the algebraic integers form a ring. Denote this ring by AI. None of the known proofs is really easy. The one that we are going to present first uses the fundamental theorem of symmetric polynomials. Consider two algebraic numbers x and y and let $x_1, x_2, ..., x_n$ and $y_1, y_2, ..., y_m$ be the conjugates of x and y respectively. Next, look at the polynomial $f(x) = \prod_{i=1}^{n} \prod_{j=1}^{m} (X - x_i - y_j)$. We claim that it has rational coefficients. (The fact that $x + y$ is a zero of f being obvious.) This follows from the fundamental theorem of symmetric polynomials applied twice. Let $R = \mathbb{Z}[y_1, y_2, ..., y_m]$ be the ring considered in the statement of the Theorem 9.2. Because the coefficients of f are symmetric polynomials in $x_1, x_2, ..., x_n$, it follows that every coefficient of f is of the form $B(\sigma_1, \sigma_2, ..., \sigma_n, y_1, y_2, ..., y_m)$, where σ_i are the symmetric sums in $x_1, x_2, ..., x_n$, and B is a polynomial with rational (respectively integer, if x, y are algebraic integers) coefficients. But the coefficients of f are also symmetric in $y_1, y_2, ..., y_m$, so by taking $R = \mathbb{Z}[\sigma_1, \sigma_2, ..., \sigma_n]$ in Theorem 9.2, we deduce that A is a polynomial with rational (or integer) coefficients in the symmetric sums in $x_1, x_2, ..., x_n$ and $y_1, y_2, ..., y_m$. Thus f has rational coefficients if x, y are algebraic and f has integer coefficients if x, y are algebraic integers.

□

There is also a solution which uses only the most elementary linear algebra! Indeed, we claim that a complex number z is an algebraic integer if and only if there exists a finitely generated commutative subring of $\mathbb{C}$ containing z. Indeed, if z is an algebraic integer, the division algorithm immediately shows that $\mathbb{Z}[z]$ is a finitely generated commutative subring of $\mathbb{C}$. Now, suppose that R is a commutative subring of $\mathbb{C}$ which is finitely generated and contains z. Take $v_1, v_2, ..., v_n$ that generate R and observe that the numbers $zv_1, zv_2, ..., zv_n$ are in R, thus they are linear combinations with integer coefficients of $v_1, v_2, ..., v_n$. Let $zv_i = a_{i1}v_1 + a_{i2}v_2 + ... + a_{in}v_n$ for some integers a_{ij} and let A be the matrix with entries a_{ij}. The above system of equations can be written as $(zI_n - A)v = o$, where v is the vector whose coordinates are $v_1, v_2, ..., v_n$. Because v is not zero, the last relation implies $\det(zI_n - A) = 0$ and thus z is a root of the characteristic polynomial of A, (which is unitary and has integer coefficients), because so does A. This proves the claim. Now, consider two algebraic integers x, y. By the previous characterization and the fact that clearly y is an algebraic integer over $\mathbb{Z}[x]$, it follows that $\mathbb{Z}[x,y] = (\mathbb{Z}[x])[y] = \mathbb{Z}[x]v_1 + \cdots + \mathbb{Z}[x]v_m$, and since x is an algebraic integer there exist $u_1, .., u_p$ such that $\mathbb{Z}[x] = \mathbb{Z}u_1 + \cdots + \mathbb{Z}u_p$. Therefore $\mathbb{Z}[x,y] \subseteq \sum\limits_{1\le k\le p, 1\le l\le m} \mathbb{Z}u_k v_l$. Because $\mathbb{Z}[x+y]$ and $\mathbb{Z}[xy]$ are subsets of $\mathbb{Z}[x,y]$, by applying the characterization again it follows that $x+y$ and xy are algebraic integers. Note however (and it is very important) that the set of algebraic integers is not a field (the following theorem will make this statement obvious), while the set of algebraic numbers is a field: if $P(x) = 0$ for some non-zero polynomial with integer coefficients P, then $Q\left(\frac{1}{x}\right) = 0$, where $Q(X) = X^{\deg(P)} \cdot P\left(\frac{1}{X}\right)$.

The next result is also very important, and we will see some of its applications in the following examples.

Theorem 9.5. *The only rational numbers which are also algebraic integers are the rational integers.*

Proof. The proof of this result is much easier. Indeed, suppose that $x = \frac{p}{q}$ is a rational number (with $\gcd(p,q) = 1$) which is also a zero of the monic polynomial with integer coefficients $f(X) = X^n + a_{n-1}X^{n-1} + \cdots + a_1X + a_0$. Then $p^n + a_{n-1}p^{n-1}q + \cdots + a_1pq^{n-1} + a_0q^n = 0$. Therefore q divides p^n and since $\gcd(q, p^n) = 1$, we must have $q = \pm 1$, which shows that x is a rational integer. Clearly, any rational integer x is an algebraic integer.

□

Here is a very nice and difficult problem that appeared in AMM in 1998, and which is a consequence of these results. We prefer to give two solutions, one using the previous results and another one using linear algebra. A variant of this problem was given in 2004 at a Team Selection Test in Romania, and it turned out to be a surprisingly difficult problem.

Example 6. Consider the sequence $(x_n)_{n\geq 0}$ defined by $x_0 = 4$, $x_1 = x_2 = 0$, $x_3 = 3$ and $x_{n+4} = x_{n+1} + x_n$. Prove that for any prime p the number x_p is a multiple of p.

AMM

Solution 1. Naturally, we start by considering the characteristic polynomial of the recursive relation: $X^4 - X - 1$. It is easy to see that it cannot have a double zero. Using the theory of linear recursive sequences, it follows that the general term of the sequence is of the form $Ar_1^n + Br_2^n + Cr_3^n + Dr_4^n$ for some constants A, B, C, D. Here r_i are the distinct zeros of the caracteristic polynomial. Because this polynomial has no rational zero, it is natural to suppose that $Ar_1^n + Br_2^n + Cr_3^n + Dr_4^n$ is symmetric in r_1, r_2, r_3, r_4 and thus $A = B = C = D$. Because $x_0 = 4$, we should take $A = B = C = D = 1$. Now, let us see whether we can prove that $x_n = r_1^n + r_2^n + r_3^n + r_4^n$ for all n. Using Viete's formulae, we can check that this holds for n less than 4. But since $r_i^{n+4} = r_i^{n+1} + r_i^n$, an inductive argument shows that the formula is true for any n. Hence we need to prove that p divides $r_1^p + r_2^p + r_3^p + r_4^p$ for any prime number p. This follows from the more general result (which is also a generalization of Fermat's little theorem):

Theorem 9.6. *Let f be a monic polynomial with integer coefficients and let $r_1, r_2, ..., r_n$ be its zeros (not necessarily distinct). Then $A = (r_1 + r_2 + \cdots + r_n)^p - (r_1^p + r_2^p + \cdots + r_n^p)$ is a rational integer divisible by p for any prime number p.*

Proof. Theorem 9.2 shows that A is a rational integer because it is a symmetric polynomial in $r_1, r_2, ..., r_n$, and thus a polynomial with integer coefficients in the coefficients of f. The difficulty is to prove that it is a multiple of p. First of all, let us prove by induction that if $a_1, a_2, ..., a_n$ are algebraic integers then $\frac{1}{p} \cdot ((a_1 + a_2 + \cdots + a_n)^p - (a_1^p + a_2^p + \cdots + a_n^p))$ is also an algebraic integer. For $n = 2$, this follows from the binomial formula $\frac{1}{p} \cdot ((a+b)^p - a^p - b^p) = \sum_{i=1}^{p-1} \frac{1}{p} \cdot \binom{p}{i} \cdot a^{p-i} b^i$. Indeed, $\frac{1}{p} \cdot \binom{p}{i}$ is an integer, and we obtain a sum of products of algebraic integers, which is an algebraic integer. Now, if the assertion is true for $n-1$, consider $a_1, a_2, ..., a_n$ algebraic integers. By the inductive hypothesis, $(a_1 + a_2 + \cdots + a_{n-1})^p - (a_1^p + a_2^p + \cdots + a_{n-1}^p) \in p \cdot AI$. The case $n = 2$ shows that $(a_1 + a_2 + \cdots + a_n)^p - (a_1 + a_2 + \cdots + a_{n-1})^p - a_n^p \in p \cdot AI$. Therefore, $(a_1 + a_2 + \cdots + a_n)^p - (a_1^p + a_2^p + \cdots + a_n^p) \in p \cdot AI$ (as being the sum of the above expressions), which is exactly what we needed to finish the inductive step. Now, finishing the proof of the theorem is easy: we know that $\frac{1}{p} \cdot ((a_1 + a_2 + \cdots + a_n)^p - (a_1^p + a_2^p + \cdots + a_n^p))$ is a rational number which is also an algebraic integer. By theorem 4, it must be a rational integer.

□

Solution 2. Let us consider the matrix

$$A = \begin{pmatrix} 0 & 0 & 0 & 1 \\ 1 & 0 & 0 & 1 \\ 0 & 1 & 0 & 0 \\ 0 & 0 & 1 & 0 \end{pmatrix}$$

and let $\mathrm{Tr}(X)$ be the sum of the entries of the main diagonal of the matrix X. We will first prove that $x_n = \mathrm{Tr}(A^n)$ (here $A^0 = I_4$). This is going to be the easy part of the solution. Indeed, for $n = 1, 2, 3$ it is not difficult to verify it. Now, assume that the statement is true for all $i = 1, 2, \ldots, n-1$ and prove that it is also true for n. This follows from

$$x_n = x_{n-4} + x_{n-3} = \mathrm{Tr}(A^{n-4}) + \mathrm{Tr}(A^{n-3}) = \mathrm{Tr}(A^{n-4}(A + I_4)) = \mathrm{Tr}(A^n).$$

We have used here the relation $A^4 = A + I_4$, which can be easily verified by a simple computation. Hence the claim is proved.
Now, let us prove an important result–that is, $\mathrm{Tr}(A^p) \equiv \mathrm{Tr}(A) \pmod{p}$ for any integral matrix and any prime p. The proof is not trivial at all. A

possible advanced solution is to start by considering the matrix $\overline{A}$ obtained by reducing all entries of A modulo p, then by working in a field in which the characteristic polynomial of A has all its zeros $\lambda_1, \lambda_2, \ldots, \lambda_n$. This field clearly has characteristic p (it contains Z_p) and so we have (using the binomial formula and the fact that all coefficients $\binom{p}{k}$, $1 \le k \le p-1$ are multiples of p)

$$\operatorname{Tr}(A^p) = \sum_{i=1}^{n} \lambda_i^p = \left(\sum_{i=1}^{n} \lambda_i\right)^p = (\operatorname{Tr} A)^p,$$

from where the conclusion is immediate, using Fermat's little theorem.
But there is a beautiful elementary solution. Let us consider two integral matrices A, B, and write

$$(A+B)^p = \sum_{A_1,\ldots,A_p \in \{A,B\}} A_1 A_2 \ldots A_p.$$

Observe that for any A, B we have $\operatorname{Tr}(AB) = \operatorname{Tr}(BA)$, and, by induction, for any $X_1, X_2, \ldots, X_n$ and any cyclic permutation σ,

$$\operatorname{Tr}(X_1 X_2 \ldots X_n) = \operatorname{Tr}(X_{\sigma(1)} X_{\sigma(2)} \cdots X_{\sigma(n)}).$$

Now, note that in the sum $\sum\limits_{A_1,\ldots,A_p \in \{A,B\}} A_1 A_2 \ldots A_p$ we can form $\dfrac{2^p - 2}{p}$ groups of p-cycles and that we have two more terms, A^p and B^p. Thus

$$\sum_{A_1,\ldots,A_p \in \{A,B\}} \operatorname{Tr}(A_1 A_2 \ldots A_p) \equiv \operatorname{Tr}(A^p) + \operatorname{Tr}(B^p)$$

modulo p (you have already noticed that Fermat's little theorem comes in handy once again), since the sum of $\operatorname{Tr}(A_1 A_2 \ldots A_p)$ is a multiple of p in any cycle. Thus we have proved that

$$\operatorname{Tr}(A+B)^p \equiv \operatorname{Tr}(A^p) + \operatorname{Tr}(B^p) \pmod{p}$$

and by an immediate induction we also have

$$\operatorname{Tr}(A_1 + \cdots + A_k)^p \equiv \operatorname{Tr}(A_1^p) + \cdots + \operatorname{Tr}(A_k^p) \pmod{p}.$$

Next, consider the matrices E_{ij} that have 1 in the position (i, j) and 0 elsewhere. For these matrices we have $\mathrm{Tr}(A^p) \equiv \mathrm{Tr}(A) \pmod{p}$ and by using the above result we can write (using Fermat's little theorem one more time):

$$\mathrm{Tr} A^p = \mathrm{Tr}\left(\sum_{i,j} a_{ij} E_{ij}\right)^p$$

$$\equiv \sum_{i,j} \mathrm{Tr}(a_{ij}^p E_{ij}^p) \equiv \sum_{i,j} a_{ij} \mathrm{Tr} E_{ij} = \mathrm{Tr} A \pmod{p}.$$

The result is proved, and with it the fact that x_p is a multiple of p.

The example we are about to discuss next generated a whole mathematical theory and even an important area of research in transcendental number theory. Let us start by introducing a definition: for a complex polynomial

$$f(X) = a_n X^n + a_{n-1} X^{n-1} + \cdots + a_1 X + a_0 = a_n (X - x_1) \cdot (X - x_2) \cdots (X - x_n)$$

define the Mahler measure of f to be

$$M(f) = |a_n| \cdot \max(1, |x_1|) \cdots \max(1, |x_n|).$$

You can immediately see that $M(fg) = M(f) \cdot M(g)$ for all polynomials f and g. Using complex analysis, we can prove the following identity:

$$M(f) = e^{\int_0^1 \ln |f(e^{2i\pi t})| dt}.$$

The next problem shows that a monic polynomial with integer coefficients and Mahler measure 1 has all of its zeros roots of unity. That is, the only algebraic integers all of whose conjugates lie on the unit disk of the complex plane are roots of the unity. This result is the celebrated Kronecker's theorem.

Example 7. Let f be a monic polynomial with integer coefficients such that $f(0) \neq 0$ and $M(f) = 1$. Then for each zero z of f there exists an n such that $z^n = 1$.

[Kronecker]

Solution. What you are going to read now is one of those mathematical jewels that you do not come across every day, so enjoy the following proof. Let $f(X) = (X - x_1) \cdot (X - x_2) \cdots (X - x_n)$ be the factorization of f in $\mathbf{C}[X]$. Consider now the polynomials $f_k(X) = (X - x_1^k) \cdot (X - x_2^k) \cdots (X - x_n^k)$. The coefficients of these polynomials are symmetric polynomials in $x_1, x_2, ..., x_n$, and since all symmetric fundamental sums of $x_1, x_2, ..., x_n$ are integers, all f_k have integer coefficients (we used Theorem 9.2 here). What is really awesome is that there is a uniform bound on the coefficients of f_k. Indeed, because all x_i have absolute values at most 1, all symmetric fundamental sums in $x_1^k, x_2^k, ..., x_n^k$ have absolute values at most $\binom{n}{[\frac{n}{2}]}$. Therefore, all coefficients of all polynomials f_k are integers between $-\binom{n}{[\frac{n}{2}]}$ and $\binom{n}{[\frac{n}{2}]}$. This shows that there are two identical polynomials among $f_1, f_2, f_3, ...$ Let $i > j$ be such that $f_i = f_j$. Consequently, there is a permutation σ of $1, 2, ..., n$ such that $x_1^i = x_{\sigma(1)}^j, ..., x_n^i = x_{\sigma(n)}^j$. An easy induction shows that $x_1^{i^r} = x_{\sigma^r(1)}^j$ for all $r \geq 1$. Because $\sigma^{n!}(1) = 1$, we deduce that $x_1^{i^{n!}-j} = 1$ and so x_1 is a root of the unity. Clearly, we can similarly prove that $x_2, x_3, ..., x_n$ are roots of the unity. After this example, a natural question appears: are there algebraic integers on the unit circle that are not roots of unity? The answer is yes, as the following example shows. Actually, part a) was known much before its publication in AMM. We invite the reader to take a look at the last chapter of this book for a proof of this more general result. Burnside proved a much more general result, which is left as exercise in the **Problems for Training** section, as a lemma in his famous theorem stating that any group whose cardinality is of the form $p^a q^b$ for some primes p, q and some positive integers a, b is solvable.

Example 8. a) If a is a root of unity whose real part is an algebraic integer, then $a^4 = 1$.
b) There are algebraic integers of absolute value 1 and which are not roots of the unity.

[H. S. Shapiro] AMM 4656

Solution. The proof of a) is very ingenious. Let $b = \text{Re}(a) = \frac{a+a^{-1}}{2}$ be the real part of a, and consider $a_1, a_2, ..., a_k$ the conjugates of a. We claim that the conjugates of b are distinct numbers among $\mathrm{Re}(a_1), \mathrm{Re}(a_2), ..., \mathrm{Re}(a_k)$. Indeed, the polynomial $\prod_{j=1}^{k}\left(X - \frac{a_j + a_j^{-1}}{2}\right)$ has b as a zero and its coefficients are symmetric polynomials in a_j (because $a_j^N = 1$ for a suitable N), and rational by the theorem of symmetric polynomials. Thus all conjugates of b are among the zeros of this polynomial. On the other hand, if $a^4 \neq 1$ then $a_j^4 \neq 1$ for all j and so $0 < |\mathrm{Re}(a_j)| < 1$, which means that the absolute value of the product of all conjugates of b is smaller than 1. Let h be the minimal polynomial of b over $\mathbb{Q}$. Because b is an algebraic integer, h has integer coefficients, thus $h(0)$ is an integer. But $|h(0)|$ is also the absolute value of the product of all conjugates of b, which is smaller than 1. It follows that $h(0) = 0$, and because h is irreducible in $\mathbb{Q}[X]$, it follows that $h(X) = X$ and so $b = 0$, which is impossible if $a^4 \neq 1$. Now b) is not so difficult. We will take a a zero of a polynomial of the form $(X+1)^4 - uX^2$ for some integer u. We need to have $|a| = 1$ and also $\mathrm{Re}(a)$ needs to be an algebraic integer. If we also manage to ensure that $a^4 \neq 1$, then we are done by a). You can easily check that by taking $u = 8$ all conditions are satisfied, and so $\sqrt{2} - 1 + i\sqrt{2\sqrt{2} - 2}$ is an algebraic integer on the unit circle which is not a root of the unity.

Some more comments on the previous examples are needed. First of all, it is not difficult to deduce from this result that the only monic polynomials with integer coefficients whose Mahler measure is 1 are products of X and some cyclotomic polynomials. A famous conjecture of Lehmer says that there exists a constant $c > 1$ such that if a polynomial with integer coefficients has Mahler measure greater than 1, then its Mahler measure is actually greater than c. The polynomial with least Mahler measure found up to now is $X^{10} + X^9 - X^7 - X^6 - X^5 - X^4 - X^3 + X + 1$, whose Mahler measure is about 1.176. For some upper bounds of the Mahler measure in terms of the coefficients of the polynomial, we refer the reader to example 16 of chapter **Pigeonhole Principle Revisited**.

Showing that a sum of square roots of positive integers is not a rational number is not difficult as long as the number of square roots is less than 3. Otherwise, this is much more complicated. Actually, one can prove the very beautiful result that if $a_1, ..., a_n$ are positive integers such that $\sqrt{a_1} + \cdots + \sqrt{a_n}$ is a rational number, then all a_i are perfect squares. The following problem claims much less, but is still not simple. We will see how easy it becomes in the framework of the above results.

Example 9. Prove that the number

$$\sqrt{1001^2+1} + \sqrt{1002^2+1} + \cdots + \sqrt{2000^2+1}$$

is irrational.

Chinese TST 2005

Solution. Let us suppose that the number is rational. Because it is a sum of algebraic integers, it is also an algebraic integer. By theorem 4, it follows that $\sqrt{1001^2+1} + \sqrt{1002^2+1} + \cdots + \sqrt{2000^2+1}$ is a rational integer. Hence

$$\sqrt{1001^2+1} + \sqrt{1002^2+1} + \cdots + \sqrt{2000^2+1} - (1001 + 1002 + \cdots + 2000)$$

is a rational integer. But this cannot hold, because

$$\sqrt{1001^2+1} + \sqrt{1002^2+1} + \cdots + \sqrt{2000^2+1} - (1001 + 1002 + \cdots + 2000) =$$

$$\frac{1}{1001 + \sqrt{1001^2+1}} + \frac{1}{1002 + \sqrt{1002^2+1}} + \cdots + \frac{1}{2000 + \sqrt{2000^2+1}}$$

is greater than 0 and smaller than 1.

The following example is very elegant, and you can easily check that the result is sharp:

Example 10. Let x, y be complex numbers such that the expression $\frac{x^n - y^n}{x - y}$ is an integer for some 4 consecutive positive integers n. Prove that it is an integer for any positive integer n.

[Clark Kimberling] AMM E 2998

Solution. Let a_n be the given expression and let $S = x + y$, $P = xy$. Observe that $a_{n+2} - Sa_{n+1} + Pa_n = 0$ for all n. Also, it is not difficult to prove that $a_{n+1}a_{n-1} - a_n^2 = -P^{n-1}$. Thus if $a_{n-1}, a_n, a_{n+1}, a_{n+2}$ are all integers, so are P^{n-1} and P^n. Thus P is an algebraic integer which is also rational (because $P = \frac{P^n}{P^{n-1}}$), so that P is an integer. On the other hand, it is immediate to prove by induction that $a_n = f_n(S)$ for some monic polynomial f_n with integer coefficients, of degree $n-1$. This shows that S is a zero of the monic polynomial with integer coefficients $f_n(X) - a_n$, so S is an algebraic integer. Because $S = \frac{a_{n+2}+Pa_n}{a_{n+1}}$, S is also rational. Thus S is an integer, and in this case it is obvious that all terms of the sequence are integers, by the recursive relation.

Here is a beautiful and difficult problem, where properties of algebraic integers come to the spotlight.

Example 11. Let $a_1, a_2, ..., a_k$ be positive real numbers such that $\sqrt[n]{a_1} + \sqrt[n]{a_2} + \cdots + \sqrt[n]{a_k}$ is a rational number for all $n \geq 2$. Prove that $a_1 = a_2 = \cdots = a_k = 1$.

Solution. First of all, we will prove that $a_1, a_2, ..., a_k$ are algebraic numbers and that $a_1 \cdot a_2 \cdots a_k = 1$. Take an integer $N > k$ and put

$$x_1 = \sqrt[N!]{a_1}, x_2 = \sqrt[N!]{a_2}, ..., x_k = \sqrt[N!]{a_k}.$$

Then clearly $x_1^j + x_2^j + \cdots + x_k^j$ is rational for all $1 \leq j \leq N$. Using Newton's formulae, we can easily deduce that all symmetric fundamental sums of $x_1, x_2, ..., x_k$ are rational numbers. Hence $x_1, x_2, ..., x_k$ are algebraic numbers, and so $a_1 = x_1^{N!}, a_2 = x_2^{N!}, ..., a_k = x_k^{N!}$ are algebraic numbers as well. Also, by the argument above, we know that

$$x_1 \cdot x_2 \cdots x_k = \sqrt[N!]{a_1 \cdot a_2 \cdots a_k}$$

is rational, and this happens for all $N > k$. This implies immediately that $a_1 \cdot a_2 \cdots a_k = 1$. Now let $f(x) = b_r X^r + b_{r-1}X^{r-1} + \cdots + b_0$ be a polynomial with integer coefficients which vanishes at $a_1, a_2, ..., a_k$. Clearly, $b_r a_1, ..., b_r a_k$ are algebraic integers. But then

$$b_r(\sqrt[n]{a_1} + \sqrt[n]{a_2} + \cdots + \sqrt[n]{a_k}) = \sqrt[n]{b_r^{n-1}} \cdot (\sqrt[n]{b_r a_1} + \sqrt[n]{b_r a_2} + \cdots + \sqrt[n]{b_r a_k})$$

is also an algebraic integer. Because it is also a rational number it follows that it is a rational integer. Consequently, $(b_r(\sqrt[n]{a_1} + \sqrt[n]{a_2} + \cdots + \sqrt[n]{a_k}))_{n \geq 1}$ is a sequence of positive integers. Because it converges to kb_r, it eventually becomes equal to kb_r (from a rank). Thus there is n such that $\sqrt[n]{a_1} + \sqrt[n]{a_2} + \cdots + \sqrt[n]{a_k} = k$. Since $a_1 \cdot a_2 \cdots a_k = 1$, the AM-GM inequality implies $a_1 = a_2 = \cdots = a_k = 1$ and the problem is solved.

9.2 Practice problems

1. Let a, b, c be relatively prime nonzero integers. Prove that for any relatively prime integers u, v, w satisfying $au + bv + cw = 0$, there are integers m, n, p such that

$$a = nw - pv, \quad b = pu - mw, \quad c = mv - nu.$$

Octavian Stănăşila, Romanian TST 1989

2. Prove that for any integers $a_1, a_2, \ldots, a_n$

$$\frac{\operatorname{lcm}(a_1, a_2, \ldots, a_n)}{a_1 a_2 \cdots a_n} \prod_{1 \le i < j \le n} (a_j - a_i)$$

is an integer divisible by $1!2! \cdots (n-2)!$. Moreover, we cannot replace $1!2! \cdots (n-2)!$ by any other multiple of $1!2! \cdots (n-2)!$.

3. Let A, B, C be lattice points such that the angles of triangle ABC are rational multiples of π. Prove that triangle ABC is right and isosceles.

4. Let α be a rational number with $0 < \alpha < 1$ and $\cos(3\pi\alpha) + 2\cos(2\pi\alpha) = 0$. Prove that $\alpha = \frac{2}{3}$.

IMO Shortlist 1991

5. (a) Let P, R be polynomials with rational coefficients with $P \neq 0$. Prove that there exists a non-zero polynomial $Q \in \mathbb{Q}[X]$ such that $P(X) | Q(R(X))$

   (b) Let P, R be polynomials with integer coefficients and suppose that P is monic. Prove that there exists a monic polynomial $Q \in \mathbb{Z}[X]$ such that $P(X) | Q(R(X))$

Iran 2006

6. Let k and n be positive integers and let $P(X)$ be a polynomial of degree n with coefficients in the set $\{-1, 0, 1\}$. Suppose that $(X-1)^k \mid P(X)$ and that there is a prime q such that

$$\frac{q}{\ln q} < \frac{k}{\ln(n+1)}.$$

Prove that the complex roots of unity of order q are roots of P.

IMC 2001

7. Prove that none of the numbers $\sqrt{n+1} - \sqrt{n}$ for positive integers n can be the written in the form $2\cos\left(\frac{2k\pi}{m}\right)$ for some integers k, m.

Chinese Olympiad

8. (a) Let $a_1, a_2, \ldots, a_m, b_1, b_2, \ldots, b_n$ be complex numbers such that

$$f_1(X) = \prod_{i=1}^{m}(X - a_i), f_2(X) = \prod_{i=1}^{n}(X - b_i) \in \mathbb{Z}[X]$$

and for which there are $g_1, g_2 \in \mathbb{Z}[X]$ such that $f_1g_1 + f_2g_2 = 1$. Prove that

$$\left|\prod_{i=1}^{m}\prod_{j=1}^{n}(a_i - b_j)\right| = 1.$$

(b) If a_i, b_i are integers and

$$\left|\prod_{i=1}^{m}\prod_{j=1}^{n}(a_i - b_j)\right| = 1,$$

prove that there are polynomials $g_1, g_2 \in \mathbb{Z}[X]$ such that $f_1g_1 + f_2g_2 = 1$.

Ibero-American Olympiad

9. Let p be a prime and let $a_1, a_2, \ldots, a_{p+1}$ be real numbers such that no matter how we eliminate one of them, the rest of the numbers can be divided into at least two nonempty pairwise disjoint subsets each having the same arithmetic mean. Prove that $a_1 = a_2 = \cdots = a_{p+1}$.

Marius Rădulescu, Romanian TST 1994

10. Let a, b be two positive rational numbers such that for some $n \geq 2$ the number $\sqrt[n]{a} + \sqrt[n]{b}$ is rational. Prove that $\sqrt[n]{a}$ is also rational.

Marius Cavachi, Gazeta Matematică

11. Let m, n be relatively prime numbers and $x > 1$ be a real number such that $x^m + \frac{1}{x^m}$ and $x^n + \frac{1}{x^n}$ are integers. Prove that $x + \frac{1}{x}$ is also an integer.

Darij Grinberg, Peter Scholze

12. Let a, b, c be integers. Define the sequence $(x_n)_{n\geq 0}$ by $x_0 = 4$, $x_1 = 0$, $x_2 = 2c$, $x_3 = 3b$ and $x_{n+3} = ax_{n-1} + bx_n + cx_{n+1}$. Prove that for any prime p and any positive integer m, the number x_{p^m} is divisible by p.

Călin Popescu, Romanian TST 2004

13. Let $\theta \in (0, \pi/2)$ be an angle such that $\cos\theta$ is irrational. Suppose that $\cos k\theta$ and $\cos(k+1)\theta$ are rational for some positive integer k. Prove that $\theta = \pi/6$.

USA TST 2007

14. Find the least positive integer n such that $\cos\frac{\pi}{n}$ cannot be written in the form $p + \sqrt{q} + \sqrt[3]{r}$ with $p, q, r \in \mathbb{Q}$.

O. Mushkarov, N. Nikolov, Bulgaria

15. Let $s_1, s_2, \ldots$ and $t_1, t_2, \ldots$ be two infinite nonconstant sequences of rational numbers such that $(s_i - s_j)(t_i - t_j)$ is an integer for all $i, j \geq 1$. Prove that there exists a rational number r such that $(s_i - s_j)r$ and $\frac{t_i - t_j}{r}$ are integers for all i, j.

USAMO 2009

16. Let p be a prime and let $n_1, n_2, \ldots, n_k$ be integers. Define

$$S = \left| \sum_{j=1}^{k} \cos \frac{2\pi n_j}{p} \right|.$$

Prove that either $S = 0$ or $S \geq k \left(\frac{1}{2k}\right)^{\frac{p-1}{2}}$.

Holden Lee

17. Let k be a positive integer and let $a_1, \ldots, a_k$ and $b_1, \ldots, b_k$ be two sequences of rational numbers with the property: for any irrational numbers $x_1, x_2, \ldots, x_k > 1$ there exist positive integers $n_1, n_2, \ldots, n_k$ and $m_1, \ldots, m_k$ such that

$$a_1 \lfloor x_1^{n_1} \rfloor + a_2 \lfloor x_2^{n_2} \rfloor + \cdots + a_k \lfloor x_k^{n_k} \rfloor = b_1 \lfloor x_1^{m_1} \rfloor + b_2 \lfloor x_2^{m_2} \rfloor + \cdots + b_k \lfloor x_k^{m_k} \rfloor.$$

Prove that $a_i = b_i$ for all i.

Gabriel Dospinescu, Mathlinks Contest

18. Prove that if $p_1, p_2, \ldots, p_n$ are distinct primes and if

$$a_1 \sqrt{p_1} + a_2 \sqrt{p_2} + \cdots + a_n \sqrt{p_n} = 0$$

for some rational numbers $a_1, a_2, \cdots, a_n$, then $a_i = 0$ for all i.

Besicovitch's theorem

19. Let the sequence $(a_n)_n$ be defined by $a_0 = 2$ and $a_{n+1} = 2a_n^2 - 1$. Prove that if $p > 2$ divides a_n, then 2^{n+3} divides $p^2 - 1$.

IMO Shortlist 2003

20. Let p, q be prime numbers and r a positive integer such that $q \mid p-1$, q does not divide r and $p > r^{q-1}$. Let $a_1, a_2, \ldots, a_r$ be integers such that $a_1^{\frac{p-1}{q}} + a_2^{\frac{p-1}{q}} + \cdots + a_r^{\frac{p-1}{q}}$ is a multiple of p. Prove that at least one of the a_i's is a multiple of p.

AMM

21. Let $a_1, a_2, \ldots, a_n$ be positive rational numbers and let $k_1, k_2, \ldots, k_n$ be integers greater than 1. If $a_1^{1/k_1} + a_2^{1/k_2} + \cdots + a_n^{1/k_n}$ is a rational number, then any term of the previous sum is also a rational number.

22. Let $a_1, a_2, \ldots, a_n$ be complex numbers such that $a_1^m + a_2^m + \cdots + a_n^m$ is an integer for all positive integers m. Prove that $(X - a_1)(X - a_2) \cdots (X - a_n) \in \mathbb{Z}[X]$.

Chinese IMO training program

23. (a) Suppose that $a_1, a_2, \ldots, a_k$ are rational numbers and $\zeta_1, \zeta_2, \ldots, \zeta_k$ are roots of unity such that $a_1\zeta_1 + a_2\zeta_2 + \cdots + a_k\zeta_k = 0$. Moreover, suppose that $\sum_{i \in I} a_i\zeta_i \neq 0$ for any proper subset I of $\{1, 2, \ldots, k\}$. Prove that $\zeta_i^m = \zeta_j^m$ for all i, j, where m is the product of primes smaller than or equal to k.

    (b) Let z be a complex number. Prove that there are at most $2^{4k^2} \cdot k^k$ k-tuples $(\zeta_1, \zeta_2, \ldots, \zeta_k)$ of roots of unity with the following property: there exist rational numbers $a_1, a_2, \ldots, a_k$ such that $z = \sum_{i=1}^k a_i\zeta_i$ and $z \neq \sum_{i \in I} a_i\zeta_i$ for any proper subset I of $\{1, 2, \ldots, k\}$.

Mann's theorem

24. Say a sequence of integers $(a_n)_n$ is linearly recursive if there exist integers r and $x_1, x_2, \ldots, x_r$ such that

$$a_{n+r} = x_1 a_{n+r-1} + x_2 a_{n+r-2} + \cdots + x_r a_n$$

for all n. Prove that if $(a_n)_n$ is linearly recursive and n divides a_n for all n, then $\frac{a_n}{n}$ is also linearly recursive.

Polya

Chapter 10

10.1 Theory and examples

Another topic with old tricks... you will probably say. Yet we might spend time on a problem just because we ignore obvious clues or basic aspects of it. This is why we think that talking about these "old fashioned tricks" is not because of lack of imagination, but rather an imperious need. In this note we combine some classical arithmetic properties of polynomials. This is just an introduction to this field, but some basic things should become second nature, and among them there will be some problems we discuss further. As usual, we keep some chestnuts for the end of the chapter, hoping that the hard-core solver will appreciate these extremely difficult problems.
Recall that if $f \in \mathbb{Z}[X]$ and a, b are integers, then $a-b$ divides $f(a)-f(b)$. This is the essential result which we will use relentlessly. Here are two applications:

Example 1. Let f, g be relatively prime polynomials with integer coefficients. Define the sequence $a_n = \gcd(f(n), g(n))$. Prove that this sequence is periodic.

AMM

Solution. As we have seen in previous problems, there exist polynomials F, G with integer coefficients and a positive integer A such that $fF+gG = A$. Thus a_n is a divisor of A for all n. Actually, we will prove that A is a period for the sequence $(a_n)_{n\geq 1}$. Let us prove that $a_n | a_{n+A}$. We know that $f(n+A) = f(n)$ (mod A), and since a_n divides A and $f(n)$, it will also divide $f(n+A)$. Similarly, a_n divides $g(n+A)$ and so $a_n | a_{n+A}$. But the same relations show that a_{n+A} divides a_n and so $a_n = a_{n+A}$.

Example 2. Let $p \in \mathbb{Z}[x]$ be such that deg $p > 1$, and let $A = \{p(n) | n \in \mathbb{Z}\}$. Prove that there exists an infinite arithmetical sequence none of whose terms can be expressed in the form $f(x)$ for some integer x.

Solution. We will argue by contradiction: suppose that for all $d > 2$ and all n at least one of the numbers $f(x)$ with x integer gives remainder n when divided by d. This means that for all n and d, the numbers $p(n), p(n+1), ..., p(n+d-1)$ give all remainders mod d. Indeed, because $n, n+1, ..., n+d-1$ are a complete system mod d, it follows that for any x, $p(x)$ gives the same remainder mod d as one of $p(n), p(n+1), ..., p(n+d-1)$. In particular, any residue mod d appears as a residue mod d of one of the numbers $p(n), p(n+1), ..., p(n+d-1)$. Because $\deg(p) > 1$, there exists n such that $d = p(n+1) - p(n) > 2$. In this case, $p(n) = p(n+1) \pmod{d}$ and so the numbers $p(n), p(n+1), ..., p(n+d-1)$ give at most $d-1$ distinct remainders mod d, which is a contradiction.

We continue with an important result, due to Schur, that appears in many variations in contests. Even though in the topic **At the Border of Analysis and Number Theory** we prove an even more general result based on a nice analytical argument, we prefer to present here a purely arithmetical proof.

Example 3. Let $f \in \mathbb{Z}[X]$ be a non-constant polynomial. Then the set of prime numbers dividing at least one nonzero number among

$$f(1), f(2), \ldots, f(n), \ldots$$

is infinite.

[Schur]

Solution. First, suppose that $f(0) = 1$ and consider the numbers $f(n!)$. For sufficiently large n, they are nonzero integers. Moreover, $f(n!) \equiv 1 \pmod{n!}$ and so if we pick a prime divisor of each of the numbers $f(n!)$, the conclusion follows (since in particular any such prime divisor is greater than n). Now, if $f(0) = 0$, everything is clear because in this case n divides $f(n)$ for all n. Suppose that $f(0) \neq 0$ and consider the polynomial $g(x) = \dfrac{f(xf(0))}{f(0)}$. Clearly, $g \in \mathbb{Z}[X]$ and $g(0) = 1$. Applying now the first part of the solution, the problem is solved.

This result has, as we have already said, important consequences. Here is a nice application.

Example 4. Suppose that $f, g \in \mathbb{Z}[X]$ are monic nonconstant irreducible polynomials such that for all sufficiently large n, $f(n)$ and $g(n)$ have the same set of prime divisors. Then $f = g$.

Solution. Indeed, by Gauss's lemma, the two polynomials are irreducible in $\mathbb{Q}[X]$. In addition, if they are not equal, then the above remark and the fact that they have the same leading coefficient implies that the two polynomials are relatively prime in $\mathbb{Q}[X]$. Using Bezout's theorem we conclude that there is a nonzero integer N and $P, Q \in \mathbb{Z}[X]$ such that $fP + gQ = N$. This shows, that for n large enough, all prime factors of $f(n)$ divide N. But, of course, this contradicts Schur's result.

The result of Example 2 remains true if we assume the same property is valid for infinitely many numbers n. Yet the proof uses some highly non-elementary results of Erdős. The interested reader will find rich literature on this field.
A refinement of Schur's theorem is discussed in the following example. The key additional ingredient is the Chinese remainder theorem.

Example 5. Let $f \in \mathbb{Z}[X]$ be a non constant polynomial and let n, k be positive integers. Prove that there exists a positive integer a such that each of the numbers $f(a), f(a+1), \ldots, f(a+n-1)$ has at least k distinct prime divisors.

Bulgarian Olympiad

Solution. Let us consider an array of distinct prime numbers $(p_{ij})_{1 \le i,j \le k}$ such that $f(x_{ij}) \equiv 0 \pmod{p_{ij}}$ for some positive integers x_{ij}. This is just a direct consequence of Schur's theorem. Now, using the Chinese remainder theorem, we can find a positive integer a such that $a + i - 1 \equiv x_{ij} \pmod{p_{ij}}$ for all indices i and j. Using the fundamental result mentioned in the beginning (namely that $f(a) - f(b)$ is always divisible by $a - b$), it follows that each of the numbers $f(a), f(a+1), \dots, f(a+n-1)$ has at least k distinct prime divisors.

We continue with two more difficult examples of problems whose solutions are based on combinations of Schur's theorem with various classical arguments.

Example 6. For integral m, let $p(m)$ be the greatest prime positive divisor of m. By convention we set $p(1) = p(-1) = 1$ and $p(0) = \infty$. Find all polynomials f with integer coefficients such that the sequence $(p(f(n^2)) - 2n)_{n \geq 0}$ is bounded above.

[Titu Andreescu, Gabriel Dospinescu] USAMO 2006

Solution. When searching for the possible answer, one should start with easy examples. Here, the quadratic polynomials might give an insight. Indeed, observe that if u is an odd integer then the polynomial $f(X) = 4X - u^2$ is a solution to the problem. This suggests that any polynomial of the form $c(4X - a_1^2)(4X - a_2^2)...(4X - a_k^2)$ is a solution if c is a nonzero integer and $a_1, a_2, ..., a_k$ are odd integers. Indeed, any prime divisor p of $f(n^2)$ is either a divisor of c (and thus in a finite set) or a divisor of some $(2n - a_j)(2n + a_j)$. In this case $p - 2n \leq \max(a_1, a_2, ..., a_k)$ and so f is a solution of the problem. We deal now with the much more difficult part: showing the converse. Take f a polynomial that satisfies the conditions of the problem, and suppose that $p(f(n^2)) - 2n \leq 2A$ for some constant A. Using Schur's theorem for the polynomial $f(X^2)$, we deduce the existence of a sequence of different prime numbers p_j and nonnegative integers k_j such that $p_j | f(k_j^2)$. Define the sequence $r_j = \min(k_j \pmod{p_j}, \quad p_j - k_j \pmod{p_j})$ and observe that p_j divides $f(r_j^2)$ and also that $0 \leq r_j \leq \frac{p_j - 1}{2}$. Hence $1 \leq p_j - 2r_j \leq A$ and so the sequence $(p_j - 2r_j)_{j \geq 1}$ must take some value a_1 infinitely many times. Let $p_j - 2r_j = a_1$ for j in an infinite set X. Then, if $m = \deg(f)$, we have $p_j | 4^m \cdot f((\frac{p_j - a_1}{2})^2)$ for all $j \in X$ and also the polynomial $4^m \cdot f((\frac{x - a_1}{2})^2)$ has integer coefficients. This shows that p_j divides $4^m \cdot f(\frac{a_1^2}{4})$ for infinitely many j. Hence $\frac{a_1^2}{4}$ is a root of f. Because $f(n^2)$ does not vanish, a_1 must be odd. This means that there exists a polynomial g with integer coefficients and a rational number r such that $f(X) = r(4X - a_1^2)g(X)$. Of course, g has the same property as f, and applying the previous arguments finitely many times we deduce that f must be of the form $c(4X - a_1^2)(4X - a_2^2)...(4X - a_k^2)$ for a certain rational number c and odd integers $a_1, a_2, ..., a_k$. But do not forget that all coefficients of f are integers! Therefore the denominator of c is a divisor of both 4^m and $a_1^2 a_2^2 ... a_k^2$, thus it is 1. This shows that c is an integer and the solution finishes here.

The next problem, which uses Schur's theorem, also needs a classical result, a very particular case of Hensel's lemma. Let us first state and prove this result and then concentrate on the following problem. So, let us first prove the following:

Lemma 10.1 (Hensel's lemma). *Let f be a polynomial with integer coefficients, p a prime number and n an integer such that p divides $f(n)$ and p does not divide $f'(n)$. Then there exists a sequence $(n_k)_{k\geq 1}$ of integers such that $n_1 = n$, p^k divides $n_{k+1} - n_k$ and p^k divides $f(n_k)$.*

Proof. The proof is surprisingly simple. Indeed, let us suppose that we have found i and search for $n_{i+1} = n_i + b \cdot p^i$ such that p^{i+1} divides $f(n_{i+1})$. Because $2i \geq i + 1$, using the binomial formula yields

$$f(n_i + b \cdot p^i) \equiv f(n_i) + bp^i f'(n_i) \pmod{p^{i+1}}.$$

Let $f(n_i) = cp^i$ for some integer c. Because $n_i \equiv n \pmod{p}$, we have $f'(n_i) \equiv f'(n) \pmod{p}$ and so $f'(n_i)$ is invertible modulo p. Let m be the inverse of $f'(n_i)$ modulo p. It is enough to choose $b = -mc$ in order to finish the inductive step.

□

We can now discuss a difficult problem used for the preparation of the Iranian IMO team:

Example 7. Find all polynomials f with integer coefficients such that $n|m$ whenever $f(n)|f(m)$.

[Mohsen Jamali] Iranian TST

Solution. [Adrian Zahariuc] With this preparation, the solution will be short, which does not mean that the problem is easy (as we already said). First of all, observe that for a nonconstant polynomial with integer coefficients such that $f(0) \neq 0$ and for any k there are infinitely many prime numbers p such that $p^k|f(n)$ for some integer n. Indeed, by working with an irreducible divisor of f, we can assume that f is irreducible. Thus f and f' are relatively prime in the ring of polynomials with rational coefficients. Bézout's theorem shows in this case that there exist integer polynomials S, Q and an integer $A \neq 0$ such that $Sf + Qf' = A$. Therefore, if p is a sufficiently large prime such that $p|f(n)$ for some n (the existence of infinitely many such primes follows from Schur's theorem), p will not divide $f'(n)$, and we can apply Hensel's lemma to finish the proof of this result.
Next, observe that $X|f(X)$. Indeed, we have $f(n)|f(n + f(n))$ for all n, so $n|n + f(n)$ for all n, which easily implies $f(0) = 0$. So, let us write $f(X) = X^k g(X)$ with $g(0) \neq 0$. Assume that g is nonconstant. By the previous result, there exists a prime p such that $p > |g(0)|$ and $p^k|g(m)$ for some integer m. Clearly, p does not divide $g(p)$, so by the Chinese remainder theorem there exists an integer n such that $n \equiv m \pmod{p^k}$ and $n \equiv p \pmod{g(p)}$. Thus $p^k|g(n)$ and $g(p)|g(n)$, and thus $f(p)|f(n)$. This implies that $p|n$, and this is impossible, because it would follow that $p|g(0)$. Thus g is constant and the answer is: all polynomials of the form aX^n.

Here is another application of Hensel's lemma. The example below is quite a difficult problem, especially because examples of small degree cannot be found:

Example 8. Is there a polynomial f with integer coefficients that has no rational zeros, but has a zero modulo any positive integer?

Kömal

Solution. The answer is yes, but it is not obvious why such polynomials exist. A very difficult theorem of Chebotarev implies that such polynomials with small degree (smaller than 5) do not exist. It can be proved that there are such polynomials of degree 5, but the example we have chosen has degree 6: define $f(X) = (X^2+3)(X^2-13)(X^2+39)$. We will prove that for any n there exists m such that $n|f(m)$. Observe first of all that it is enough to prove this if n is a power of a prime. Indeed, if we can find m_1, m_2 such that $n_1|f(m_1)$ and $n_2|f(m_2)$ for some relatively prime integers n_1, n_2 then by taking m such that $m = m_1 \pmod{n_1}$ and $m = m_2 \pmod{n_2}$ (which is possible by the Chinese remainder theorem) we have an m such that $n_1 n_2 | f(m)$. Let us now deal with powers of 2. We will prove by induction the existence of a sequence x_n such that $2^n | x_n^2 + 39$. For $n = 1$ we take $x_1 = 1$, for $n = 2$ we take $x_2 = 1$, as well as $x_3 = 1$. Now assume that $x_n^2 + 39 = 2^n \cdot k$ for some integer k and $n \geq 3$. Then $(2^{n-1}x + x_n)^2 + 39 = 2^n(xx_n + k) \pmod{2^{n+1}}$. If k is even we define $x_{n+1} = x_n$. Otherwise, we define $x = 1$ and so $x_{n+1} = x_n + k$. In either case, $2^{n+1} | x_{n+1}^2 + 39$. Now, let us deal with powers of 3 and 13; actually, this case follows immediately from Hensel's lemma applied to the polynomials $X^2 - 13$ and X^2+3, with $n = 1$ and $n = 6$ respectively. Finally, take p a prime number different from $2, 3, 13$ and observe that the identity $\left(\frac{-39}{p}\right) \cdot \left(\frac{13}{p}\right) \cdot \left(\frac{-3}{p}\right) = 1$ (where $\left(\frac{x}{p}\right)$ denotes the Legendre symbol) implies that one of the numbers $\left(\frac{-39}{p}\right), \left(\frac{13}{p}\right)$ and $\left(\frac{-3}{p}\right)$ equals 1. This shows the existence of an integer m such that for some a equal to $3, -13, 39$ we have $p | m^2 + a$. It is now enough to apply Hensel's lemma for the polynomial $X^2 + a$ in order to obtain a sequence x_n such that $p^n | x_n^2 + a$ for all n. This shows that f has a root modulo p^n for any prime p and any positive integer n, and by the remark in the beginning of the solution this polynomial is a solution of the problem.

Example 9. Find all polynomials f with integer coefficients and the following property: for any relatively prime positive integers a, b, the sequence $(f(an + b))_{n \geq 0}$ contains an infinite number of terms, any two of which are relatively prime.

[Gabriel Dospinescu]

Solution. Clearly, constant polynomials can be eliminated. We will prove that the only polynomials with this property are those of the form X^n and $-X^n$, with n a positive integer. Because changing f with its opposite does not modify the property of the polynomial, we can assume that the leading coefficient of f is positive. Hence there exists a constant M such that $f(n) > 2$ for all $n > M$. From now on, we consider only $n > M$. Let us prove that we have $\gcd(f(n), n) \neq 1$ for any such n. Suppose that there is an $n > M$ such that $\gcd(f(n), n) = 1$. The sequence $(f(n + kf(n)))_{k \geq 0}$ would contain at least two relatively prime numbers. Let them be s and r. Because $f(n) \mid kf(n) = kf(n) + n - n \mid f(kf(n) + n) - f(n)$, we have $f(n) \mid f(n + kf(n))$ for any positive integer k. It follows that s and r are both multiples of $f(n) > 2$, which is impossible. We have shown that $\gcd(f(n), n) \neq 1$ for any $n > M$. Thus for any prime $p > M$ we have $p | f(p)$ and so $p | f(0)$. Because any nonzero integer has a finite number of divisors, we conclude that $f(0) = 0$. Hence there is a polynomial q with integer coefficients such that $f(X) = Xq(X)$. It is clear that q has positive leading coefficient and the same property as f. Repeating the above argument, we infer that if q is nonconstant, then $q(0) = 0$ and $q(X) = Xh(X)$. Because f is nonconstant, the above argument cannot be repeated infinitely many times, and thus one of the polynomials g and h must be constant. Consequently, there are positive integers n, k such that $f(X) = kX^n$. But since the sequence $(f(2n + 3))_{n \geq 0}$ contains at least two relatively prime integers, we must have $k = 1$. We obtain that f is of the form X^n. Because f is a solution if and only if $-f$ is a solution, we infer that any solution of the problem is a polynomial of the form $\pm X^n$.

Now let us prove that the polynomials of the form X^n, $-X^n$ are solutions. It suffices to prove it for X^n and even for X; but this follows from Dirichlet's theorem. There is another more elementary approach. Suppose that $x_1, x_2, \ldots, x_p$ are chosen such that the numbers $ax_i + b$ are pairwise relatively prime. We prove that we can add x_{p+1} so that $ax_1 + b, ax_2 + b, \ldots, ax_{p+1} + b$ are pairwise relatively prime. Clearly, $ax_1 + b, ax_2 + b, \ldots, ax_p + b$ are relatively prime to a, so we can apply the Chinese remainder theorem to find an x_{p+1} greater than $x_1, x_2, \ldots, x_p$, such that $x_{p+1} \equiv (1 - b)a_i^{-1} \pmod{ax_i + b}$, $i \in \{1, 2, \ldots, p\}$, where a_i^{-1} is a's inverse in $\mathbb{Z}^*_{ax_i+b}$. Then $\gcd(ax_{p+1} + b, ax_i + b) = 1$ for $i \in \{1, 2, \ldots, p\}$ and thus we can add x_{p+1}.

Example 10. Find all polynomials f with integer coefficients such that $f(n)|n^{n-1}-1$ for all sufficiently large n.

[Gabriel Dospinescu]

Solution. Clearly, $f(X)=X-1$ is a solution, so let us consider an arbitrary solution and write it in the form $f(X)=(X-1)^r g(X)$ with $r \geq 0$ and $g \in \mathbb{Z}[X]$ with $g(1) \neq 0$. Thus there exists M such that $g(n)|n^{n-1}-1$ for all $n > M$. We will prove that g is constant. Assuming the contrary, we may assume without loss of generality that the leading coefficient of g is positive. Thus there is $k > M$ such that $g(n) > 2$ and $g(n)|n^{n-1}-1$ for all $n > k$. Now, since $n+g(n)-n|g(n+g(n))-g(n)$, we deduce that $g(n)|g(n+g(n))$ for all n. In particular, for all $n > k$ we have

$$g(n)|g(n+g(n))|(n+g(n))^{n+g(n)-1}-1$$

and $g(n)|n^{n-1}-1$. Of course, this implies that

$$g(n)|n^{n+g(n)-1}-1=(n^{n-1}-1)n^{g(n)}+n^{g(n)}-1,$$

that is $g(n)|n^{g(n)}-1$ for all $n > k$. Now, let us consider a prime number $p > k$ and let us look at the smallest prime divisor q of $g(p+1) > 2$. We clearly have $q|g(p+1)|(p+1)^{g(p+1)}-1$ and $q|(p+1)^{q-1}-1$. Since $\gcd(g(p+1), q-1)=1$ (by minimality) and

$$\gcd((p+1)^{g(p+1)}-1,(p+1)^{q-1}-1)=(p+1)^{\gcd(g(p+1),q-1)}-1=p,$$

it follows that we actually have $p=q$. This shows that $p|g(p+1)$ and thus (again using the fundamental result) $p|g(1)$. Because this occurs for any prime number $p > k$, we must have $g(1)=0$. This contradiction shows that g is indeed constant.

Let $g(X)=c$. Thus $c|2^{n(2^n-1)}-1$ for all $n > M$. Given that $\gcd(2^a-1, 2^b-1)=2^{\gcd(a,b)}-1$, in order to show that $|c|=1$, it suffices to exhibit $k < m < n$ such that $\gcd(m(2^m-1), n(2^n-1))=1$. This is easy to achieve. Indeed, it suffices to take a prime number m greater than M, k and to choose

a prime number n greater than $m(2^m - 1)$. A simple argument shows that $\gcd(m(2^m - 1), n(2^n - 1)) = 1$ and so $|c| = 1$.
Finally, let us prove that $r \leq 2$. Assuming the contrary, we deduce that

$$(n-1)^3 | n^{n-1} - 1 \Leftrightarrow (n-1)^2 | n^{n-2} + n^{n-3} + \cdots + n + 1$$

for all sufficiently large n, and since

$$n^{n-2} + n^{n-3} + \cdots + n + 1 =$$

$$= (n-1)[n^{n-3} + 2n^{n-4} + \cdots + (n-3)n + (n-2) + 1],$$

we obtain $n - 1 | n^{n-3} + 2n^{n-4} + \cdots + (n-3)n + (n-2) + 1$ for all sufficiently large n, which is clearly impossible, since

$$n^{n-3} + 2n^{n-4} + \cdots + (n-3)n + (n-2) + 1 \equiv 1 + 2 + \cdots + (n-2) + 1$$

$$\equiv \frac{(n-1)(n-2)}{2} + 1 \pmod{n-1}.$$

Hence $r \leq 2$. The relation

$$n^{n-1} - 1 = (n-1)^2[n^{n-3} + 2n^{n-4} + \cdots + (n-3)n + (n-2) + 1]$$

shows that $(n-1)^2 | n^{n-1} - 1$ for all $n > 1$ and allows us to conclude that all solutions are the polynomials $\pm(X-1)^r$, with $r \in \{0, 1, 2\}$.

After reading the solution of the following problem, you might think that the problem is very simple. Actually, it is extremely difficult. There are many possible approaches that fail and the time spent for solving such a problem can be significant.

Example 11. Let $f \in \mathbb{Z}[X]$ be a nonconstant polynomial, and let $k \geq 2$ be an integer such that $\sqrt[k]{f(n)} \in \mathbb{Z}$ for all positive integers n. Then there exists a polynomial $g \in \mathbb{Z}[X]$ such that $f = g^k$.

Solution. Let us assume the contrary, and let us factor $f = p_1^{k_1} \dots p_s^{k_s} g^k$ where $1 \le k_i < k$, and p_i are different irreducible polynomials in $\mathbb{Q}[X]$. Suppose that $s \ge 1$ (which is the same as negating the conclusion). Because p_1 is irreducible in $\mathbb{Q}[X]$, it is relatively prime with $p_1' p_2 \dots p_s$ and thus (using Bezout's theorem and multiplication by integers) there exist polynomials Q, R with integer coefficients and a positive integer c such that

$$Q(x)p_1(x) + R(x)p_1'(x)p_2(x)\dots p_s(x) = c.$$

Now, using the result from Example 1, we can take a prime number $q > |c|$ and a number n such that $q | p_1(n) \neq 0$. We have of course $q | p_1(n+q)$ (since $p_1(n+q) \equiv p_1(n) \pmod{q}$). The choice $q > |c|$ ensures that q does not divide $p_2(n)\dots p_s(n)$ and so $v_q(f(n)) = v_q(p_1(n)) + kv_q(g(n))$. But the hypothesis implies that $k \mid v_q(f(n))$, so $v_q(p_1(n)) \ge 2$. In a similar manner we obtain $v_q(p_1(n+q)) \ge 2$. Yet, using the binomial formula,

$$p_1(n+q) \equiv p_1(n) + qp_1'(n) \pmod{q^2}.$$

Hence we must have $q | p_1(n)$, which contradicts the fact that $q > |c|$ and

$$Q(x)p_1(x) + R(x)p_1'(x)p_2(x)\dots p_s(x) = c.$$

This contradiction shows that $s \ge 1$ is false and the result follows.

The next problem was given at the USA TST 2005 and uses a nice combination of arithmetic considerations and complex number computations. We take advantage of many arithmetical properties of polynomials in this problem, although the problem itself is not so difficult (if we find a good way to solve it, of course...).

Example 12. A polynomial $f \in \mathbb{Z}[X]$ is called special if for any positive integer $k > 1$, the sequence $f(1), f(2), f(3), \dots$ contains numbers which are relatively prime to k. Prove that for any $n > 1$, at least 71% of all monic polynomials of degree n with coefficients in the set $\{1, 2, \dots, n!\}$ are special.

[Titu Andreescu, Gabriel Dospinescu] USA TST 2005

Solution. Of course, before counting such polynomials, it would be better to find an easier characterization for them.
Let $p_1, p_2, \ldots, p_r$ be all the prime numbers not exceeding n, and consider the sets $A_i = \{f \in M|\ p_i|f(m),\ \forall\ m \in \mathbb{N}^*\}$, where M is the set of monic polynomials of degree n with coefficients in the set $\{1, 2, \ldots, n!\}$. We will prove that the set T of special polynomials is exactly $M \setminus \bigcup_{i=1}^{r} A_i$. Clearly, $T \subset M \setminus \bigcup_{i \le r} A_i$. The converse, however, is not that easy. Let us suppose that $f \in \mathbb{Z}[X]$ belongs to $M \setminus \bigcup_{i=1}^{r} A_i$ and let p be a prime number greater than n. Because f is monic, Lagrange's theorem ensures that we can find m such that p is not a divisor of $f(m)$. It follows that for any prime number q at least one of the numbers $f(1), f(2), f(3), \ldots$ is not a multiple of q. Let $k > 1$ and let $q_1, q_2, \ldots, q_s$ be its prime divisors. Then we can find $u_1, \ldots, u_s$ such that q_i does not divide $f(u_i)$. Using the Chinese remainder theorem, there is a positive integer x such that $x \equiv u_i \pmod{q_i}$. Consequently, $f(x) \equiv f(u_i) \pmod{q_i}$ and thus q_i does not divide $f(x)$, so $\gcd(f(x), k) = 1$. The equality of the two sets is now proved.
Using a raw estimation, we obtain

$$|T| = |M| - \left| \bigcup_{i=1}^{r} A_i \right| \ge |M| - \sum_{i=1}^{r} |A_i|.$$

Let us now compute $|A_i|$. Actually, we will show that $|A_i| = \dfrac{(n!)^n}{p_i^{p_i}}$. Let f be a monic polynomial in A_i,

$$f(X) = X^n + a_{n-1}X^{n-1} + \cdots + a_1 X + a_0.$$

Then, for any $m > 1$,

$$0 \equiv f(m) \equiv a_0 + (a_1 + a_p + a_{2p-1} + a_{3p-2} + \ldots)m$$

$$+(a_2 + a_{p+1} + a_{2p} + \ldots)m^2 + \cdots + (a_{p-1} + a_{2p-2} + a_{3p-3} + \ldots)m^{p-1} \pmod{p},$$

where, for simplicity, we put $p = p_i$. Again, using Lagrange's theorem it follows that $p \mid a_0,\quad p \mid a_1 + a_p + a_{2p-1} + \cdots, \cdots, p \mid a_{p-1} + a_{2p-2} + \cdots$. We are going to use this later, but a small observation is still needed. Let us count the number of s-tuples $(x_1, x_2, \ldots, x_s) \in \{1, 2, \ldots, n!\}^s$ such that $x_1 + x_2 + \cdots + x_s \equiv u \pmod{p}$, where u is fixed. Let

$$\varepsilon = \cos\frac{2\pi}{p} + i\sin\frac{2\pi}{p}$$

and observe that

$$0 = (\varepsilon + \varepsilon^2 + \cdots + \varepsilon^{n!})^s$$

$$= \sum_{k=0}^{p-1} \varepsilon^k \left|\{(x_1, x_2, \ldots, x_s) \in \{1, 2, \ldots, n!\}^s \mid x_1 + \cdots + x_s \equiv k \pmod{p)}\}\right|.$$

A simple argument related to the irreducibility of the polynomial $1+X+X^2+\cdots+X^{p-1}$ shows that all numbers that appear in the above sum are equal, and that their sum is $(n!)^s$, thus each number equals $\dfrac{(n!)^s}{p}$.

We are now ready to finish the proof. Assume that among the numbers $a_1, a_p, a_{2p-1}, \ldots$ there are exactly v_1 numbers, and so on, finally there are v_{p-1} numbers among $a_{p-1}, a_{2p-2}, \ldots$. Using the above observations, it follows that

$$|A_i| = \frac{n!}{p} \cdot \frac{(n!)^{v_1}}{p} \cdots \frac{(n!)^{v_{p-1}}}{p} = \frac{(n!)^n}{p^p}.$$

Hence

$$|T| \geq (n!)^n - \sum_{p \text{ prime}} \frac{(n!)^n}{p^p}.$$

But

$$\frac{1}{5^5} + \frac{1}{7^7} + \cdots < \frac{1}{5^5}\left(1 + \frac{1}{5} + \frac{1}{5^2} + \ldots\right) < \frac{1}{1000}$$

and so the percent of special polynomials is at least

$$100\left(1 - \frac{1}{4} - \frac{1}{27} - \frac{1}{1000}\right) = 75 - \frac{100}{27} - \frac{1}{10} > 71.$$

Just a few more observations about this problem. The authors discovered (after the problem was submitted and given in the TST) that this question was the object of Jan Turk's article *The fixed divisor of a polynomial* published in the fourth issue of the American Mathematical Monthly, 1986. In this article, with a completely different idea and technique, much more involved and precise estimations are obtained. For instance, the author proves that the probability for a random polynomial with integer coefficients to be special is $\prod_p \left(1 - \frac{1}{p^p}\right)$, which is approximately 0.722. This shows that even though our estimations were very elementary, they were not far from reality. We invite the reader to read this fascinating article.

Example 13. Suppose that a polynomial f with integer coefficients has no double zeros. Then for any positive integer r there exists an n such that in the prime decomposition of $f(n)$ there are at least r distinct prime divisors, all of them with exponent 1.

Iranian Olympiad

Solution. Already for $r = 1$ the problem is in no way obvious. So let's not attack the general case directly, but rather concentrate first on the case $r = 1$. Suppose the contrary, that is for all n the prime divisors of $f(n)$ have exponent at least 2. Because f has no double zero, $\gcd(f, f') = 1$ in $\mathbb{C}[X]$ and thus also in $\mathbb{Q}[X]$ (because of the division algorithm and Euclid's algorithm). Using Bèzout's theorem in $\mathbb{Q}[X]$, we can find polynomials P, Q with integer coefficients such that $P(n)f(n) + Q(n)f'(n) = c$ for some positive integer c. Using the result in the first example, we can take $q > c$ a prime divisor of some $f(n)$. Our hypothesis ensures that $q^2 | f(n)$. But then, also, $q | f(n + q)$ and so $q^2 | f(n+q)$. Using Newton's binomial formula, we deduce immediately that $f(n + q) \equiv f(n) + qf'(n) \pmod{q^2}$. We finally find $q | f'(n)$ and so $q | c$, which is impossible, since our choice was $q > c$. Thus the case $r = 1$ is proved. Let us now try to prove the property by induction and suppose it is true for r. Of course, the existence of P, Q such that $P(n)f(n) + Q(n)f'(n) = c$ for some positive integer c did not depend on r, so we keep the above notations. By the inductive hypothesis, there is n such that at least r prime divisors of $f(n)$ have exponent 1. Let these prime factors be $p_1, p_2, \ldots, p_r$. But it is clear that $n + kp_1^2 p_2^2 \ldots p_r^2$ has the same property: all prime divisors $p_1, p_2, \ldots, p_r$ have exponent 1 in the decomposition of $f(n+kp_1^2 p_2^2 \ldots p_r^2)$. Because at most a finite number among them can be zeros of f, we may assume from the beginning that n is not a zero of f. Consider now the polynomial $g(X) = f(n+(p_1 \ldots p_r)^2 X)$, which is obviously nonconstant. Thus using again the result in Example 1, we find a prime number $q > \max\{|c|, p_1, \ldots, p_r, |p(n)|\}$ and a number u such that $q|g(u)$. If $v_q(g(u)) = 1$, victory is ours, since a trivial verification shows that $q, p_1, \ldots, p_r$ are different prime numbers whose exponents in $f(n+(p_1 \ldots p_r)^2 u)$ are all 1. The difficult case is when $v_q(g(u)) \geq 2$. In this case, we will consider the number

$$N = n + u(p_1 \ldots p_r)^2 + uq(p_1 \ldots p_r)^2.$$

Let us prove that in the decomposition of $f(N)$, all prime numbers $q, p_1, \ldots, p_r$ have exponent 1. For any p_i, this is true since $f(N) \equiv f(n) \pmod{(p_1 \ldots p_r)^2}$. Using once again the binomial formula, we obtain

$$f(N) \equiv f(n + (p_1 \ldots p_r)^2 u) + uq(p_1 \ldots p_r)^2 f'(N) \pmod{q^2}.$$

Now, if $v_q(f(n)) \geq 2$, then since $v_q(f(n + (p_1 \ldots p_r)^2 u)) = v_q(g(u)) \geq 2$, we have $q|u(p_1 \ldots p_r)^2 f'(N)$. Recall that the choice was $q > \max\{|c|, p_1, \ldots, p_r, |p(n)|\}$ so necessarily $q|u$ (if $q|f'(N) \Rightarrow q|(f(N), f'(N))|c \Rightarrow q \leq |c|$, contradiction). But since $q|g(u)$, we have $q|g(0) = f(n)$. Fortunately, we ensured that n is not a zero of our polynomial and also that $q > \max\{|c|, p_1, \ldots, p_r, |p(n)|\}$ so the last divisibility cannot hold. This finishes the inductive step and solves the problem.

Did you like Erdős's Corner in chapter **Look at the Exponent**? We repeat the experience, with a series of difficult problems related to prime divisors of polynomials. When we say difficult, we say however solvable, because one should know that most of the problems concerning quantitative estimates for prime divisors of polynomials are still unsolved and will probably remain so for very long time. Let us recall a few terrible results that have been obtained so far, of course without proofs. Let $P(n)$ be the greatest prime divisor of n. Even the fact that $P(f(n))$ tends to ∞ for any polynomial f of degree at least

2 is a very difficult result (even the case $\deg(f) = 2$ requires the Thue-Siegel theorems). An extremely difficult theorem of Erdős shows that the largest prime divisor of $f(1)f(2)...f(n)$ is greater than $n \cdot e^{(\ln n)^c}$ for some absolute constant $c > 0$. All these results require very deep results in algebraic and analytic number theory. Another is the famous open question of prime-producing polynomials: any polynomial f without a fixed divisor should produce prime numbers infinitely many times. All these questions are far beyond the known results. But, of course, we will discuss just a few results with elementary (more or less) solutions.

The first problem investigates Schur's theorem for a family of polynomials. The following solution was suggested to us by Vesselin Dimitrov. The beauty of the result can be easily seen when studying the second part of the problem, where we prove by elementary means a result that usually was proved using Galois theory. Even though we haven't found the first article studying this problem, we did find one signed by T. Nagell, so we will call this Nagell's theorem.

Example 14. a) Let $f_1, f_2, \cdots, f_n$ be nonconstant polynomials with integer coefficients. Prove that there are infinitely many primes numbers p with the property that $f_1, f_2, \cdots, f_n$ have a zero in $\mathbb{Z}/p\mathbb{Z}$ (that is, there exist integers $k_1, k_2, \cdots, k_n$ such that $p | f_i(k_i)$ for all i).
b) Prove that for any nonconstant polynomial f with integer coefficients and any positive integer k there are infinitely many primes of the form $1 + qk$ that divide at least one of the numbers $f(1), f(2), f(3), ...$

Nagell's theorem

Solution. For $n = 1$, a) is just Schur's theorem. Actually, the idea is to reduce the study to this special case, by proving the existence of polynomials $g_1, g_2, \ldots, g_n$ such that $f_i(g_i(X))$ have a common nontrivial divisor. This is not immediate, however. Let us see what we are asking for: of course, if there exists a common nontrivial divisor, it must have a complex root z, so first of all we should see whether we can find g_i with rational coefficients and some z such that $f_i(g_i(X))$ have common root z. In this case, $g_i(z)$ would be all the zeros of f_i, so it is more than natural to start by fixing some roots $x_1, x_2, \ldots, x_n$ of $f_1, f_2, \ldots, f_n$ respectively and trying to find some z and some g_i with $g_i(z) = x_i$. And now, a very useful theorem from algebraic number theory (but whose proof is completely elementary) helps us: actually, any finite extension of the field of rational numbers is generated by one element. T hat is, if $a_1, a_2, \ldots, a_k$ are algebraic numbers (over the field of rational numbers), then there exists an algebraic number α such that $\mathbb{Q}(a_1, a_2, \ldots, a_k) = \mathbb{Q}(\alpha)$. We will leave the proof of this theorem as a beautiful exercise for the reader (in case you do not manage to solve it alone, any introductory book to algebraic number theory gives a proof of this result). Now, x_i are clearly algebraic, since they are roots of f_i. Thus there exists some algebraic number z for which $\mathbb{Q}(x_1, x_2, \ldots, x_n) = \mathbb{Q}(z)$. By multiplying z by a suitable integer, we may assume that z is actually an algebraic integer. This means that each x_i can be written in the form $g_i(z)$ for some polynomial g_i with rational coefficients. Of course, there exists some integer N for which $h_i = Ng_i$ have integer coefficients and there exists some large d for which $F_i(X) = N^d f_i\left(\frac{h_i(X)}{N}\right)$ also has integer coefficients. Now, all F_i are divisible by P, the minimal polynomial of z in $\mathbb{Q}[X]$. Because z is an algebraic integer, P is a monic polynomial with integer coefficients, and thus primitive. From Gauss's lemma, it follows that F_i are divisible by P in $\mathbb{Z}[X]$. Finally, let us apply Schur's theorem to this polynomial. There are infinitely many prime $p > N$ for which F has a zero n_p in $\mathbb{Z}/p\mathbb{Z}$. Fix such a prime p and note that $x = n_p$. Let $f_i(X) = A_s X^s + A_{s-1} X^{s-1} + \cdots + A_0$. We know that p divides

$$A_s N^{d-s} h_i(x)^s + A_{s-1} N^{d-s+1} h_i(x)^{s-1} + \cdots + A_0 N^d.$$

Of course, p is relatively prime to N, so p will actually divide

$$A_s h_i(x)^s + A_{s-1} N h_i(x)^{s-1} + \cdots + A_0 N^s.$$

Thus, if N' is the inverse of N in $\mathbb{Z}/p\mathbb{Z}$, $N' h_i(x)$ is a zero of f_i modulo p. Since i was arbitrary, it follows that all f_i have a zero in $\mathbb{Z}/p\mathbb{Z}$ for any such prime p. The conclusion follows.

Part b) is actually a fairly immediate consequence of a). The idea is that for $n > 1$, any prime divisor of $\phi_n(a)$, the nth cyclotomic polynomial, is either congruent to 1 modulo p or divides n. The proof of this result is not very difficult. Indeed, consider p such a prime divisor. Then $p|a^n - 1$ and thus, if d is the order of a modulo p, we have $d|n$ and $d|p-1$. Clearly, if $d = n$, we are done, so assume that $d < n$. Then since $p|a^d - 1 = \prod_{k|d} \phi_k(a)$, there exists a divisor k of d such that $p|\phi_k(a)$. However, $X^n - 1$ is the product of all cyclotomic polynomials whose orders divide n, so it is a multiple of $\phi_k(X) \cdot \phi_n(X)$. Therefore, $X^n - 1$ will have a as a double root in $\mathbb{Z}/p\mathbb{Z}$. This is impossible unless $p|n$, because in this case a would be a root of nX^{n-1} and thus $p|n$ (since p is not a divisor of a). This proves the claim. Now, using a) for the polynomials $\phi_k(X)$ and $f(X)$, we know there are infinitely many primes p such that both these polynomials have roots in the field with p elements. But the observation made in the beginning of b) shows that only finitely many of these prime numbers are not congruent to 1 modulo k. Thus, infinitely many are of the form $1 + kq$ and the proof finishes here.

The next example concerns the very classical problem of square free numbers among polynomial values. More generally, one defines k-free numbers as nonzero integers which are not divisible by any k-th power of a prime. One can prove (the idea is exactly the same as in the problem that we will discuss) that if f is a primitive polynomial of degree d and if f is not the d-th power of a linear polynomial, then a positive proportion of positive integers n have the property that $f(n)$ is d-free. A more difficult result was proved by Erdős: under some natural conditions imposed on f, there are infinitely many n for which $f(n)$ is $d-$free. Needless to say, the proof if highly nontrivial. We will discuss a closely related problem concerning square free numbers of a special form.

The next result is a lot stronger than the one proved by Laurentiu Panaitopol, stating that there are infinitely many triples of consecutive numbers, all square free. The solution is adapted from a beautiful argument due to Ravi Boppana. Before passing to this problem, let us give a definition: we say that a set A of positive integers has positive density if there exists a constant $c > 0$ such that for all sufficiently large x there are at least cx elements of A less than x.

Example 15. Prove that the set of positive integers n such that

$$\frac{1}{2}n(n+1)(n+2)(n^2+1)$$

is square free has positive density.

[Vesselin Dimitrov]

Solution. Let us search for such numbers of the form $n = 180k+1$ for some positive integer k. By this choice, $\frac{1}{2}n(n+1)(n+2)(n^2+1)$ is not divisible by 4 or 9 or 25. So we can ignore the prime factors 2, 3 and 5. Let p be a prime greater than 5. There is exactly one $k \pmod{p^2}$ such that $n = 180k+1$ is divisible by p^2, exactly one $k \pmod{p^2}$ such that $n+1$ is divisible by p^2, and exactly one $k \pmod{p^2}$ such that $n+2$ is divisible by p^2. Also, there are at most two $k \pmod{p^2}$ such that n^2+1 is divisible by p^2. Indeed, if $p^2 | a^2+1$ and $p^2 | b^2+1$, then $p^2 | (a-b)(a+b)$. Then $p^2 | a-b$ or $p^2 | a+b$ (otherwise, p divides $a-b$ and $a+b$, thus it divides a too, which is clearly impossible). Altogether there are at most five $k \pmod{p^2}$ such that one of n, $n+1$, $n+2$, or n^2+1 is divisible by p^2. Let N be a large positive integer. By the previous observation, there are at most $5\left\lceil \frac{N}{p^2} \right\rceil$ values of k between 1 and N such that n, $n+1$, $n+2$, or n^2+1 is divisible by p^2. If $p > 180N+1$, then p is too large for n, $n+1$, $n+2$, or n^2+1 to be divisible by p^2. Altogether the number of k between 1 and N such that one of n, $n+1$, $n+2$, or n^2+1 is not square free is at most $\sum_{p=7}^{180N+1} 5\left\lceil \frac{N}{p^2} \right\rceil$.
We can bound the last sum by

$$\sum_{p=7}^{180N+1}\left(5+\frac{5N}{p^2}\right) \le 5\pi(180N+1) + 5N\sum_{p\ge 7}\frac{1}{p^2}$$

and since

$$\sum_{p\ge 7}\frac{1}{p^2} \le \sum_{m\ge 3}\frac{1}{(2m+1)(2m-1)} \le \frac{1}{2}\left(\frac{1}{5}-\frac{1}{7}+\frac{1}{7}-\frac{1}{9}+\cdots\right) = \frac{1}{10},$$

we infer that the number of "bad" k is at most $\frac{N}{2} + o(N)$. We used here the classical fact that $\pi(x) = o(x)$, where $\pi(x) = \sum_{p \le x} 1$ is the counting function of the prime numbers (for a proof of this result, see the chapter **At the Border between Analysis and Number Theory**).

Therefore, the number of $1 \le k \le N$ for which all numbers $n, n+1, n+2, n^2+1$ (where $n = 180k+1$) are squarefree is at least $\frac{N}{2} + o(N)$. For any such number k, $\frac{1}{2}n(n+1)(n+2)(n^2+1)$ is squarefree (the only common prime divisors of two numbers among $n, n+1, n+2, n^2+1$ are $2, 3, 5$ and we saw that the choice of n ensures that $4, 9, 25$ are not divisors of $\frac{1}{2}n(n+1)(n+2)(n^2+1)$). Thus, the number of $n < 181N$ such that $\frac{1}{2}n(n+1)(n+2)(n^2+1)$ is squarefree is at least $\frac{N}{2} + o(N)$, which means that the set of n for which $\frac{1}{2}n(n+1)(n+2)(n^2+1)$ is squarefree has positive density.

10.2 Practice problems

1. Let $f_1, f_2, \ldots, f_k$ be nonconstant polynomials with integer coefficients. Prove that for infinitely many n all numbers $f_1(n), f_2(n), \ldots, f_k(n)$ are composite.

2. Let $f \in \mathbb{Z}[X]$ and $n > 3$. Prove that there are no integers $x_1, x_2, \ldots, x_n$ such that $f(x_i) = x_{i-1}$, $i = 1, 2, \ldots, n$, indices being taken mod n.

3. Let $f \in \mathbb{Z}[X]$ be a polynomial of degree $n \geq 2$. Prove that the polynomial $f(f(X)) - X$ has at most n integer zeros.

Gh. Eckstein, Romanian TST

4. Find all integers $n > 1$ for which there is a polynomial $f \in \mathbb{Z}[X]$ with the property: for any integer k one has $f(k) = 0 \pmod{n}$ or $f(k) = 1 \pmod{n}$ and both these equations have solutions.

5. Find all polynomials f with integer coefficients such that $f(n) \mid 2^n - 1$ for all positive integer n.

Polish Olympiad

6. Let p be a prime and let $f \in \mathbb{Z}[X]$ be a polynomial such that the numbers $f(0), f(1), \ldots, f(p^2 - 1)$ give distinct remainders when divided by p^2. Prove that the numbers $f(0), f(1), \ldots, f(p^3 - 1)$ give distinct remainders when divided by p^3.

Putnam 2008

7. Is there a nonconstant polynomial $f \in \mathbb{Z}[X]$ and an integer $a > 1$ such that the numbers $f(a), f(a^2), f(a^3), \ldots$ are pairwise relatively prime?

St Petersburg 1998

8. Let $f \in \mathbb{Z}[X]$ be a nonconstant polynomial. Prove that the sequence $f(3^n) \pmod{n}$ is not bounded.

9. Find all polynomials f with integer coefficients, having the following property: there exists k such that for all primes p, $f(p)$ has at most k prime factors.

10. Find all polynomials $f \in \mathbb{Z}[X]$ with the property that for any relatively prime integers m, n, the numbers $f(m), f(n)$ are also relatively prime.

Iran TST

11. Let f be a polynomial with integer coefficients and let $a_0 = 0$ and $a_n = f(a_{n-1})$ for all $n \geq 1$. Prove that $(a_n)_{n\geq 0}$ is a Mersenne sequence, that is $\gcd(a_m, a_n) = a_{\gcd(m,n)}$ for all positive integers m and n.

Romanian TST

12. Find all integers k such that if a polynomial with integer coefficients f satisfies $0 \leq f(0), f(1), ..., f(k+1) \leq k$ then $f(0) = f(1) = \cdots = f(k) = f(k+1)$.

IMO 1997 Shortlist

13. (a) Let f be a polynomial with real coefficients. Prove the equivalence of the following two assertions:

    i. for any integer n one has $f(n) \in \mathbb{Z}$;

    ii. There exist integers n and $a_0, a_1, a_2, \ldots, a_n$ such that

$$f(X) = a_0 + a_1 X + a_2 \frac{X(X-1)}{2} + \cdots + a_n \cdot \frac{X(X-1)\cdots(X-n+1)}{n!}.$$

    (b) Let n be a positive integer. What is the least degree of a monic polynomial f with integer coefficients such that $n \mid f(k)$ for any integer k?

14. Let f be a polynomial with rational coefficients such that $f(n) \in \mathbb{Z}$ for all $n \in \mathbb{Z}$. Prove that for any integers m, n the number $\text{lcm}[1, 2, \ldots, \deg(f)] \cdot \frac{f(m)-f(n)}{m-n}$ is an integer.

MOSP 2001

15. Let f be a polynomial of degree d such that $f(\mathbb{Z}) \subset \mathbb{Z}$ and $\frac{f(n)-f(m)}{m-n} \in \mathbb{Z}$ for all $0 \leq m, n \leq d$. Prove that $\frac{f(n)-f(m)}{m-n} \in \mathbb{Z}$ for all integers $m \neq n$.

Holden Lee

16. Let P be a polynomial with integer coefficients such that $P(0) = 0$ and

$$\gcd(P(0), P(1), P(2), \ldots,) = 1.$$

Show there are infinitely many n such that

$$\gcd(P(n) - P(0), P(n+1) - P(1), P(n+2) - P(2), \ldots) = n.$$

USA TST 2010

17. Let d, r be positive integers with $d \geq 2$. Prove that for any non-constant polynomial $f \in \mathbb{R}[X]$ of degree smaller than r, the numbers $f(0), f(1), \ldots, f(d^r - 1)$ can be divided into d groups such that the sum of the numbers in each group is the same.

J. O. Shallit, AMM E 3032

18. Let $(a_n)_{n \geq 1}$ be an increasing sequence of positive integers such that for some polynomial $f \in \mathbb{Z}[X]$ we have $a_n \leq f(n)$ for all n. Suppose also that $m - n \mid a_m - a_n$ for all distinct positive integers m, n. Prove that there exists a polynomial $g \in \mathbb{Q}[X]$ such that $a_n = g(n)$ for all n.

USAMO 1995

19. (a) Prove that for each positive integer n there is a polynomial $f \in \mathbb{Z}[X]$ such that $f(1) < f(2) < \cdots < f(n)$ are primes.
   (b) As above, but now the numbers are powers of 2, not primes.
   (c) Let $a > 1$ be an integer and let n be a positive integer. Prove that there exists a polynomial f of degree n, having integer coefficients such that $f(0), f(1), \ldots, f(n)$ are pairwise distinct positive integers, all of the form $2a^k + 3$ for some integer k.

Chinese TST 2004

20. Let $p > 5$ be a prime and let a, b, c be integers such that p does not divide any of the numbers $a - b, b - c, c - a$. Let i, j, k be nonnegative integers such that $i + j + k$ is divisible by $p - 1$ and such that for all integers x, the number

$$(x-a)(x-b)(x-c)[(x-a)^i(x-b)^j(x-c)^k - 1]$$

is divisible by p. Prove that each of i, j, k is divisible by $p - 1$.

Kiran Kedlaya and Peter Shor, USA TST 2009

21. Consider all sequences $(f(1) \pmod{n}, f(3) \pmod{n}, \ldots, f(1023) \pmod{n})$, where $n = 1024$ and f is an arbitrary polynomial with integer coefficients. Prove that among these sequences there are at most 2^{35} that are permutations of the numbers $1, 3, 5, \ldots, 1023$.

USA TST 2007

22. Prove that for all n there exists a polynomial f with integer coefficients and degree not exceeding n such that 2^n divides $f(x)$ for all even integers x and 2^n divides $f(x) - 1$ for all odd integers x.

P. Hajnal, Komal

23. Let n be an even positive integer. Find the least positive integer k for which one can find polynomials with integer coefficients f, g such that

$$f(X)(X+1)^n + g(X)(X^n+1) = k.$$

IMO Shortlist 1996

24. Let $f \in \mathbb{Z}[X]$ be a polynomial of degree at least 2, with positive leading coefficient. Show that there are infinitely many n such that $f(n!)$ is composite.

IMO Shortlist 2005

25. Suppose that n is a positive integer not divisible by the cube of a prime number. Consider all sequences $(x_1, x_2, \ldots, x_n)$ with $x_i \in \mathbb{Z}/n\mathbb{Z}$. For how many of these can we find a polynomial f with integer coefficients such that $f(i) \pmod{n} = x_i$ for all i?

USA TST 2008

26. Prove the existence of a number $c > 0$ with the following property: for any prime p, there are at most $cp^{2/3}$ positive integers n such that p divides $n! + 1$.

Chinese TST 2009

27. Find all polynomials $f \in \mathbb{Z}[X]$ such that $f(p) \mid 2^p - 2$ for any prime number p.

Gabriel Dospinescu, Peter Scholze

28. Let $f \in \mathbb{Z}[X]$ be a polynomial and let a be an integer. Consider the sequence $a_0 = a, a_{n+1} = f(a_n)$. If $a_n \to \infty$ and if the set of prime divisors of $\{a_n\}$ is finite, prove that $f(X) = AX^d$ for some A, d.

Tuymaada Olympiad 2003

Chapter 11

11.1 Theory and examples

Almost everyone knows the Chinese Remainder Theorem, which is a remarkable tool in number theory. But does everyone know the analogous form for polynomials? Stated like this, this question may seem impossible to answer. Then, let us make it easier and also reformulate it: is it true that given some pairwise distinct real numbers $x_0, x_1, x_2, \ldots, x_n$ and some arbitrary real numbers $a_0, a_1, a_2, \ldots, a_n$, we can find a polynomial f with real coefficients such that $f(x_i) = a_i$ for $i \in \{0, 1, \ldots, n\}$? The answer turns out to be positive, and a possible solution to this question is based on Lagrange's interpolation formula. It says that an example of such polynomial is

$$f(x) = \sum_{i=0}^{n} a_i \prod_{\substack{0 \leq j \leq n \\ j \neq i}} \frac{x - x_j}{x_i - x_j}. \tag{11.1}$$

(In what follows along this unit, a product like the above one will be written, for simplicity, just as $\prod_{j \neq i} \frac{x - x_j}{x_i - x_j}$.)

Indeed, it is immediate to see that $f(x_i) = a_i$ for $i \in \{0, 1, \ldots, n\}$. Also, the above expression shows that this polynomial has degree less than or equal to n. Is this the only polynomial with this supplementary property? Yes, and the proof is not difficult at all. Just suppose we have another polynomial g of degree less than or equal to n and such that $g(x_i) = a_i$ for $i \in \{0, 1, \ldots, n\}$. Then the polynomial $g - f$ also has degree less than or equal to n and vanishes at $x_0, x_1, \ldots, x_n$. Thus, it must be null, and the uniqueness is proved.

What is Lagrange's interpolation theorem good for? We will see in the following problems that it helps us to immediately find the value of a polynomial in a certain point if we know the values in some given points. And the reader may have already noticed that this follows directly from the formula (1), which shows that if we know the value in $1 + \deg f$ points, then we can find the value in any other point without solving a complicated linear system. Also, we will see that it helps in establishing some inequalities and bounds for certain special polynomials, and will even help us in finding and proving some beautiful identities. Now, let us begin the journey through some nice examples of problems where this idea can be used. As promised, we will first see how we can rapidly compute the value in a certain point for some polynomials. This was one of the favorites problems in the old Olympiads, as the following example illustrates.

Example 1. Let $F_1 = F_2 = 1$, $F_{n+2} = F_n + F_{n+1}$ and let f be a polynomial of degree 990 such that $f(k) = F_k$ for $k \in \{992, \ldots, 1982\}$. Show that $f(1983) = F_{1983} - 1$.

[Titu Andreescu] IMO 1983 Shortlist

Solution. So, we have $f(k+992) = F_{k+992}$ for $k = 0, 1, ..., 990$ and we need to prove that $f(992+991) = F_{1983} - 1$. This simple observation shows that we don't have to bother too much with $k + 992$, since we could work as well with the polynomial $g(x) = f(x+992)$, which also has degree 990. Now, the problem becomes: if $g(k) = F_{k+992}$, for $k = 0, 1, ..., 990$, then $g(991) = F_{1983} - 1$. But we know how to compute $g(991)$. Indeed, looking again at the interpolation formula (as we name equation (1)), we find that

$$g(991) = \sum_{k=0}^{990} g(k)\binom{991}{k}(-1)^k = \sum_{k=0}^{990}\binom{991}{k}F_{k+992}(-1)^k$$

which shows that we need to prove the identity

$$\sum_{k=0}^{990}\binom{991}{k}F_{k+992}(-1)^k = F_{1983} - 1.$$

We know that

$$F_n = \frac{a^n - b^n}{\sqrt{5}},$$

where $a = \dfrac{\sqrt{5}+1}{2}$ and $b = \dfrac{1-\sqrt{5}}{2}$. Bearing this in mind, we can of course try a direct approach:

$$\sum_{k=0}^{990} \binom{991}{k} F_{k+992}(-1)^k$$

$$= \frac{1}{\sqrt{5}} \left[\sum_{k=0}^{990} \binom{991}{k} a^{k+992}(-1)^k - \sum_{k=0}^{990} \binom{991}{k} b^{k+992}(-1)^k \right].$$

But using the binomial theorem, the above sums vanish:

$$\sum_{k=0}^{990} \binom{991}{k} a^{k+992}(-1)^k = a^{992} \sum_{k=0}^{990} \binom{991}{k} (-a)^k = a^{992}[(1-a)^{991} + a^{991}].$$

Since $a^2 = a+1$, we have

$$a^{992}[(1-a)^{991} + a^{991}] = a(a-a^2)^{991} + a^{1983} = -a + a^{1983}.$$

Since in all this argument we have used only the fact that $a^2 = a+1$ and since b also satifies this relation, we find that

$$\sum_{k=0}^{990} \binom{991}{k} F_{k+992}(-1)^k = \frac{1}{\sqrt{5}}(a^{1983} - b^{1983} - a + b)$$

$$= \frac{a^{1983} - b^{1983}}{\sqrt{5}} - \frac{a-b}{\sqrt{5}} = F_{1983} - 1.$$

And this is how, with the help of a precious formula and with some smart computations, we could solve this problem and also find a nice property of the Fibonacci numbers.

The following example is a very nice problem proposed for IMO 1997. Here, the steps following the use of Lagrange's interpolation formula are even better hidden in some congruences. It is the typical example of a good Olympiad problem: no matter how much the contestant knows in that field, one may have great difficulties in solving it.

Example 2. Let f be a polynomial with integer coefficients, and let p be a prime such that $f(0) = 0$, $f(1) = 1$ and $f(k)$ is congruent to either 0 or 1 modulo p, for all positive integers k. Show that the degree of f is at least $p-1$.

IMO 1997 Shortlist

Solution. Such a problem should be solved indirectly, arguing by contradiction. So, let us suppose that $\deg f \le p-2$. Then, using the Interpolation Formula, we find that

$$f(x) = \sum_{k=0}^{p-1} f(k) \prod_{j \neq k} \frac{x-j}{k-j}.$$

Now, since $\deg f \le p-2$, the coefficient of x^{p-1} in the right-hand side of the identity must be zero. Consequently, we have

$$\sum_{k=0}^{p-1} \frac{(-1)^{p-k-1}}{k!(p-1-k)!} f(k) = 0.$$

From here we have one more step. Indeed, let us write the above relation in the form

$$\sum_{k=0}^{p-1} (-1)^k \binom{p-1}{k} f(k) = 0$$

and let us take this equality modulo p. Since

$$k!\binom{p-1}{k} = (p-k)(p-k+1)\ldots(p-1) \equiv (-1)^k k! \pmod p$$

we find that

$$\binom{p-1}{k} \equiv (-1)^k \pmod p$$

and so

$$\sum_{k=0}^{p-1} (-1)^k \binom{p-1}{k} f(k) \equiv \sum_{k=0}^{p-1} f(k) \pmod p.$$

Thus,

$$\sum_{k=0}^{p-1} f(k) \equiv 0 \pmod{p},$$

which is impossible, since $f(k) \equiv 0, 1 \pmod{p}$ for all k and not all of the numbers $f(k)$ have the same remainder modulo p (for example, $f(0)$ and $f(1)$). This contradiction shows that our assumption was wrong and the conclusion follows.

It's time now to see how some formidable identities are simple consequences of the Lagrange interpolation formula, although in these problems polynomials do not appear at first sight.

Example 3. Let $a_1, a_2, \ldots, a_n$ be pairwise distinct positive integers. Prove that for any positive integer k the number $\sum_{i=1}^{n} \frac{a_i^k}{\prod_{j \neq i}(a_i - a_j)}$ is an integer.

United Kingdom

Solution. Just by looking at the expression, we recognize the Lagrange Interpolation formula for the polynomial $f(x) = x^k$. But we may have some problems when the degree of this polynomial is greater than or equal to n. This can be solved by working with the remainder of f modulo $g(x) = (x-a_1)(x-a_2)\ldots(x-a_n)$. So, let us proceed by writing $f(x) = g(x)h(x)+r(x)$, where r is a polynomial of degree at most $n-1$. This time we don't have to worry since the formula works, and we obtain

$$r(x) = \sum_{i=1}^{n} r(a_i) \prod_{j \neq i} \frac{x - a_j}{a_i - a_j}.$$

Now, we need three observations. The first one is $r(a_i) = a_i^k$, the second one is that the polynomial r has integer coefficients, and the third one is that

$\displaystyle\sum_{i=1}^{n} \frac{a_i^k}{\prod\limits_{j\neq i}(a_i - a_j)}$ is just the coefficient of x^{n-1} in the polynomial

$$\sum_{i=1}^{n} r(a_i) \prod_{j\neq i} \frac{x - a_j}{a_i - a_j}.$$

Combining these observations, we find that $\displaystyle\sum_{i=1}^{n} \frac{a_i^k}{\prod\limits_{j\neq i}(a_i - a_j)}$ is the coefficient of x^{n-1} in r, which is an integer. Thus, not only have we solved the problem, but we have also found a rapid way to compute the sums of the form

$$\sum_{i=1}^{n} \frac{a_i^k}{\prod_{j\neq i}(a_i - a_j)}.$$

The following two problems concern some combinatorial sums. If the first one is relatively easy to prove using a combinatorial argument (it is a very good exercise for the reader to find this argument), for the second problem a combinatorial approach is much more difficult. But we will see that both are immediate consequences of the interpolation formula.

Example 4. Let $f(x) = \displaystyle\sum_{k=0}^{n} a_k x^{n-k}$. Prove that for any non-zero real number h and any real number A we have

$$\sum_{k=0}^{n} (-1)^{n-k} \binom{n}{k} f(A + kh) = a_0 \cdot n! \cdot h^n.$$

Solution. Since this polynomial has degree at most n, we have no problems in applying the interpolation formula

$$f(x) = \sum_{k=0}^{n} f(A + kh) \prod_{j\neq k} \frac{x - A - jh}{(k - j)h}.$$

Now, let us identify the leading coefficients in both polynomials that appear in the equality. We find that

$$a_0 = \sum_{k=0}^{n} f(A+kh)\frac{1}{\prod_{j\neq k}[(k-j)h]} = \frac{1}{n!h^n}\sum_{k=0}^{n}(-1)^{n-k}\binom{n}{k}f(A+kh),$$

which is exactly what we had to prove. Simple and elegant! Notice that the above problem implies the well-known combinatorial identities

$$\sum_{k=0}^{n}(-1)^k\binom{n}{k}k^p = 0$$

for all $p \in \{0, 1, 2, \ldots, n-1\}$ and $\sum_{k=0}^{n}(-1)^{n-k}\binom{n}{k}k^n = n!$. (Notice that the identity remains valid for $h = 0$, too!)

As we promised, we will discuss a much more difficult problem. The reader might say after reading the solution: but this is quite natural! Yes, it is natural for someone who knows the Lagrange Interpolation formula very well and especially for someone who thinks that using it could lead to a solution. Unfortunately, this isn't always so easy.

Example 5. Prove the identity

$$\sum_{k=0}^{n}(-1)^{n-k}\binom{n}{k}k^{n+1} = \frac{n(n+1)!}{2}.$$

Solution. We take the polynomial $f(x) = x^n$. (Why don't we take the polynomial $f(x) = x^{n+1}$? Simply because $(-1)^{n-k}\binom{n}{k}$ appears when writing the formula for a polynomial of degree at most n.) We write the Interpolation Formula

$$x^n = \sum_{k=0}^{n} k^n \frac{x(x-1)\cdots(x-k-1)(x-k+1)\cdots(x-n)}{(n-k)!k!}(-1)^{n-k}$$

Now, we identify the coefficient of x^{n-1} in both terms. We find that

$$0=\sum_{k=0}^{n}(-1)^{n-k}\binom{n}{k}k^n(1+2+\cdots+n-k).$$

And now the problem is solved, since we found that

$$\sum_{k=0}^{n}(-1)^{n-k}\binom{n}{k}k^{n+1}=\frac{n(n+1)}{2}\sum_{k=0}^{n}(-1)^{n-k}\binom{n}{k}k^n$$

and we also know that

$$\sum_{k=0}^{n}(-1)^{n-k}\binom{n}{k}k^n=n!$$

from the previous problem.

If the Lagrange interpolation formula was good only to establish identities and to compute values of polynomials, it would not have been such a great discovery. Of course this is not the case—it plays a fundamental role in analysis. However, we are not going to enter this field, and we prefer to concentrate on another elementary aspect of this formula and see how it can help us establish some remarkable inequalities. And some of them will be really tough.
We begin with a really difficult inequality, originally published by H.S. Shapiro in the American Mathematical Monthly, in which the interpolation formula is really well hidden. But denominators can give valuable indications from time to time.

Example 6. Prove that for any real numbers $x_1, x_2, \ldots, x_n \in [-1,1]$ the following inequality is true:

$$\sum_{i=1}^{n}\frac{1}{\prod_{j\neq i}|x_j-x_i|}\geq 2^{n-2}.$$

[H. S. Shapiro] Iranian Olympiad

Solution. The presence of $\prod_{j\neq i}|x_j - x_i|$ is the only hint to this problem. But even if we know it, how do we choose the polynomial? The answer is simple: we will choose it to be arbitrary, and only in the end we will decide which one is optimal. So, let us proceed by taking $f(x) = \sum_{k=0}^{n-1} a_k x^k$ an arbitrary polynomial of degree $n-1$. Then we have

$$f(x) = \sum_{k=1}^{n} f(x_k) \prod_{j\neq k} \frac{x - x_j}{x_k - x_j}.$$

Combining this with the triangle inequality, we get

$$|f(x)| \leq \sum_{k=1}^{n} |f(x_k)| \prod_{j\neq k} \left| \frac{x - x_j}{x_k - x_j} \right|.$$

Only now comes the beautiful idea, which is in fact the main step. From the above inequality we find that

$$\left| \frac{f(x)}{x^{n-1}} \right| \leq \sum_{k=1}^{n} \frac{|f(x_k)|}{\prod_{j\neq k} |x_k - x_j|} \left| \prod_{j\neq k} \left(1 - \frac{x_j}{x}\right) \right|$$

and since this is true for all non-zero real numbers x, we may take the limit when $x \to \infty$ and the result is pretty nice:

$$|a_{n-1}| \leq \sum_{k=1}^{n} \frac{|f(x_k)|}{\prod_{j\neq k} |x_k - x_j|}.$$

This is the right moment to decide what polynomial to take. We need a polynomial f such that $|f(x)| \leq 1$ for all $x \in [-1, 1]$ and such that the leading coefficient is 2^{n-2}. This time our mathematical culture will decide. And it says that Chebyshev polynomials are the best, since they are the polynomials with the minimum deviation on $[-1, 1]$ (the reader will wait just a few seconds and will see a beautiful proof of this remarkable result using Lagrange's interpolation theorem). So, we take the polynomial defined by $f(\cos x) = \cos(n-1)x$. It is easy to see that such a polynomial exists, has degree $n-1$, and leading coefficient 2^{n-2}, so this choice solves our problem.

Note also that the inequality $|a_{n-1}| \le \sum_{k=1}^{n} \frac{|f(x_k)|}{\prod_{j\neq k} |x_k - x_j|}$ can be proved by identifying the leading coefficients in the identity

$$f(x) = \sum_{k=1}^{n} f(x_k) \prod_{j\neq k} \frac{x - x_j}{x_k - x_j}$$

and then using the triangle inequality.

The following example is a fine combination of ideas. The problem is not simple at all, since many possible approaches fail. Yet, in the framework of the previous problems and with the experience of Lagrange's interpolation formula, it is not so hard after all.

Example 7. Let $f \in \mathbb{R}[X]$ be a polynomial of degree n with leading coefficient 1, and let $x_0 < x_1 < x_2 < \cdots < x_n$ be some integers. Prove that there exists $k \in \{0, 1, \ldots, n\}$ such that

$$|f(x_k)| \ge \frac{n!}{2^n}.$$

Crux Matematicorum

Solution. Naturally (but would this be naturally without having discussed so many related problems before?), we start with the identity

$$f(x) = \sum_{k=0}^{n} f(x_k) \prod_{j\neq k} \frac{x - x_j}{x_k - x_j}.$$

Now, repeating the argument in the previous problem and using the fact that the leading coefficient is 1, we find that

$$\sum_{k=0}^{n} \frac{|f(x_k)|}{\prod_{j\neq k} |x_k - x_j|} \geq 1.$$

It is time to use the fact that we are dealing with integers. This will allow us to find a good lower bound for $\prod_{j\neq k} |x_k - x_j|$ This is easy, since

$$\prod_{j\neq k} |x_k - x_j| = |(x_k - x_0)(x_k - x_1)\cdots(x_k - x_{k-1})(x_{k+1} - x_k)\cdots(x_n - x_k)|$$

$$\geq k(k-1)(k-2)\cdots 2\cdot 1\cdot 1\cdot 2\cdots(n-k) = k!(n-k)!.$$

And yes, we are done, since using these inequalities, we deduce that

$$\sum_{k=0}^{n} \frac{|f(x_k)|}{k!(n-k)!} \geq 1.$$

Now, since

$$\sum_{k=0}^{n} \frac{1}{k!(n-k)!} = \frac{1}{n!}\sum_{k=0}^{n} \binom{n}{k} = \frac{2^n}{n!},$$

it follows that

$$|f(x_k)| \geq \frac{n!}{2^n}$$

for some $0 \leq k \leq n$.

The following example is an answer to a conjecture of F. J. Dyson (1962). The elegant proof presented here, based on an identity obtained by Lagrange's interpolation formula, is due to I. J. Good (1970):

Example 8. Let $a_1, a_2, ..., a_n$ be nonnegative integers and let $f(a_1, a_2, ..., a_n)$ be the constant term of the "polynomial"

$$\prod_{\substack{1\le i,j\le n\\ i\ne j}} \left(1-\frac{x_i}{x_j}\right)^{a_i}.$$

Prove that

$$f(a_1, a_2, ..., a_n) = \frac{(a_1+a_2+\cdots+a_n)!}{a_1!a_2!...a_n!}.$$

Solution. Define

$$g(a_1, a_2, ..., a_n) = \frac{(a_1+a_2+\cdots+a_n)!}{a_1!a_2!...a_n!}.$$

We will prove by induction on $a_1+a_2+\cdots+a_n$ that $f(a_1, a_2, ..., a_n) = g(a_1, a_2, ..., a_n)$. If $a_1 = a_2 = \cdots = a_n = 0$ the claim is obviously true. Now, observe that

$$g(a_1, a_2, ..., a_n) = g(a_1-1, a_2, ..., a_n) + \cdots + g(a_1, a_2, ..., a_n - 1)$$

if all a_i are positive and

$$g(a_1, a_2, ..., a_{k-1}, 0, a_{k+1}, ..., a_n) = g(a_1, a_2, ..., a_{k-1}, a_{k+1}, ..., a_n)$$

if $a_k = 0$. Therefore it would be enough to prove the same relation for f. If $a_k = 0$ it is clear that

$$f(a_1, a_2, ..., a_{k-1}, 0, a_{k+1}, ..., a_n) = f(a_1, a_2, ..., a_{k-1}, a_{k+1}, ..., a_n),$$

so assume that all a_i are positive. In order to prove that

$$f(a_1, a_2, ..., a_n) = f(a_1-1, a_2, ..., a_n) + \cdots + f(a_1, a_2, ..., a_n - 1)$$

it is enough to prove the identity

$$\prod_{\substack{1\le i,j\le n\\ i\ne j}} \left(1-\frac{x_i}{x_j}\right)^{a_i} = \sum_{i=1}^{n} \prod_{j\ne i} \left(1-\frac{x_i}{x_j}\right)^{a_i} \cdot \prod_{j\ne i} \left(1-\frac{x_i}{x_j}\right)^{-1},$$

which reduces of course to

$$1=\sum_{i=1}^{n}\prod_{j\neq i}\left(1-\frac{x_i}{x_j}\right)^{-1}.$$

But this is just Lagrange's interpolation formula written for the polynomial $f(X)=1$ with nodes $x_1, x_2, ..., x_n$ and evaluated at $x=0$.

We will discuss one more problem before embarking on a more detailed study of Chebyshev polynomials and their properties. This was given in the Romanian Mathematical Olympiad and is a very nice application of Lagrange's interpolation formula. It is sufficient to say that it follows trivially using a little bit of integration theory and Fourier series.

Example 9. Prove that for any polynomial f of degree n and with leading coefficient 1 there exists a point z such that

$$|z|=1 \text{ and } |f(z)|\geq 1.$$

[Marius Cavachi] Romanian Olympiad

Solution. Of course, the idea is always the same, but this time it is necessary to find the good points for which we should write the interpolation formula. As we did before, we will be blind at the beginning and we will try to find these points in the end. Until then, let us call them $x_0, x_1, x_2, \ldots, x_n$ and write

$$\sum_{k=0}^{n}\frac{|f(x_k)|}{\prod\limits_{j\neq k}|x_k-x_j|}\geq 1.$$

This inequality was already proved in Example 6. Now, consider the polynomial

$$g(x)=\prod_{i=0}^{n}(x-x_i).$$

We have then

$$|g'(x_i)| = \left|\prod_{j\neq i}(x_i - x_j)\right|.$$

Now, we would like, if possible, to have $|x_i| = 1$ and also $\sum_{k=0}^{n} \frac{1}{|g'(x_k)|} \leq 1$. In this case it would follow from $\sum_{k=0}^{n} \frac{|f(x_k)|}{\prod_{j\neq k} |x_k - x_j|} \geq 1$ that at least one of the numbers $|f(x_k)|$ is greater than or equal to 1 and the problem would be solved. Thus, we should find a monic polynomial g of degree $n+1$ with all roots of modulus 1 and such that $\sum_{k=0}^{n} \frac{1}{|g'(x_k)|} \leq 1$. This is trivial: it suffices, of course, to consider $g(x) = x^{n+1} - 1$. The conclusion follows.
We have an explanation to give: we said the problem follows trivially with a little bit of integration theory tools. Indeed, if we write $f(x) = \sum_{k=0}^{n} a_k x^k$ then one can check with a trivial computation that

$$a_k = \frac{1}{2\pi}\int_0^{2\pi} f(e^{it})e^{-ikt}dt$$

and from here the conclusion follows since we will have

$$2\pi = \left|\int_0^{2\pi} f(e^{it})e^{-int}dt\right| \leq \int_0^{2\pi} |f(e^{it})|dt \leq 2\pi \max_{|z|=1} |f(z)|.$$

Of course, knowing this already in 10th grade (since the problem was given to 10th grade students) is not something common...

Before passing to the next more computational problem (which does not mean less interesting, of course), let us recall some properties of the Chebyshev's polynomials of the first kind. They are defined by $T_n(x) = \cos(n \arccos(x))$, or, equivalently, $T_n(\cos x) = \cos(nx)$. You can easily check by induction, using the obvious relation $T_{n+1}(x) = 2xT_n(x) - T_{n-1}(x)$ that this gives you a polynomial of degree n, having leading coefficient 2^{n-1} and all of whose coefficients are integers. Among hundreds of interesting and useful properties of these polynomials, let us state a few, the proof of which is left as a very useful exercise for the interested reader.

Theorem 11.1. *The polynomials T_n have the following properties:*

- *An explicit formula for T_n is*

$$T_n(x) = \frac{(x+\sqrt{x^2-1})^n + (x-\sqrt{x^2-1})^n}{2}.$$

- *The polynomials T_n and T_m commute, that is $T_n(T_m(x)) = T_m(T_n(x))$ for all m, n and all x.*

- *The generating function of these polynomials is given by:*

$$\sum_{n\geq 1} T_n(x)z^n = \frac{z(x-z)}{1-2zx+z^2}$$

for all $|z| < 1$ and $|x| < 1$.

- *They form an orthogonal system on the interval $[-1,1]$ for the weight $v(x) = \frac{1}{\sqrt{1-x^2}}$; that is for all $i \neq j$ positive integers the following relation holds*

$$\int_{-1}^{1} \frac{T_i(x)T_j(x)}{\sqrt{1-x^2}}dx = 0.$$

The following problems will be based on a very nice identity that will allow us to prove some classical results about norms of polynomials, to find the polynomials having minimal deviation on $[-1,1]$, and also to establish some new inequalities. In order to do all this, we need two quite technical lemmas, which are not difficult to establish, but very useful.

Lemma 11.2. *If we let* $t_k = \cos\dfrac{k\pi}{n}$, $0 \le k \le n$, *then*

$$f(x) = \prod_{k=0}^{n}(x - t_k) = \frac{\sqrt{x^2-1}}{2^n}[(x+\sqrt{x^2-1})^n - (x-\sqrt{x^2-1})^n].$$

Proof. The proof is simple. Indeed, if we consider

$$g(x) = \frac{\sqrt{x^2-1}}{2^n}[(x+\sqrt{x^2-1})^n - (x-\sqrt{x^2-1})^n],$$

using the binomial formula we can establish immediately that it is a polynomial. Moreover, from the obvious fact that $\lim\limits_{x\to\infty}\dfrac{g(x)}{x^{n+1}} = 1$, we deduce that it is actually a monic polynomial of degree $n+1$. The fact that $g(t_k) = 0$ for all $0 \le k \le n$ is easily verified using de Moivre's formula. All these prove the first lemma. □

A little bit more computational is the second lemma.

Lemma 11.3. *The following relations are true:*

i) $\displaystyle\prod_{j\ne k}(t_k - t_j) = \frac{(-1)^k n}{2^{n-1}}$ *if* $1 \le k \le n-1$;

ii) $\displaystyle\prod_{j=1}^{n}(t_0 - t_j) = \frac{n}{2^{n-2}}$;

iii) $\displaystyle\prod_{j=0}^{n-1}(t_n - t_j) = \frac{(-1)^n n}{2^{n-2}}$.

Proof. Simple computations, left to the reader, allow us to write:

$$f'(x) = \frac{n}{2^n}[(x+\sqrt{x^2-1})^n + (x-\sqrt{x^2-1})^n] +$$

$$+\frac{x}{2^n\sqrt{x^2-1}}[(x+\sqrt{x^2-1})^n - (x-\sqrt{x^2-1})^n].$$

Using this formula and de Moivre's formula we easily deduce i).

To prove ii) and iii) it suffices to compute $\lim_{x\to 1} f'(x)$ and $\lim_{x\to -1} f'(x)$, using the above formula; we let the reader to do the dirty job. □

Of course, you hope that all these computations have an honorable purpose; and you're right, since these lemmas will allow us to prove some very nice results. The first one is a classical theorem of Chebyshev, about minimal deviation of polynomials on $[-1,1]$.

Example 10. (Chebyshev's theorem) Let $f \in \mathbb{R}[X]$ be a monic polynomial of degree n. Then

$$\max_{x\in[-1,1]} |f(x)| \geq \frac{1}{2^{n-1}}$$

and this bound cannot be improved.

<u>**Solution.**</u> Using again the observation from Problem 7, we obtain the identity:

$$\sum_{k=0}^{n} f(t_k) \prod_{j\neq k} \frac{1}{t_k - t_j} = 1.$$

Thus, we have

$$1 \leq \max_{0\leq k\leq n} |f(t_k)| \sum_{k=0}^{n} \left| \prod_{j\neq k} (t_k - t_j) \right|^{-1}.$$

Now, it suffices to apply Lemma 2 to conclude that we actually have

$$\sum_{k=0}^{n} \left| \prod_{j\neq k} (t_k - t_j) \right|^{-1} = 2^{n-1}.$$

This shows that $\max_{x\in[-1,1]} |f(x)| \geq \frac{1}{2^{n-1}}$ and so the result is proved. To prove that this bound is optimal, it suffices to use the polynomial T_n. Then the polynomial $\frac{1}{2^{n-1}}T_n$ is monic of degree n and

$$\max_{x\in[-1,1]}\left|\frac{1}{2^{n-1}}T_n(x)\right| = \frac{1}{2^{n-1}}. \tag{11.2}$$

There are many other proofs of this result, a lot of them are much easier, but we chose this one because it shows the power of Lagrange interpolation theory. Not to mention that the use of the two lemmas allowed us to prove that the inequality presented in Example 7 is actually the best.

Some years ago, Walther Janous presented in Crux the following as an open problem. It is true that it is a very difficult one, but here is a very simple solution using the results already achieved.

Example 11. Suppose that $a_0, a_1, \ldots, a_n$ are real numbers such that for all $x \in [-1, 1]$ we have

$$|a_0 + a_1x + \cdots + a_nx^n| \leq 1.$$

Then for all $x \in [-1, 1]$ we also have

$$|a_n + a_{n-1}x + \cdots + a_0x^n| \leq 2^{n-1}.$$

[Walther Janous] Crux Matematicorum

Solution. Actually, we are going to prove a stronger result, that is:

Lemma 11.4. *Denote, for $f \in \mathbb{R}[X]$,*

$$\|f\| = \max_{x\in[-1,1]} |f(x)|.$$

Then for any polynomial $f \in \mathbb{R}[X]$ of degree n the following inequality is satisfied:

$$|f(x)| \le |T_n(x)| \cdot \|f\| \text{ for all } |x| \ge 1.$$

Proof. Using Lagrange's interpolation formula and the triangle inequality, we deduce that for all $u \in [-1,1]$, $u \neq 0$, we have:

$$\left|f\left(\frac{1}{u}\right)\right| \le \frac{1}{|u|^n}\|f\| \sum_{k=0}^{n} \prod_{j \neq k} \frac{1 - t_j u}{|t_k - t_j|}.$$

The brilliant idea is to use the Lagrange interpolation formula again, this time for the polynomial T_n. We shall then have (also for $u \in [-1,1], u \neq 0$)

$$\left|T_n\left(\frac{1}{u}\right)\right| = \frac{1}{|u|^n}\left|\sum_{k=0}^{n}(-1)^k \prod_{j \neq k} \frac{1 - ut_j}{t_k - t_j}\right| = \frac{1}{|u|^n}\sum_{k=0}^{n} \prod_{j \neq k} \frac{1 - ut_j}{|t_k - t_j|}$$

(the last identity being ensured by lemma 11.2). By combining the two results, we obtain

$$\left|f\left(\frac{1}{u}\right)\right| \le \left|T_n\left(\frac{1}{u}\right)\right| \|f\| \text{ for all } u \in [-1,1], u \neq 0$$

and the conclusion follows. □

Coming back to the problem and considering the polynomial $f(x) = \sum_{k=0}^{n} a_k x^k$, the hypothesis says that $\|f\| \le 1$ and so by the lemma we have

$$|f(x)| \le |T_n(x)| \text{ for all } |x| \ge 1.$$

We will then have for all $x \in [-1,1]$, $x \neq 0$:

$$|a_n + a_{n-1}x + \cdots + a_0 x^n| = \left|x^n f\left(\frac{1}{x}\right)\right| \le \left|x^n T_n\left(\frac{1}{x}\right)\right|.$$

It suffices to prove that

$$\left|x^n T_n\left(\frac{1}{x}\right)\right| \le 2^{n-1},$$

which can be also written as

$$(1+\sqrt{1-x^2})^n + (1-\sqrt{1-x^2})^n \le 2^n.$$

But this inequality is very easy to prove: just set $a = \sqrt{1-x^2} \in [0,1]$ and observe that $h(a) = (1-a)^n + (1+a)^n$ is a convex function on $[0,1]$, thus its superior bound is attained at 0 or 1 and there the inequality is trivially verified. Therefore we have

$$|a_n + a_{n-1}x + \cdots + a_0 x^n| \le 2^{n-1}$$

and the problem is solved.

Since we are here, why not continuing with some classical, but very important results of Riesz, Bernstein and Markov? We unified these results in a single example because they have the same idea, and moreover they follow one from another. We must mention that a) is a result of M. Riesz, while b) was obtained by S. Bernstein and finally c) is a famous theorem of A. Markov, the real equivalent of an even more celebrated Bernstein's complex theorem (whose proof you will surely enjoy: it is among the training problems).

Example 12. a) Let P be a polynomial with real coefficients of degree at most $n-1$ such that $\sqrt{1-x^2}|P(x)| \le 1$ for all $x \in [-1,1]$. Prove that $|P(x)| \le n$ for all $x \in [-1,1]$.
b) Let

$$f(x) = \sum_{k=0}^{n} (a_k \cos(kx) + b_k \sin(kx))$$

be a trigonometric polynomial of degree n with real coefficients. Suppose that $|f(x)| \le 1$ for all real numbers x. Prove that $|f'(x)| \le n$ for all real numbers x.
c) Prove that if P has degree n and real coefficients and moreover $|P(x)| \le 1$ for all $x \in [-1,1]$ then $|P'(x)| \le n^2$ for all $x \in [-1,1]$.

Solution. a) Let us write the Lagrange interpolation formula for P with the points $x_1, x_2, ..., x_n$, where $x_j = \cos(\frac{(2j-1)\pi}{2n})$ are the zeros of the nth Chebyshev's polynomial T_n. We obtain the important identity

$$P(x) = \frac{1}{n} \cdot \sum_{i=1}^{n} (-1)^{i-1} \cdot \sqrt{1 - x_i^2} \cdot P(x_i) \cdot \frac{T_n(x)}{x - x_i}.$$

Take now $x \in [-1, 1]$. Observe that if $x \in [x_n, x_1] = [-x_1, x_1]$ then by the hypothesis $|P(x)| \leq \frac{1}{\sqrt{1-x^2}} \leq \frac{1}{\sqrt{1-x_1^2}} \leq n$, the last inequality being equivalent to $\sin(\frac{\pi}{2n}) \geq \frac{1}{n}$, which is clear by a convexity argument. So, assume that $x \geq x_1$, the case $x \leq -x_1$ being identical. In this case the triangle inequality applied to the previous identity shows that

$$|P(x)| \leq \frac{1}{n} \cdot \sum_{i=1}^{n} \frac{T_n(x)}{x - x_i}.$$

But the last sum is exactly $\frac{1}{n} T_n'(x)$. Because $T_n(\cos u) = \cos(nu)$, we have $T_n'(\cos u) = n\frac{\sin(nu)}{\sin u}$. However, an easy induction shows that $|\sin(nu)| \leq n \leq |\sin u|$ for all u and all positive integers n. This implies that $|T_n'(x)| \leq n^2$ for all $x \in [-1, 1]$. Combining this with the inequality $|P(x)| \leq \frac{1}{n} \cdot T_n'(x)$ we deduce that $|P(x)| \leq n$ for all $x \geq x_1$. This finishes the proof of the first part.

b) First of all, let us see what happens when all a_i are zero, that is

$$f(x) = b_1 \sin x + b_2 \sin(2x) + \cdots + b_n \sin(nx).$$

Observe that in exactly the same way as you could have proved the existence of the polynomial T_n (that is, by induction), you can prove the existence of a polynomial R_n of degree $n - 1$ such that $R_n(\cos x) = \frac{\sin(nx)}{\sin x}$. Therefore there exists a polynomial P of degree at most $n - 1$, with real coefficients, and such that $\frac{f(x)}{\sin x} = P(\cos x)$. Observe that this polynomial satisfies the conditions of a), because $|\sin x \cdot P(\cos x)| \leq 1$ for all real x. Therefore we can apply a) to deduce that $|P(x)| \leq n$ for all $x \in [-1, 1]$, that is $|f(x)| \leq n \cdot |\sin x|$ for all x. Dividing by x and letting $x \to 0$ we deduce that $|f'(0)| \leq n$. Now, let us come back to the general problem and fix a real number x_0. Define $g(x) = \frac{f(x+x_0)-f(x-x_0)}{2}$. Using standard trigonometric formulae, we deduce that $g(x)$ is of the form $c_1 \sin x + c_2 \sin(2x) + \cdots + c_n \sin(nx)$ for some real numbers c_j. The triangle inequality also ensures that $|g(x)| \leq 1$ for all real numbers x. Thus, by the result that we have just obtained, we must have $|g'(0)| \leq n$. Because $g'(0) = f'(x_0)$ and because x_0 was arbitrary, b) is proved.

c) Let us consider this time $f(x) = P(\cos x)$. An immediate induction based on the most elementary product formulae for trigonometric functions shows that $(\cos x)^j$ is a trigonometric polynomial of degree at most n. Thus by b) we must have $|f'(x)| \leq n$ for all x. This means that $|\sin x \cdot P'(\cos x)| \leq n$ for all x, which shows that the polynomial $\frac{1}{n}P'$ satisfies the conditions of a). Thus it has values not exceeding n on $[-1,1]$, which means that P' does not exceed n^2 on $[-1,1]$, and this finishes the proof of this beautiful theorem.

We end this topic with a very difficult problem, which refines an older one given in a Japanese mathematical Olympiad in 1994. The problem has a nice story: given initially in an old Russian Olympiad, it asked to prove that

$$\max_{x\in[0,2]} \prod_{i=1}^{n} |x-a_i| \leq 108^n \max_{x\in[0,1]} \prod_{i=1}^{n} |x-a_i|$$

for any real numbers $a_1, a_2, \ldots, a_n$. The Japanese problem asked only to prove the existence of a constant that could replace 108. A brute force choice of points in the Lagrange interpolation theorem gives a better bound of approximately 12 for this constant. Recent work by Alexandru Lupaş reduces this bound to $1 + 2\sqrt{6}$. In the following, we present the optimal bound.

Example 13. For any real numbers $a_1, a_2, \ldots, a_n$, the following inequality holds:

$$\max_{x\in[0,2]} \prod_{i=1}^{n} |x-a_i| \leq \frac{(3+2\sqrt{2})^n + (3-2\sqrt{2})^n}{2} \max_{x\in[0,1]} \prod_{i=1}^{n} |x-a_i|.$$

[Gabriel Dospinescu]

Solution. Let us denote

$$\|f\|_{[a,b]} = \max_{x\in[a,b]} |f(x)|$$

for a polynomial f and let, for simplicity,

$$c_n = \frac{(3+2\sqrt{2})^n + (3-2\sqrt{2})^n}{2}.$$

We thus need to prove that $\|f\|_{[0,2]} \le c_n \|f\|_{[0,1]}$ where

$$f(x) = \prod_{i=1}^{n}(x-a_i).$$

We shall prove that this inequality is true for any polynomial f, which allows us to suppose that $\|f\|_{[0,1]} = 1$. We shall prove that for all $x \in [1,2]$ we have $|f(x)| \le c_n$. Let us fix $x \in [1,2]$ and consider the numbers $x_k = \frac{1+t_k}{2}$, where t_k's are as in Lemma 11.2. Using the Lagrange interpolation formula, we deduce that

$$|f(x)| \le \sum_{k=0}^{n}\left|\prod_{j\neq k}\frac{x-x_j}{x_k-x_j}\right| = \sum_{k=0}^{n}\prod_{j\neq k}\frac{x-x_j}{|x_k-x_j|}$$

$$\le \sum_{k=0}^{n}\prod_{j\neq k}\frac{2-x_j}{|x_k-x_j|} = \sum_{k=0}^{n}\prod_{j\neq k}\frac{3-t_j}{|t_k-t_j|}.$$

Using Lemma 11.3, we can write

$$\sum_{k=0}^{n}\prod_{j\neq k}\frac{3-t_j}{|t_k-t_j|} = \frac{2^{n-1}}{n}\sum_{k=1}^{n-1}\prod_{j\neq k}(3-t_j) +$$

$$+\frac{2^{n-2}}{n}\left[\prod_{j=0}^{n-1}(3-t_j) + \prod_{j=1}^{n}(3-t_j)\right].$$

Based on the expression of the derivative from the proof of Lemma 11.3, we obtain:

$$\sum_{k=1}^{n-1}\prod_{j\neq k}(3-t_j)+\prod_{j=0}^{n-1}(3-t_j)+\prod_{j=1}^{n}(3-t_j)=$$

$$=\frac{n}{2^n}[(3+2\sqrt{2})^n+(3-2\sqrt{2})^n]+\frac{3}{2^{n+1}\sqrt{2}}[(3+2\sqrt{2})^n-(3-2\sqrt{2})^n].$$

All we have to do now is to compute

$$\prod_{j=0}^{n-1}(3-t_j)+\prod_{j=1}^{n}(3-t_j)=6\prod_{j=1}^{n-1}(3-t_j).$$

But, according to Lemma 11.2, we deduce immediately that

$$\prod_{j=1}^{n-1}(3-t_j)=\frac{1}{2^{n+1}\sqrt{2}}[(3+2\sqrt{2})^n-(3-2\sqrt{2})^n].$$

Putting all these observations together and making a small computation, that we leave to the reader, we easily deduce that $|f(x)|\leq c_n$. This proves that $\|f\|_{[0,2]}\leq c_n\|f\|_{[0,1]}$ and solves the problem.

11.2 Practice problems

1. A polynomial p of degree n satisfies $p(k) = 2^k$ for all $0 \le k \le n$. Find its value at $n+1$.

   Murray Klamkin

2. A polynomial f of degree n satisfies

$$f(k) = \frac{1}{\binom{n+1}{k}}$$

   for all $0 \le k \le n$. Find $f(n+1)$.

   Titu Andreescu, IMO Shortlist 1981

3. Prove that for any real number a we have the following identity

$$\sum_{k=0}^{n} (-1)^k \binom{n}{k} (a-k)^n = n!.$$

   Tepper's identity

4. Prove that

$$\sum_{k=0}^{n} \frac{x_k^{n+1}}{\prod\limits_{j \ne k} (x_k - x_j)} = \sum_{k=0}^{n} x_k$$

   and compute

$$\sum_{k=0}^{n} \frac{x_k^{n+2}}{\prod\limits_{j \ne k} (x_k - x_j)}.$$

5. Let a, b, c be real numbers and let $f(x) = ax^2 + bx + c$ such that $\max\{|f(\pm 1)|, |f(0)|\} \leq 1$. Prove that if $|x| \leq 1$ then

$$|f(x)| \leq \frac{5}{4} \text{ and } \left|x^2 f\left(\frac{1}{x}\right)\right| \leq 2.$$

Spain 1996

6. Find the greatest possible value of the expression $a^2 + b^2 + c^2$ if $|ax^2 + bx + c| \leq 1$ for all $x \in [-1, 1]$.

Laurenţiu Panaitopol

7. Define $F(a, b, c) = \max_{x \in [0,3]} |x^3 - ax^2 - bx - c|$. What is the least possible value of this function over $\mathbb{R}^3$?

Chinese TST 2001

8. Let $a, b, c, d \in \mathbb{R}$ such that $|ax^3 + bx^2 + cx + d| \leq 1$ for all $x \in [-1, 1]$. What is the greatest possible value of $|c|$? For which polynomials is the maximum attained?

Gabriel Dospinescu

9. Let $f \in \mathbb{R}[X]$ be a polynomial of degree n that satisfies $|f(x)| \leq 1$ for all $x \in [0, 1]$. Prove that

$$\left|f\left(-\frac{1}{n}\right)\right| \leq 2^{n+1} - 1.$$

Komal

10. Let $a \geq 3$ be a real number and p be a real polynomial of degree n. Prove that

$$\max_{i=0,1,\ldots,n+1} |a^i - p(i)| \geq 1.$$

India 1998

11. Let $a, b, c, d \in \mathbb{R}$ such that $|ax^3 + bx^2 + cx + d| \leq 1$ for all $x \in [-1, 1]$. Prove that

$$|a| + |b| + |c| + |d| \leq 7.$$

IMO Shortlist 1996

12. Let $A = \left\{ p \in \mathbb{R}[X] \mid \ \deg p \leq 3, \ |p(\pm 1)| \leq 1, \ \left| p\left(\pm \frac{1}{2}\right) \right| \leq 1 \right\}.$

Find $\sup\limits_{p \in A} \max\limits_{|x| \leq 1} |p''(x)|$.

IMC 1998

13. Prove the identity

$$\sum_{k=1}^{n} \frac{(-1)^{k-1}}{k} \binom{n}{k} (n-k)^n = n^n \sum_{k=2}^{n} \frac{1}{k}.$$

Peter Ungar, AMM E 3052

14. Let $n \geq 3$ and let $f, g \in \mathbb{R}[X]$ be polynomials such that the points

$$(f(1), g(1)), (f(2), g(2)), \ldots, (f(n), g(n))$$

are the vertices of a regular n-gon in counterclockwise order. Prove that $\max(\deg f, \deg g) \geq n - 1$.

Putnam 2008

15. Let f be a complex polynomial of degree n such that $|f(x)| \le 1$ for all $x \in [-1, 1]$. Prove that for all k and all real numbers x such that $|x| \ge 1$, $|f^{(k)}(x)| \le |T_n^{(k)}(x)|$. Prove that Chebyshev's theorem is a consequence of this result.

W. W. Rogosinski

16. Let f and g be polynomials in one variable with real coefficients and let a be a polynomial in two variables such that

$$f(x) - f(y) = a(x, y)(g(x) - g(y))$$

for all x, y. Prove that there is a polynomial h such that $f(x) = h(g(x))$.

Russia 2004

17. a) Prove that for any polynomial f having degree at most n, the following identity is satisfied:

$$xf'(x) = \frac{n}{2} f(x) + \frac{1}{n} \sum_{k=1}^{n} f(xz_k) \frac{2z_k}{(1 - z_k)^2},$$

where z_k are the roots of the polynomial $X^n + 1$.
b) Deduce Bernstein's inequality: $\|f'\| \le n\|f\|$ where $\|f\| = \max_{|x|\le 1} |f(x)|$.

P. J. O'Hara, AMM

18. Let f be a complex polynomial of degree at most n and let $z_0, z_1, \dots, z_d$ be the zeros of the polynomial $X^{d+1} - 1$, where $d > n$. Define $||f|| = \max_{|z|=1} |f(z)|$.

   (a) Prove that if there exist $n + 1$ pairwise distinct zeros $x_0, x_1, \dots, x_n$ among $z_0, z_1, \dots, z_d$ such that $|f(x_i)| \le \frac{1}{2^d}$, then $||f|| < 1$.

   (b) Deduce that

$$||f|| \cdot ||g|| \le 4^{\deg(f)+\deg(g)} \cdot ||fg||.$$

Gelfond

Chapter 12

12.1 Theory and examples

It is probably time to see the contribution of non-elementary mathematics in combinatorics. It is quite difficult to imagine that behind a simple game such as football, for example, or behind a day-to-day situation such as handshakes, there exists such a complicated machinery. But this sometimes happens, as will be soon demonstrated. In the beginning of the discussion, the reader does not need any special knowledge, just imagination and the most basic properties of matrices, but, as soon as we advance, things may change. Anyway, the most important fact is not the knowledge, but the ideas and, as we will see, it is not always easy to discover that non-elementary fact that hides behind a completely elementary problem. Now that we have clarified what is the purpose of the unit, we can begin.

The first problem we are going to discuss is not classical, but it is relatively easy and shows how a very nice application of linear-algebra can solve elementary problems.

Example 1. Let $n \geq 3$ and let A_n, B_n be the sets of all even, respectively odd, permutations of the set $\{1, 2, \ldots, n\}$. Prove that

$$\sum_{\sigma \in A_n} \sum_{i=1}^{n} |i - \sigma(i)| = \sum_{\sigma \in B_n} \sum_{i=1}^{n} |i - \sigma(i)|.$$

[Nicolae Popescu] Gazeta Matematică

Solution. Writing the difference

$$\sum_{\sigma \in A_n} \sum_{i=1}^{n} |i - \sigma(i)| - \sum_{\sigma \in B_n} \sum_{i=1}^{n} |i - \sigma(i)|$$

as

$$\sum_{\sigma \in S_n} \varepsilon(\sigma) \sum_{i=1}^{n} |i - \sigma(i)|,$$

where

$$\varepsilon(\sigma) = \begin{cases} +1, & \text{if} \quad \sigma \in A_n \\ -1, & \text{if} \quad \sigma \in B_n \end{cases}$$

reminds us about the formula

$$\det A = \sum_{\sigma \in S_n} \varepsilon(\sigma) a_{1\sigma(1)} a_{2\sigma(2)} \cdots a_{n\sigma(n)}.$$

We have taken here $S_n = A_n \cup B_n$. But we have no product in our sum! This is why we take an arbitrary positive number x and consider the matrix $A = (x^{|i-j|})_{1 \le i,j \le n}$. We have

$$\det A = \sum_{\sigma \in S_n} (-1)^{\varepsilon(\alpha)} x^{|1-\sigma(1)|} \cdots x^{|n-\sigma(n)|} =$$

$$= \sum_{\sigma \in A_n} x^{\sum\limits_{i=1}^{n} |i-\sigma(i)|} - \sum_{\sigma \in B_n} x^{\sum\limits_{i=1}^{n} |i-\sigma(i)|}$$

This is how we obtain the identity

$$\begin{vmatrix} 1 & x & x^2 & \dots & x^{n-2} & x^{n-1} \\ x & 1 & x & \dots & x^{n-3} & x^{n-2} \\ x^2 & x & 1 & \dots & x^{n-4} & x^{n-3} \\ \dots & \dots & \dots & \dots & \dots & \dots \\ x^{n-1} & x^{n-2} & \dots & \dots & x & 1 \end{vmatrix}$$

$$= \sum_{\substack{\sigma \in S_n \\ \sigma \text{ even}}} x^{\sum\limits_{i=1}^{n} |i-\sigma(i)|} - \sum_{\substack{\sigma \in S_n \\ \sigma \text{ odd}}} x^{\sum\limits_{i=1}^{n} |i-\sigma(i)|}. \tag{12.1}$$

Anyway, we do not have the desired difference yet. The most natural way is to differentiate the last relation, which is nothing other than a polynomial identity, and then to take $x = 1$. Before doing that, let us observe that the polynomial

$$\begin{vmatrix} 1 & x & x^2 & \dots & x^{n-2} & x^{n-1} \\ x & 1 & x & \dots & x^{n-3} & x^{n-2} \\ x^2 & x & 1 & \dots & x^{n-4} & x^{n-3} \\ \dots & \dots & \dots & \dots & \dots & \dots \\ x^{n-1} & x^{n-2} & \dots & \dots & x & 1 \end{vmatrix}$$

is divisible by $(x-1)^2$. This can be easily seen by subtracting the first line from the second and the third one and taking from each of these lines $x - 1$ as a common factor. Thus, the derivative of this polynomial is a polynomial divisible by $x - 1$, which shows that after we differentiate (12.1) and take $x = 1$, the left-hand side vanishes, while the right-hand side becomes

$$\sum_{\sigma \in A_n} \sum_{i=1}^{n} |i - \sigma(i)| - \sum_{\sigma \in B_n} \sum_{i=1}^{n} |i - \sigma(i)|.$$

This completes the proof.

Here is another nice application of this idea. You probably know how many permutations do not have a fixed point. The question that arises is how many of them are even. Using determinants provides a direct answer to the question.

Example 2. Find the number of even permutations of the set $\{1, 2, \dots, n\}$ that do not have fixed points.

Solution. Let C_n and D_n be the sets of even and odd permutations of the set $\{1, 2, \dots, n\}$, that do not have any fixed points, respectively. You may recall how to find the sum $|C_n| + |D_n|$: using the inclusion-exclusion principle, it is not difficult to establish that it is equal to

$$n!\left(1 - \frac{1}{1!} + \frac{1}{2!} - \dots + \frac{(-1)^n}{n!}\right).$$

Hence if we manage to compute the difference $|C_n| - |D_n|$, we will be able to answer to the question. Write

$$|C_n| - |D_n| = \sum_{\substack{\sigma \in A_n \\ \sigma(i) \neq i}} 1 - \sum_{\substack{\sigma \in B_n \\ \sigma(i) \neq i}} 1$$

using A_n, B_n from Example 1, observe that this reduces to computing the determinant of the matrix $T = (t_{ij})_{1\le i,j\le n}$, where

$$t_{ij} = \begin{cases} 1, & \text{if} \quad i \ne j \\ 0, & \text{if} \quad i = j \end{cases}$$

That is,

$$|C_n| - |D_n| = \begin{vmatrix} 0 & 1 & 1 & \dots & 1 \\ 1 & 0 & 1 & \dots & 1 \\ \dots & \dots & \dots & \dots & \dots \\ 1 & 1 & 1 & \dots & 0 \end{vmatrix}.$$

But computing this determinant is not difficult. Indeed, we add all columns to the first and factor $n-1$, then we subtract the first column from each of the other columns. The result is $|C_n| - |D_n| = (-1)^{n-1}(n-1)$, and the conclusion is:

$$|C_n| = \frac{1}{2}\left[n!\left(1 - \frac{1}{2!} + \frac{1}{3!} - \dots + \frac{(-1)^{n-2}}{(n-2)!}\right) + (-1)^{n-1}(n-1)\right].$$

In the following problems we will focus on a very important combinatorial tool, that is the incidence matrix. Suppose we have a set $X = \{x_1, x_2, \dots, x_n\}$ and $X_1, X_2, \dots, X_k$ a family of subsets of X. Now, define the matrix $A = (a_{ij})_{\substack{1\le i\le n \\ 1\le j\le k}}$, where

$$a_{ij} = \begin{cases} 1, & \text{if} \quad x_i \in X_j \\ 0, & \text{if} \quad x_i \notin X_j \end{cases}$$

This is the incidence matrix of the family $X_1, X_2, \dots, X_k$ and the set X. In many situations, computing the product $A^T \cdot A$ helps translate the conditions and the conclusion of certain problems. From this point, we turn on this machinery, and solving the problem is on its way.

Let us discuss first a classical problem. It appeared at the USAMO 1979, Tournament of the Towns 1985 and in the Bulgarian Spring Mathematical Competition 1995. This says something about the classical character and beauty of this problem.

Example 3. Let $A_1, A_2, \ldots, A_{n+1}$ be distinct subsets of the set $\{1, 2, \ldots, n\}$, each having exactly three elements. Prove that there are two subsets among them that have exactly one common element.

Solution. We argue by contradiction and suppose that $|A_i \cap A_j| \in \{0, 2\}$ for all $i \neq j$. Now, let $T = (t_{ij})_{\substack{1 \le i \le n \\ 1 \le j \le n+1}}$ be the incidence matrix of the family $A_1, A_2, \ldots, A_{n+1}$ and compute the product

$$ {}^tT \cdot T = \begin{pmatrix} \sum_{k=1}^{n} t_{k,1}^2 & \sum_{k=1}^{n} t_{k,1}t_{k,2} & \ldots & \sum_{k=1}^{n} t_{k,1}t_{k,n+1} \\ \ldots & \ldots & \ldots & \ldots \\ \sum_{k=1}^{n} t_{k,n+1}t_{k,1} & \sum_{k=1}^{n} t_{k,n+1}t_{k,2} & \ldots & \sum_{k=1}^{n} t_{k,n+1}^2 \end{pmatrix}. $$

But $\sum_{k=1}^{n} t_{ki}^2 = |A_i| = 3$ and $\sum_{k=1}^{n} t_{ki}t_{kj} = |A_i \cap A_j| \in \{0, 2\}$.
Thus, considered in the field $\mathbb{Z}/2\mathbb{Z}$, we have

$$ \overline{{}^tT \cdot T} = \begin{pmatrix} \widehat{1} & \widehat{0} & \ldots & \widehat{0} & \widehat{0} \\ \ldots & \ldots & \ldots & \ldots & \ldots \\ \widehat{0} & \widehat{0} & \ldots & \widehat{0} & \widehat{1} \end{pmatrix}, $$

where $\overline{X}$ is the matrix having as elements the residues classes of the elements of the matrix T. Because $\det \overline{X} = \overline{\det X}$, the previous relation shows that $\det {}^tT \cdot T$ is odd, hence nonzero. This means that ${}^tTT$ is an invertible matrix of order $n+1$, thus $\operatorname{rank}({}^tT \cdot T) = n+1$ which contradicts the inequality $\operatorname{rank}({}^tT \cdot T) \le \operatorname{rank}(T) \le n$. This shows that our assumption is wrong and there indeed exist indices $i \neq j$ such that $|A_i \cap A_j| = 1$.

The following problem is very difficult to solve by elementary means, but the solution using Linear Algebra is straightforward.

Example 4. Let n be even and let $A_1, A_2, \ldots, A_n$ be distinct subsets of the set $\{1, 2, \ldots, n\}$, each of them having an even number of elements. Prove that among these subsets there are two having an even number of common elements.

Solution. Indeed, if T is the incidence matrix of the family $A_1, A_2, \ldots, A_n$, we obtain as in the previous problem the following relation

$$
{}^tT \cdot T = \begin{pmatrix} |A_1| & |A_1 \cap A_2| & \ldots & |A_1 \cap A_n| \\ \ldots & \ldots & \ldots & \ldots \\ |A_n \cap A_1| & |A_n \cap A_2| & \ldots & |A_n| \end{pmatrix} \tag{12.2}
$$

Now, let us suppose that all the numbers $|A_i \cap A_j|$ are odd and interpret the above relation in the field $\mathbb{Z}/2\mathbb{Z}$. We find that

$$
\overline{{}^tT \cdot T} = \begin{pmatrix} \widehat{0} & \widehat{1} & \ldots & \widehat{1} & \widehat{1} \\ \ldots & \ldots & \ldots & \ldots & \ldots \\ \widehat{1} & \widehat{1} & \ldots & \widehat{1} & \widehat{0} \end{pmatrix},
$$

which means again that $\det {}^tT \cdot T$ is odd. Indeed, if we work in $\mathbb{Z}/2\mathbb{Z}$, we obtain

$$
\begin{vmatrix} \widehat{0} & \widehat{1} & \ldots & \widehat{1} & \widehat{1} \\ \ldots & \ldots & \ldots & \ldots & \ldots \\ \widehat{1} & \widehat{1} & \ldots & \widehat{1} & \widehat{0} \end{vmatrix} = \widehat{1}.
$$

The technique used is exactly the same as in the previous example. Note that this is the moment when we use the hypothesis that n is even. Now, since $\det {}^tT \cdot T = \det^2 T$, we obtain that $\det T$ is also an odd number. Hence we should try to prove that in fact $\det T$ is an even number and the problem will be solved. Just observe that the sum of elements of the column i of T is $|A_i|$, hence an even number. Thus, if we add all lines to the first, we will obtain only even numbers on the first line. Because the value of the determinant does not change under this operation, it follows that $\det T$ is an even number. But a number cannot be both even and odd, so our assumption is wrong and the problem is solved.

Working in a simple field such as $\mathbb{Z}/2\mathbb{Z}$ can allow us to find quite interesting solutions. For example, we will discuss the following problem, used in the preparation of the Romanian IMO team in 2004.

Example 5. The squares of an $n \times n$ table are colored white and black. Suppose that there exists a nonempty set of rows A such that any column of the table has an even number of white squares that also belong to A. Prove that there exists a nonempty set of columns B such that any row of the table contains an even number of white squares that also belong to B.

[Gabriel Dospinescu]

Solution. This is just the combinatorial translation of the well-known fact that a matrix T is invertible in a field if and only if its transpose is also invertible in that field. But this is not so easy to see. In each white square we write the number 1 and in each black square we put a 0. We thus obtain a binary matrix $T = (t_{ij})_{1 \le i,j \le n}$. From now on, we work only in $\mathbb{Z}/2\mathbb{Z}$. Suppose that A contains the rows $a_1, a_2, \ldots, a_k$. It follows that $\sum_{i=1}^{k} t_{a_i,j} = 0$ for all $j = 1, 2, ..., n$. Now, let us take

$$x_i = \begin{cases} 1, & \text{if } \ i \in A \\ 0, & \text{if } \ i \notin A \end{cases}$$

It follows that the system

$$\begin{cases} t_{11}z_1 + t_{21}z_2 + \cdots + t_{n1}z_n = 0 \\ t_{12}z_1 + t_{22}z_2 + \cdots + t_{n2}z_n = 0 \\ \dots\dots\dots\dots\dots\dots\dots\dots\dots\dots \\ t_{1n}z_1 + t_{2n}z_2 + \cdots + t_{nn}z_n = 0 \end{cases}$$

has the nontrivial solution $(x_1, x_2, \ldots, x_n)$. Thus, $\det T = 0$ and consequently $\det {}^tT = 0$. But this means that the system

$$\begin{cases} t_{11}y_1 + t_{12}y_2 + \cdots + t_{1n}y_n = 0 \\ t_{21}y_1 + t_{22}y_2 + \cdots + t_{2n}y_n = 0 \\ \dots\dots\dots\dots\dots\dots\dots\dots\dots\dots \\ t_{n1}y_1 + t_{n2}y_2 + \cdots + t_{nn}y_n = 0 \end{cases}$$

also has a nontrivial solution in $\mathbb{Z}/2\mathbb{Z}$. Now, we take $B = \{i \mid y_i \neq 0\}$ and we clearly have $B \neq \emptyset$ and $\sum_{x \in B} u_{ix} = 0,\ i = 1, 2, ..., n$. But this means that any line of the table contains an even number of white squares that also belong to B, and the problem is solved.

The cherry on the cake is the following very difficult problem, where just knowing the trick of computing ${}^tA \cdot A$ does not suffice. It is true that it is one of the main steps, but there are many more things to do after we compute ${}^tA \cdot A$. And if for these first problems we have used only intuitive or well-known properties of the matrices and fields, this time we need a more sophisticated arsenal: the properties of the characteristic polynomial and the eingenvalues of a matrix. It is exactly the kind of problem that knocks you down when you feel most confident. Note that the problem can also be reformulated in a more down-to-earth way: for which m, n is there a directed graph with n vertices in which every pair of vertices is connected by exactly m paths of length 2?

Example 6. Let $S = \{1, 2, \ldots, n\}$ and let A be a family of pairs of elements in S with the following property: for any $i, j \in S$ there exist exactly m indices $k \in S$ for which $(i, k), (k, j) \in A$. Find all possible values of m and n for which this is possible.

[Gabriel Carrol]

Solution. It is not difficult to see what hides behind this problem. Indeed, if we take $T = (t_{ij})_{1 \leq i,j \leq n}$, where

$$t_{ij} = \begin{cases} 1, & \text{if } (i, j) \in A \\ 0, & \text{otherwise} \end{cases}$$

the existence of the family A reduces to

$$T^2 = \begin{pmatrix} m & m & \ldots & m \\ m & m & \ldots & m \\ \ldots & \ldots & \ldots & \ldots \\ m & m & \ldots & m \end{pmatrix}.$$

So we must find all values of m and n for which there exist a binary matrix T such that

$$T^2 = \begin{pmatrix} m & m & \dots & m \\ m & m & \dots & m \\ \dots & \dots & \dots & \dots \\ m & m & \dots & m \end{pmatrix}.$$

Let us consider

$$B = \begin{pmatrix} m & m & \dots & m \\ m & m & \dots & m \\ \dots & \dots & \dots & \dots \\ m & m & \dots & m \end{pmatrix}.$$

and find the eigenvalues of B. This is not difficult, since if x is an eingenvalue, then

$$\begin{vmatrix} m-x & m & \dots & m \\ m & m-x & \dots & m \\ \dots & \dots & \dots & \dots \\ m & m & \dots & m-x \end{vmatrix} = 0$$

If we add all columns to the first and then take the common factor $mn - x$, we obtain the equivalent form

$$(mn-x)\begin{vmatrix} 1 & m & \dots & m \\ 1 & m-x & \dots & m \\ \dots & \dots & \dots & \dots \\ 1 & m & \dots & m-x \end{vmatrix} = 0.$$

In this last determinant, we subtract from each column the first column multiplied by m and we obtain in the end the equation $x^{n-1}(mn - x) = 0$, which shows that the eigenvalues of B are precisely $\underbrace{0, 0, \dots, 0}_{n-1 \text{ times}}, mn$. But these are exactly the squares of the eigenvalues of T. Hence T has the eigenvalues $\underbrace{0, 0, \dots, 0}_{n-1}, \sqrt{mn}$, because the sum of the eigenvalues is nonnegative (being equal to the sum of the elements of the matrix situated on the main diagonal). Since $Tr(T) \in \mathbb{Z}$, we find that mn must be a perfect square. Also, because $Tr(T) \leq n$, we must have $m \leq n$.

Now, let us prove the converse. Suppose that $m \leq n$ and mn is a perfect square and write $m = du^2$, $n = dv^2$. Let us take the matrices

$$I = (\underbrace{1\ldots1}_{dv\ \text{times}}), \quad O = (\underbrace{0\ldots0}_{dv\ \text{times}}).$$

Now, let us define the circulant matrix

$$S = \begin{pmatrix} \underbrace{I\ldots I}_{u}\underbrace{O\ldots O}_{v-u} \\ O\underbrace{I\ldots I}_{u}\underbrace{O\ldots O}_{v-u-1} \\ \ldots\ldots\ldots\ldots\ldots\ldots \\ \underbrace{I\ldots I}_{u-1}\underbrace{O\ldots O}_{v-u}I \end{pmatrix} \in M_{v,n}(\{0,1\}).$$

Finally, we take

$$A = \begin{pmatrix} S \\ S \\ \ldots \\ S \end{pmatrix} \in M_n(\{0,1\}).$$

It is not difficult to see that

$$A^2 = \begin{pmatrix} m & m & \ldots & m \\ m & m & \ldots & m \\ \ldots & \ldots & \ldots & \ldots \\ m & m & \ldots & m \end{pmatrix},$$

which concludes the proof.

The last idea that we present here (but certainly these are not all the methods of higher mathematics applied to combinatorics) is the use of vector spaces. Again, we will not insist on complicated concepts from the theory of vector spaces, just the basic facts and theorems. Maybe the most useful fact is that if V is a vector space of dimension n (that is, V has a basis of cardinality n), then any $n+1$ or more vectors are linearly dependent. As a direct application, we will discuss the following problem, which is very difficult to solve by means of elementary mathematics. Try first to solve it without vectors and you will see how hard it is. The following example is classical too, but few people know the trick behind it.

Example 7. Let n be a positive integer and let $A_1, \dots, A_{n+1}$ be nonempty subsets of the set $\{1, 2, \dots, n\}$. Show that there exist nonempty and disjoint index sets $I = \{i_1, i_2, \dots, i_p\}$ and $J = \{j_1, \dots, j_q\}$ such that

$$A_{i_1} \cup A_{i_2} \cup \cdots \cup A_{i_p} = A_{j_1} \cup A_{j_2} \cup \cdots \cup A_{j_q}.$$

Chinese Olympiad

Solution. Let us assign to each subset A_k a vector $v_k \in \mathbb{R}^n$, where $v_k = (x_k^1, x_k^2, \dots, x_k^n)$ and

$$x_k^l = \begin{cases} 0, & \text{if } \ l \in A_k \\ 1, & \text{if } \ l \notin A_k \end{cases}$$

Because $\dim \mathbb{R}^n = n$, the vectors we have just constructed must be linearly dependent. So, we can find $a_1, a_2, \dots, a_{n+1} \in \mathbb{R}$, not all of them 0, such that

$$a_1 v_1 + a_2 v_2 + \cdots + a_{n+1} v_{n+1} = 0.$$

Now take $I = \{i \in \{1, 2, \dots, n+1\} \mid a_i > 0\}$ and $J = \{j \in \{1, 2, \dots, n+1\} \mid a_j < 0\}$. It is clear that I and J are nonempty and disjoint. Let us prove that $\bigcup_{i \in I} A_i = \bigcup_{j \in J} A_j$ and the solution will be complete. Take $h \in \bigcup_{i \in I} A_i$ and suppose that $h \notin \bigcup_{j \in J} A_j$. Then the vectors v_j with $j \in J$ have zero on their hth component, so the hth component of the vector $a_1 v_1 + a_2 v_2 + \cdots + a_{n+1} v_{n+1}$ is $\sum_{\substack{x \in A_i \\ i \in I}} a_i > 0$, which is impossible, since $a_1 v_1 + a_2 v_2 + \cdots + a_{n+1} v_{n+1} = 0$. This shows that $\bigcup_{i \in I} A_i \subset \bigcup_{j \in J} A_j$. The reversed inclusion can be proved in exactly the same way, so we conclude that $\bigcup_{i \in I} A_i = \bigcup_{j \in J} A_j$.

And probably an even more difficult result:

Example 8. Let S be a finite subset of $[0, 1]$ containing 0 and 1 and such that every distance that occurs between pairs of elements occurs at least twice, except for the distance 1. Prove that S contains only rational numbers.

[E.G.Strauss] Iran 1998

Solution. Let $(e_1, e_2, ..., e_n)$ be a basis of the linear space spanned by S over the field of rational numbers; this basis can be chosen such that $e_n = 1$. Now, write each element x_i of S in this basis: $x_i = a_{i1}e_1 + a_{i2}e_2 + \cdots + a_{in}e_n$. We can define an order relation on the set of these vectors, by saying that $x_i > x_j$ if there exists a position inn which the two vectors differ and x_i has a larger coordinate in the first position where they differ. This (lexicographic) order is total, so we can choose the maximal and minimal elements for it to be x_i and x_j respectively. We know that $x_i - x_j = x_k - x_l$ for some k, l. Thus using the maximality and minimality of x_i and x_j respectively, we deduce that $x_i = (0, 0, ..., 0, 1)$ and $x_j = (0, 0, ..., 0)$. Because any other vector x_r is less than x_i but greater than x_j, we deduce that all vectors have the first $n - 1$ coordinates zero, which is equivalent to the fact that all elements of S are rational.
We conclude this discussion with another problem, proposed for the TST 2004 in Romania, whose idea is also related to vector spaces.

Example 9. Thirty boys and twenty girls are training for the Team Selection Test. They observed that any two boys have an even number of common acquaintances among the girls and exactly nine boys know an odd number of girls. Prove that there exists a group of sixteen boys such that any girl attending the training is known by an even number of boys from this group.

[Gabriel Dospinescu]

Solution. Let us consider the matrix $A = (a_{ij})$ where

$$a_{ij} = \begin{cases} 1, & \text{if } B_i \text{ knows } F_j \\ 0, & \text{otherwise} \end{cases}$$

We have considered here that $B_1, B_2, \ldots, B_{30}$ are the boys and $F_1, F_2, \ldots, F_{20}$ are the girls. Now, consider the matrix $T = A \cdot A^t$. Observe that all the elements of the matrix T, except those from the main diagonal, are even (because $t_{ij} = \sum_{k=1}^{20} a_{ik}a_{jk}$ is the number of common acquaintances among the girls of the boys B_i, B_j). Each element on the main diagonal of T is precisely the number of girls known by the corresponding boy. Thus, if we consider the matrix T in $(\mathbb{Z}/2\mathbb{Z}, +, \cdot)$, it will be diagonal, with exactly nine nonzero elements on its main diagonal. From now on, we will work only in $\mathbb{Z}/2\mathbb{Z}$. We have seen that $\text{rank}(T) = 9$. Using Sylvester's inequality, we have

$$9 = \text{rank}(T) \geq \text{rank}(A) + \text{rank}\left(A^t\right) - 20 = 2 \cdot \text{rank}\left(A^t\right) - 20$$

hence $r = \text{rank}\left(A^t\right) \leq 14$. Let us consider now the linear system in $(\mathbb{Z}/2\mathbb{Z}, +, \cdot)$:

$$\begin{cases} a_{11}x_1 + a_{21}x_2 + \cdots + a_{30,1}x_{30} = 0 \\ a_{12}x_1 + a_{22}x_2 + \cdots + a_{30,2}x_{30} = 0 \\ \cdots\cdots\cdots\cdots\cdots\cdots\cdots\cdots\cdots \\ a_{1,20}x_1 + a_{2,20}x_2 + \cdots + a_{30,20}x_{30} = 0 \end{cases}$$

The set of solutions of this system is a vector space of dimension $30 - r \geq 16$. This is why we can choose a solution $(x_1, x_2, \ldots, x_{30})$ of the system, having at least 16 components equal to $\widehat{1}$. Finally, consider the set $M = \{i \in \{1, 2, \ldots, 30\} |\ x_i = \widehat{1}\}$. We have proved that $|M| \geq 16$ and also $\sum_{j \in M} a_{ji} = 0$ for all $i = 1, 2, ..., 20$. But observe that $\sum_{j \in M} a_{ji}$ is just the number of boys B_k with $k \in M$ such that B_k knows F_i. Thus, if we choose the group of those boys B_k with $k \in M$, then each girl is known by an even number of boys from this group, and the problem is solved.

A famous result of Sylvester (proved by Gallai and then by many other mathematicians) states that if A is a finite set of points in the plane such that there is no line which contains A, then there exists a line passing through exactly two points of A. The following example is a refinement of this result and the proof is almost magical:

Example 10. Prove that n distinct points, not all of them lying on a line, determine at least n distinct lines.

[Paul Erdős]

Solution. Number the points with $1, 2, ..., n$. Let X be the set of distinct lines passing through two of the n points and let A_i be the set of those lines in X that contain the point i. Then any two of the sets A_i, A_j have exactly one common element. We need to prove that their union, X has at least n elements. Suppose the contrary, namely that there are only $p < n$ such elements of X (and let $l_1, \ldots, l_p$ be these lines). Then because any homogeneous linear system with p equations and more than p unknowns has a nontrivial solution, it follows that we can assign numbers $x_1, x_2, ..., x_n$, not all 0, to the points such that the sum of the numbers on each line of X is 0. Then $\sum_{i \in l_j} x_i = 0$ for all j. Therefore

$$0 = \sum_{j=1}^{p} (\sum_{i \in l_j} x_i)^2.$$

However, observe that in the last sum every x_i^2 appears at least twice (since not all the points are on the same line), yet every product $2x_ix_j$ with $i \neq j$ appears only once (this is where we use the fact that any two sets among $A_1, A_2, ..., A_n$ have exactly one common point). We therefore obtain $x_1^2 + x_2^2 + \cdots + x_n^2 + (x_1 + x_2 + \cdots + x_n)^2 \leq 0$, which forces all x_i to be zero, which contradicts the choice of $x_1, x_2, ..., x_n$.

In the framework of the previous problem, the next example should not be very difficult to solve. However, it is worth saying that this problem has no combinatorial proof until now: this is the famous Graham-Pollak theorem. The solution, due to Tverberg, is taken from the excellent work **Proofs from The Book**.

Example 11. There exists no partition of the complete graph on n vertices with fewer than $n-1$ complete bipartite subgraphs (such that every edge belongs to exactly one subgraph).

[R. Graham, O. Pollak]

Solution. Denote by $1, 2, ..., n$ the vertices of the complete graph on n vertices and suppose that $B_1, B_2, ..., B_m$ is a partition of this graph with complete bipartite subgraphs. Every such subgraph B_k is defined by two sets of vertices L_k and R_k. Put a real number x_i in each vertex of the complete graph K_n.

The hypothesis implies that

$$\sum_{1\leq i<j\leq n} x_i x_j = \sum_{k=1}^{m} \left[\left(\sum_{i\in L_k} x_i \right) \cdot \left(\sum_{j\in R_k} x_j \right) \right].$$

The idea is (like in the previous problem) that if $m < n-1$ then we can choose the real numbers $x_1, x_2, ..., x_n$ such that not all of them are zero, $x_1 + x_2 + \cdots + x_n = 0$ and $\sum\limits_{i\in L_k} x_i = 0$ for all k. Indeed, this linear system has a nontrivial solution, because the number of unknowns exceeds the number of equations. Using the above identity and the fact that $\sum\limits_{i=1}^{n} x_i^2 = (\sum\limits_{i=1}^{n} x_i)^2 - 2 \sum\limits_{1\leq i<j\leq n} x_i x_j$, we infer that $x_1^2 + x_2^2 + \cdots + x_n^2 = 0$, which contradicts the choice of $x_1, x_2, ..., x_n$.

We end this chapter with a very tricky problem, which became classical: we found traces of it and variants in AMM, Mathematics Magazine, as well as Iranian, Russian, and German Olympiads:

Example 12. Let G be a simple graph, all of whose vertices are colored in white. A legal operation consists of choosing a vertex and changing the color of that vertex and of all of its neighbors (vertices connected to it) to the opposite color (black in white and white in black). Prove that one can make all vertices of the graph black in a finite number of legal operations.

Solution. We will assume by convention that any vertex is connected to itself, so that the adjacency matrix A of the graph (defined by $a_{ij} = 0$ if i and j are not connected and 1 otherwise) is symmetric and has only 1 on the main diagonal. The idea is to prove the existence of a set S of vertices of the graph such that any vertex of G is connected to an odd number of vertices of S. In this case, all we need is to perform legal operations on the vertices of S in order to change the color of all vertices of the graph. Now, observe that if we find a vector $v = (v_1, v_2, ..., v_n)$ with integer coefficients such that Av has all coordinates odd numbers we are done: it is enough to choose S the set of those i such that v_i is odd. Thus, the problem reduces to proving that for any binary symmetric matrix A with diagonal $(1, 1, ..., 1)$, there exists a vector v such that Av has all coordinates odd numbers. Translated in the field $F = \mathbb{Z}/2\mathbb{Z}$, this comes down to proving that for any symmetric matrix $A \in M_n(F)$, there exists a vector $v \in F^n$ such that $Av = (1, 1, ..., 1)$. By a classical argument, it is enough to show that the orthogonal vector space of $\mathrm{Im}(A)$ is a subset of the orthogonal vector space of $(1, 1, ..., 1)$. But if x is orthogonal to $\mathrm{Im}(A)$, then we must have

$$\sum_{i=1}^{n} x_i \sum_{j=1}^{n} a_{ij} y_j = 0$$

for all $y_1, ..., y_n \in F$, which means that $\sum_{i=1}^{n} a_{ij} x_i = 0$ for all j. Thus

$$\sum_{j=1}^{n} x_j \sum_{i=1}^{n} a_{ij} x_i = 0,$$

which can be also written as

$$\sum_{1\leq i<j\leq n}(a_{ij}+a_{ji})x_ix_j+\sum_{i=1}^{n}a_{ii}x_i^2=0.$$

The matrix is symmetric, so the first sum is 0. Also, we have $x_i^2 = x_i$ and $a_{ii} = 1$, so we infer that $x_1 + x_2 + \cdots + x_n = 0$, which means that x is orthogonal to v. This finishes the proof.

12.2 Practice problems

1. Consider $2n+1$ real numbers with the property that no matter how we eliminate one of them, the rest can be divided into two groups of n numbers, the sum of the numbers in the two groups being the same. Then all the numbers must be equal.

2. A handbook classifies plants by 100 attributes (each plant either has a given attribute or does not have it). Two plants are dissimilar if they differ in at least 51 attributes. Show that the handbook cannot give 51 plants all dissimilar from each other.

   Tournament of the Towns 1993

3. Let $A_1, A_2, \ldots, A_m$ be distinct subsets of a set A with $n \geq 2$ elements. Suppose that any two of these subsets have exactly one element in common. Prove that $m \leq n$.

   Fisher's inequality

4. The edges of a regular 2^n-gon are colored red and blue. A step consists of recoloring each edge which has the same color as both of its neighbors in red, and recoloring each other edge in blue. Prove that after 2^{n-1} steps all of the edges will be red and that need not hold after fewer steps.

   Iranian Olympiad 1998

5. Is there in the plane a configuration of 22 circles and 22 points on their union (ie the union of their circumferences) such that any circle contains at least 7 points and any point belongs to at least 7 circles?

   Gabriel Dospinescu, Moldova TST 2004

6. In an m by n table, real numbers are written such that for any two lines and any two columns, the sum of the numbers situated in the opposite vertices of the rectangle formed by them is equal to the sum of the numbers situated in the other two opposite vertices. Some of the numbers are erased, but the remaining ones allow us to find the erased numbers using the above property. Prove that at least $n+m-1$ numbers remained on the table.

Russian Olympiad 1971

7. A number n of teams compete in a tournament, and each team plays against any other team exactly once. In each game, 2 points are given to the winner, 1 point for a draw, and 0 points for the loser. It is known that for any subset S of teams, one can find a team (possibly in S) whose total score in the games with teams in S is odd. Prove that n is even.

D. Karpov, Russian Olympiad 1972

8. A simple graph has the property: given any nonempty set H of its vertices, there is a vertex x of the graph such that the number of edges connecting x with the points in H is odd. Prove that the graph has an even number of vertices.

Komal

9. Let $A_1, A_2, \ldots, A_n$ be subsets of $A = \{1, 2, \ldots, n\}$ such that for any nonempty subset T of A, there is an $i \in A$ such that $|A_i \cap T|$ is odd. Suppose that B_1, B_2 are subsets of A such that

$$|A_i \cap B_1| = |A_i \cap B_2| = 1$$

for all i. Prove that $B_1 = B_2$.

Gabriel Dospinescu, Mathematical Reflections

10. Light bulbs $L_1, L_2, \ldots, L_n$ are controlled by switches $S_1, S_2, \ldots, S_n$. Switch S_i changes the on/off status of light L_i and possibly the status of some other lights. Suppose that if S_i changes the status of L_j then S_j changes the status of L_i. Initially all lights are off. Is it possible to operate the switches in such a way that all the lights are on?

Uri Peled, AMM 10197

11. Let G be a graph. Prove that the set of its vertices can be partitioned in two groups (possibly empty) such that each group induces a subgraph in which all vertices have even degree.

Gallai Cycle-Cocyle partition theorem

12. At a certain mathematical conference, every pair of mathematicians are either friends or strangers. At mealtime, every participant eats in one of two large dining rooms. Each mathematician insists upon eating in a room which contains an even number of his or her friends. Prove that the number of ways that the mathematicians may be split between the two rooms is a power of two .

USAMO 2008

13. Let $n \geq 2$. Find the largest p such that for all $k \in \{1, 2, \ldots, p\}$ we have

$$\sum_{\sigma \in A_n} \left(\sum_{i=1}^{n} i\sigma(i) \right)^k = \sum_{\sigma \in B_n} \left(\sum_{i=1}^{n} i\sigma(i) \right)^k,$$

where A_n, B_n are the sets of all even and odd permutations of the set $\{1, 2, \ldots, n\}$ respectively.

Gabriel Dospinescu

14. Let s be a function defined by

$$s(a_1, a_2, , a_r) = (|a_1 - a_2|, |a_2 - a_3|, \ldots, |a_r - a_1|).$$

Prove the equivalence of the following statements:

(a) for all nonnegative integers $a_1, a_2, \ldots, a_r$, there exists n such that the n-th iterate of s evaluated at $(a_1, a_2, \ldots, a_r)$ is $(0, 0, \ldots, 0)$;

(b) r is a power of 2.

Ducci's problem

15. An $m \times n$ matrix is filled with 0s and 1s such that any two rows differ in at least $n/2$ positions. Prove that $m \leq 2n$.

Iranian Olympiad

16. Let n be a positive integer. Find the largest number k with the following property: there exist k 2^n-tuples of integers, all equal to 0 or 1 and such that $d(u, v) \geq 2^{n-1}$ for any two distinct tuples u, v. Here

$$d(u, v) = \sum_{i=1}^{2^n} |u_i - v_i|$$

if u_i, v_j are the components of u, v.

Chinese TST 2005

17. In a contest consisting of n problems, the jury defines the difficulty of each problem by assigning it a positive integral number of points (the same number of points may be assigned to different problems). Any participant who answers the problem correctly receives that number of points for the problem; any other participant receives 0 points. After the participants submitted their answers, the jury realizes that given any ordering of the participants (where ties are not permitted), it could have

defined the problems' difficulty levels to make that ordering coincide with the participants' ranking according to their total scores. Determine, in terms of n, the maximum number of participants for which such a scenario could occur.

Russian Olympiad 2001

18. For a permutation σ of $\{1, 2, \ldots, n\}$ let $\varepsilon(\sigma) = 1$ if σ is even and -1 otherwise. Let $f(\sigma)$ be the number of fixed points of σ. Proved that

$$\sum_{\sigma} \frac{\varepsilon(\sigma)}{1 + f(\sigma)} = (-1)^{n+1} \cdot \frac{n}{n+1},$$

where the sum is taken over all permutations σ of $\{1, 2, \ldots, n\}$.

Putnam 2005

19. A permutation σ of $\{1, 2, \ldots, n\}$ is called k-limited if $|\sigma(i) - i| \le k$ for all $1 \le i \le n$. Prove that the number of k-limited permutations of $\{1, 2, \ldots, n\}$ is odd if and only if $n = 0, 1 \pmod{2k+1}$.

Putnam 2008

20. Let $G_1, G_2, \ldots, G_k$ be complete bipartite subgraphs of the complete graph K_{2n} with $2n$ vertices. Assume that every edge of K_{2n} is contained in an odd number of subgraphs $G_1, G_2, \ldots, G_k$. Prove that $k \ge n$.

21. Let $A_1, A_2, \ldots, A_m$ and $B_1, B_2, \ldots, B_p$ be subsets of $\{1, 2, \ldots, n\}$ such that $A_i \cap B_j$ is an odd number for all i and j. Then $mp \le 2^{n-1}$.

Benny Sudakov

22. On an $n \times m$ sheet of paper a grid dividing the sheet into unit squares is drawn. The two sides of length n are taped together to form a cylinder.

Prove that it is possible to write a real number in each square, not all zero, so that each number is the sum of the numbers in the neighboring squares, if and only if there exist integers k, l such that $n+1$ does not divide k and

$$\cos\frac{2l\pi}{m} + \cos\frac{k\pi}{n+1} = \frac{1}{2}.$$

Ciprian Manolescu, Romanian TST 1998

23. Let $m > n+1$ and let $A_1, A_2, \ldots, A_m$ be subsets of $\{1, 2, \ldots, n\}$.Then there are disjoint sets I, J such that $\bigcup_{i\in I} A_i = \bigcup_{j\in J} A_j$ and $\bigcap_{i\in I} A_i = \bigcap_{j\in J} A_j$.

Lindstrom's theorem

24. In a society, acquaintance is mutual and even more, any two persons have exactly one common friend. Then there is a person who knows all the others.

Erdos-Renyi-Sos, Friendship theorem

25. Let $x_1, x_2, \ldots, x_n$ be real numbers and suppose that the vector space spanned by $x_i - x_j$ over the rationals has dimension m. Then the vector space spanned only by those $x_i - x_j$ for which $x_i - x_j \neq x_k - x_l$ whenever $(i, j) \neq (k, l)$ also has dimension m.

Straus's theorem

26. In a graph G with $n^2 + 1$ vertices every vertex has degree n. Moreover, any cycle has length at least 5. Prove that $n \in \{1, 2, 3, 7, 57\}$.

Hoffman-Singleton theorem

27. Let $A_1, A_2, \ldots, A_m$ and $B_1, B_2, \ldots, B_m$ be sets such that

    (a) $|A_1| = |A_2| = \cdots = |A_m| = a$ and $|B_1| = |B_2| = \cdots = |B_m| = b$.

    (b) $A_i \cap B_j$ is nonempty if and only if $i \neq j$.

    Prove that $m \leq \binom{a+b}{b}$.

Bollobas's theorem

28. Let F be a family of subsets of $\{1, 2, \ldots, n\}$ with the following property: there is no $Y \subset \{1, 2, \ldots, n\}$ with k elements such that $\{Y \cap A | A \in F\}$ is the set of all subsets of Y. Prove that

$$|F| \leq \binom{n}{0} + \binom{n}{1} + \cdots + \binom{n}{k-1}.$$

Sauer-Shelah lemma

29. Let m, n be positive integers and let S be a figure made of 1×1 squares and having the property: whenever the 1×1 squares of an $m \times n$ table are filled with real numbers whose sum is positive, the figure can be placed on the table (possibly after being rotated by a multiple of $\pi/2$, but such that its squares are contained in the table) so that the sum of the numbers in the squares covered by the figure is positive. Prove that one can place a number of such figures on the table such that each 1×1 square of the table is covered by the same number of figures.

Russia 1998

30. Let k be a positive real number. Prove that the unit square can be tiled with finitely many rectangles similar to the $1 \times k$ rectangle if and only if k is algebraic and all of its algebraic conjugates have positive real part.

Laczkovich-Szekeres' theorem

31. An $a \times b$ rectangle is divided into squares with side lengths $x_1, x_2, \ldots, x_n$. Prove that $\frac{a}{x_i}$ and $\frac{b}{x_i}$ are rational numbers.

Dehn's theorem

32. Given a rectangle R, a finite set $S \subset \mathbb{R}^+$ and a positive integer $n \in \mathbb{N}$, there are only finitely many dissections of R into n rectangles whose side ratio is in S.

Vesselin Dimitrov

Chapter 13

13.1 Theory and examples

It may seem weird, but geometry is really useful in number theory, and sometimes it can help in proving difficult results with some extremely simple arguments. In the sequel we are going to exhibit a few applications of geometry in number theory, almost all of them revolving around the celebrated Minkowski's theorem. This theorem will give a very efficient criterion for a centrally symmetric convex body to contain a nontrivial lattice point. The existence of this point has important consequences in the theory of representation of numbers by quadratic forms, and in the approximation of real numbers by rational numbers. As usual, we will present only a mere introduction to this extremely well-developed field of mathematics. You will surely have the pleasure of consulting some reference books about this fascinating area of research.

First of all, let us define the notion of convex body (or convex set; in what follows we will call bodies sets in $\mathbb{R}^n$). A subset A of $\mathbb{R}^n$ will be called a convex body if it is convex, that is A contains the segment $\{tx+(1-t)y|0 \leq t \leq 1\}$ once it contains two points x, y. A is called centrally symmetric if it is symmetric with respect to the origin, that is $-x \in A$ if $x \in A$. We will take for granted that convex bodies have volumes (this is more delicate than it seems, actually). We start by proving the celebrated Minkowski's theorem.

Theorem 13.1 (Minkowski). *Suppose that A is a bounded centrally symmetric convex body in $\mathbb{R}^n$ having volume strictly greater than 2^n. Then there is a lattice point in A different from the origin.*

Proof. The proof is surprisingly simple. Indeed, begin by making a partition of $\mathbb{R}^n$ into cubes of edge 2, having as centers the points that have all coordinates even integers. It is clear that any two such cubes have disjoint interiors and that they cover all space. That is why we can say that the volume of A is equal to the sum of the volumes of the intersections of A with each cube (because A is bounded, it is clear that the sum will be finite). But of course, one can bring any cube into the cube centered around the origin by using a translation by a vector all of whose coordinates are even. Since translations preserve volume, we will have now an agglomeration of bodies in the central cube (the one centered at the origin), and the sum of volumes of all these bodies is greater than 2^n. It follows that there are two bodies which intersect at a point X. Now, look at the cubes where these two bodies where taken from and look at the points in these cubes whose image under these translations is the point X. We have found two different points x, y in our convex body such that $x - y \in 2\mathbb{Z}^n$. But since A is centrally symmetric and convex, it follows that $\frac{x-y}{2}$ is a lattice point different from the origin and belonging to A. The theorem is proved. □

Here is a surprising result that follows directly from this theorem.

Example 1. Suppose that at each lattice point in space except for the origin one draws a ball of radius $r > 0$ (common for all the balls). Then any line that passes through the origin will intercept some ball.

Solution. Let us suppose the contrary and consider a cylinder having as axis that very line and base a circle of radius $\frac{r}{2}$. We choose it sufficiently long to ensure that it has a volume greater than 8. This is clearly a bounded centrally symmetric convex body in space and using Minkowski's theorem we deduce the existence of a nontrivial lattice point in this cylinder (or on the border or the corresponding sphere). This means that the line will intercept the ball centered around this point.

Actually, the theorem proved before admits a more general formulation:

Theorem 13.2 (Minkowski). *Let A be a convex body in $\mathbb{R}^n$ and let $v_1, v_2, \dots, v_n$ be linearly independent vectors in $\mathbb{R}^n$. Consider the fundamental parallelepiped $P = \left\{ \sum_{i=1}^{n} x_i v_i \mid 0 \le x_i \le 1 \right\}$ and denote by Vol(P) its volume. If A has a volume greater than $2^n \cdot$ Vol(P), A must contain at least one point of the lattice $L = \mathbb{Z}v_1 + \cdots + \mathbb{Z}v_n$ different from the origin.*

Proof. With all these terms, it would seem that this is extremely difficult to prove. Actually, it follows trivially from the first theorem. Indeed, by considering the linear transformation f sending v_i into $e_i = (0, 0, \ldots, 1, 0, \ldots, 0)$, one can easily see that P is sent into the "normal" cube in $\mathbb{R}^n$ (that is, the set of vectors all of whose components are between 0 and 1), and that f maps L into $\mathbb{Z}^n$. Because the transformation is linear, it will send A into a bounded centrally symmetric convex body of volume $\dfrac{\text{Vol}(A)}{\text{Vol}(P)} > 2^n$. It suffices to apply the first theorem to this bounded centrally symmetric convex body and to look at the preimage of the lattice point (in $\mathbb{Z}^n$), in order to find a nontrivial point of $A \cap L$. This finishes the proof of the second theorem. □

In the chapter **Primes and Squares** we proved that any prime number of the form $4k+1$ is the sum of two squares. Let us prove it differently, this time using Minkowski's theorem.

Example 2. Any prime number of the form $4k+1$ is the sum of two squares.

Solution. We have already proved that for any prime number of the form $4k+1$, call it p, we can find an integer a such that $p|a^2+1$. Consider then $v_1 = (p, 0)$ and $v_2 = (a, 1)$. Clearly, they are linearly independent and moreover for any point (x, y) in the lattice $L = \mathbb{Z}v_1 + \mathbb{Z}v_2$ we have $p|x^2+y^2$. Indeed, there are $m, n \in \mathbb{Z}$ such that $x = mp + na$, $y = n$ and thus $x^2+y^2 \equiv n^2(a^2+1) \equiv 0 \pmod{p}$. In addition, the area of the fundamental parallelogram is $\|v_1 \wedge v_2\| = p$. Next, consider as convex body (when the context is clear, we will no longer add bounded centrally symmetric convex body, just convex body) the disc centered at the origin and having radius $\sqrt{2p}$. Clearly, its area is strictly greater than four times the area of the fundamental parallelogram. Thus, there is a point (x, y) different from the origin that lies in this disc and also in the lattice $L = \mathbb{Z}v_1 + \mathbb{Z}v_2$. For this point we have $p|x^2+y^2$ and $x^2+y^2 < 2p$, which shows that $p = x^2+y^2$.

Proving that some Diophantine equation has no solution is a classical problem, but what can we do when we are asked to prove that some equation has solutions? Minkowski's theorem and, in general, the geometry of numbers give responses to such problems. Here is an example:

Example 3. Consider positive integers a, b, c such that $ac = b^2 + b + 1$. Prove that the equation $ax^2 - (2b+1)xy + cy^2 = 1$ has integer solutions.

Polish Olympiad

Solution. Here is a very quick approach: consider in $\mathbb{R}^2$ the set of points satisfying $ax^2 - (2b+1)xy + cy^2 < 2$. A simple computation shows that it is an elliptical disc having area $\frac{4\pi}{\sqrt{3}} > 4$. An elliptical disc is obviously a convex body and, even more, it certainly is symmetric about the origin. Thus by Minkowski's theorem there is a point in this region different from the origin. Since $ac = b^2 + b + 1$, we have for all x, y not both equal to 0 the inequality $ax^2 - (2b+1)xy + cy^2 > 0$. Thus for $(x, y) \in \mathbb{Z}^2 \setminus \{(0,0)\}$, we have $ax^2 - (2b+1)xy + cy^2 = 1$ and the existence of a solution of the given equation is proved.

The following problem (like the one above) has a quite difficult elementary solution. The solution using geometry of numbers is more natural, but it is not at all obvious how to proceed. Yet... the experience gained by solving the previous problem should ring a bell.

Example 4. Suppose that n is a positive integer for which the equation $x^2 + xy + y^2 = n$ has rational solutions. Then this equation has integer solutions as well.

Kömal

Solution. Of course, the problem reduces to: if there are integers a, b, c such that $a^2 + ab + b^2 = c^2 n$, then $x^2 + xy + y^2 = n$ has integer solutions. We will assume that a and b are nonzero (otherwise the conclusion follows trivially); and more, a classical argument allows us to assume that a and b are each relatively prime (which implies that a and b are each relatively prime to n, too). We try again to find a pair $(x, y) \in \mathbb{Z}^2 \setminus \{(0,0)\}$ such that $x^2+xy+y^2 < 2n$ and such that n divides x^2+xy+y^2. In this case we will have $x^2+xy+y^2 = n$ and the conclusion follows.

First, let us look at the region defined by $x^2 + xy + y^2 < 2n$. Again, simple computations show that it is an elliptical disc of area $\frac{4\pi}{\sqrt{3}}n$. Next, consider the lattice formed by the points (x, y) such that n divides $ax - by$. The area of the fundamental parallelogram is clearly at most n. By Minkowski's theorem, we can find $(x, y) \in \mathbb{Z}^2 \setminus \{(0,0)\}$ such that $x^2 + xy + y^2 < 2n$ and n divides $ax - by$. We claim that this yields an integer solution to the equation. Observe that $ab(x^2 + xy + y^2) = c^2xyn + (ax - by)(bx - ay)$ and so n also divides x^2+xy+y^2 (since n is relatively prime with a and b) and the conclusion follows.

Before continuing with some more difficult problems, let us recall that for any symmetric real matrix A such that

$$\sum_{1 \leq i,j \leq n} a_{ij}x_ix_j > 0$$

for all $x = (x_1, x_2, \ldots, x_n) \in \mathbb{R}^n \setminus \{0\}$ the set of points satisfying

$$\sum_{1 \leq i,j \leq n} a_{ij}x_ix_j \leq 1$$

has volume equal to $\frac{\text{Vol}(B_n)}{\sqrt{\det A}}$, where

$$\text{Vol}(B_n) = \frac{\pi^{\frac{n}{2}}}{\Gamma\left(1 + \frac{n}{2}\right)},$$

where B_n is the nth dimensional Euclidean ball (and $\Gamma(x) = \int_0^\infty e^{-t}t^{x-1}dt$ is Euler's gamma function). There are explicit formulae for $\Gamma(1 + \frac{n}{2})$ because

$\Gamma(n) = (n-1)!$ for positive integers n (so this takes care of the case n even) and

$$\Gamma\left(1+\frac{n}{2}\right) = \frac{\sqrt{\pi} \times n!}{2^{\frac{3n+1}{2}} \times ((n-1)/2)!}$$

for odd n. The proof of this result is not elementary and we invite you to read more about it in any decent book of multivariate integral calculus. In particular, you should notice that these results can be applied to previous problems to facilitate the computations of different areas and volumes. With these fact in mind, let us attack some serious problems.
If we talked about squares, why not present the beautiful classical proof of Lagrange's theorem on representations using four squares?

Example 5. Any positive integer is a sum of four perfect squares.

[Lagrange]

Solution. This is going to be much more complicated, but the idea is always the same. The main difficulty is finding the appropriate lattice and centrally symetric convex body. First of all, let us prove the result for prime numbers. Let p be an odd prime number (for the prime 2 the result is obvious) and consider the sets $A = \{a^2 | \ a \in \mathbb{Z}/p\mathbb{Z}\}$, $B = \{-b^2 - 1| \ b \in \mathbb{Z}/p\mathbb{Z}\}$. Since there are $\frac{p+1}{2}$ distinct squares in $\mathbb{Z}/p\mathbb{Z}$ (as we have already seen in previous chapters), these two sets cannot be disjoint. In particular, there are integers x and y such that $0 \le x, y \le p-1$ and $p | x^2 + y^2 + 1$. This is the observation that will enable us to find a good lattice. Consider now the vectors

$$v_1 = (p, 0, 0, 0),\ v_2 = (0, p, 0, 0),\ v_3 = (x, y, 1, 0),\ v_4 = (y, -x, 0, 1)$$

and the lattice L generated by these vectors. A simple computation (using the above formulas) allows us to prove that the volume of the fundamental parallelepiped is p^2. Moreover, one can easily verify that for each point $(t, u, v, w) \in L$ we have $p | t^2 + u^2 + v^2 + w^2$. Even more, we can also prove (by employing the non-elementary results stated before) that the volume of

the convex body $C = \{(t, u, v, w) \in \mathbb{R}^4 | \; t^2 + u^2 + v^2 + w^2 < 2p\}$ is equal to $2\pi^2 p^2 > 16\mathrm{Vol}(P)$. Thus $C \cap L$ is not empty. It suffices then to choose a point $(t, u, v, w) \in C \cap L$ and we will clearly have $t^2 + u^2 + v^2 + w^2 = p$. This finishes the proof for prime numbers.
Of course, everything would be nice if the product of two sums of four squares is always a sum of four squares. Fortunately, this is the case, but the proof is not obvious at all. It follows form the miraculous identity:

$$(a^2 + b^2 + c^2 + d^2)(x^2 + y^2 + z^2 + t^2) = (ax + by + cz + dt)^2$$

$$+(ay - bx + ct - dz)^2 + (az - bt + dy - cx)^2 + (at + bz - cy - dx)^2.$$

This is very nice, but how could one answer the eternal question: how on earth should I think of such an identity? Well, this time there is a very nice reason: instead of thinking in eight variables, let us reason only with four. Consider the numbers $z_1 = a + bi$, $z_2 = c + di$, $z_3 = x + yi$, $z_4 = z + ti$ and introduce the matrices

$$M = \begin{pmatrix} z_1 & z_2 \\ -\overline{z_2} & \overline{z_1} \end{pmatrix}, \quad N = \begin{pmatrix} z_3 & z_4 \\ -\overline{z_4} & \overline{z_3} \end{pmatrix}.$$

We have

$$\det(M) = |z_1|^2 + |z_2|^2 = a^2 + b^2 + c^2 + d^2$$

and similarly

$$\det(N) = x^2 + y^2 + z^2 + t^2.$$

It is then natural to express $(a^2 + b^2 + c^2 + d^2)(x^2 + y^2 + z^2 + t^2)$ as $\det(MN)$. But surprise! We have

$$MN = \begin{pmatrix} z_1 z_3 - z_2 \overline{z_4} & z_1 z_4 + z_2 \overline{z_3} \\ \overline{-z_1 z_4 - z_2 \overline{z_3}} & \overline{z_1 z_3 - z_2 \overline{z_4}} \end{pmatrix}$$

and so $\det(MN)$ is again a sum of four squares. The identity is now motivated.

Let us concentrate a little bit more on approximations of real numbers. We have some beautiful results of Minkowski that deserve to be presented after this small introduction to the geometry of numbers. The following one is extremely important while studying algebraic number fields.

Example 6. Let $A = (a_{ij})$ be an $n \times n$ invertible matrix with real entries, and let $c_1, c_2, \ldots, c_n$ be positive real numbers such that $c_1 c_2 \cdots c_n > |\det A|$. Then there are integers $x_1, x_2, \ldots, x_n$, not all 0, such that $\left|\sum_{j=1}^{n} a_{ij}x_j\right| < c_i$ for all $i = 1, \ldots, n$.

Minkowski's linear forms theorem

Solution. We need to prove that there exists a nonzero integer vector X that also belongs to the region $\{Y \in \mathbb{R}^n |\ |(AY)_i| < c_i,\ i = 1, \ldots, n\}$ (here $(AY)_i$ denotes the i-th coordinate of AY). But $\{Y \in \mathbb{R}^n |\ |(AY)_i| < c_i,\ i = 1, \ldots, n\}$ is exactly the image through A^{-1} of the parallelepiped $\{Z \in \mathbb{R}^n |\ -c_i < Z_i < c_i,\ i = 1, \ldots, n\}$ (which has volume $2^n c_1 \ldots c_n$). Thus $\{Y \in \mathbb{R}^n |\ |(AY)_i| < c_i,\ i = 1, \ldots, n\}$ is a centrally symetric convex body of volume $\frac{1}{|\det A|} 2^n c_1 \ldots c_n > 2^n$. By Minkowski's theorem, this body will contain a nonzero lattice point, which satisfies the conditions of the problem.

Actually, there exists a very useful sharpening of the last result: if we suppose only that $c_1 c_2 \cdots c_n \geq |\det A|$, then the integers $x_1, x_2, \ldots, x_n$, not all 0, can be chosen such that $\left|\sum_{j=1}^{n} a_{1j}x_j\right| \leq c_1$ and $\left|\sum_{j=1}^{n} a_{ij}x_j\right| < c_i$ for all $i \geq 2$. The proof is not difficult at all, once example 6 is proved. Indeed, note that if $\epsilon > 0$ then by the previous result there are integers $x_1(\epsilon), x_2(\epsilon), \ldots, x_n(\epsilon)$, not all 0 and such that $\left|\sum_{j=1}^{n} a_{ij}x_j(\epsilon)\right| < c_i$ for $i = 2, 3, ..., n$ and $\left|\sum_{j=1}^{n} a_{1j}x_j(\epsilon)\right| < c_1(1+\epsilon)$. Because the matrix A is invertible, there exist only finitely many $(x_1(\epsilon), x_2(\epsilon), ..., x_n(\epsilon))$ with these properties for fixed ϵ. Indeed, the condition says that the vector $Ax(\epsilon)$ is bounded where $x(\epsilon)$ is the vector with components $x_i(\epsilon)$. Thus the vector $x(\epsilon)$ is also bounded in $\mathbb{R}^n$. This shows that it is possible to construct a sequence ϵ_k that converges to 0 and such that $x_j = x_j(\epsilon_k)$ does not depend on k for all j. All we need is then to make $k \to \infty$ in the above inequalities.

Now, this theorem implies Dirichlet's approximation theorem (also discussed in the chapter **Density and regular distribution**: for all real numbers $a_1, a_2, ..., a_n$ and all positive integers M there exist integers $m_1, m_2, ..., m_n, p$ such that $|p| \leq M^n$ and $|m_i - pa_i| < \frac{1}{M}$ for all i). Indeed, all we need is to apply the above result to the $(n+1) \times (n+1)$ matrix

$$\begin{pmatrix} 1 & 0 & 0 & \dots & 0 & -a_1 \\ 0 & 1 & 0 & \dots & 0 & -a_2 \\ 0 & 0 & 1 & \dots & 0 & -a_3 \\ \dots & \dots & \dots & \dots & \dots & \dots \\ 0 & 0 & 0 & \dots & 1 & -a_n \\ 0 & 0 & 0 & \dots & 0 & 1 \end{pmatrix}$$

And here is a nice consequence of the previous example. Our last example of Diophantine approximation that can be obtained using Minkowski's theorem will imply the product theorem for homogeneous linear forms:

Example 7. Let $A = (a_{ij})$ be an $n \times n$ invertible matrix with real entries ($n \geq 2$). Show the existence of integers $x_1, x_2, ..., x_n$, not all 0, for which

$$\sum_{i=1}^{n} |a_{i1}x_1 + a_{i2}x_2 + \cdots + a_{in}x_n| \leq \sqrt[n]{n! \cdot |\det A|}.$$

Solution. Let us start by computing the volume of the figure $O(x, n)$ consisting of all points $(x_1, x_2, ..., x_n)$ such that $|x_1| + |x_2| + \cdots + |x_n| \leq x$. For $n = 1$ it is certainly $2x$. Now, using Fubini's theorem we can write

$$\mathrm{Vol}(O(x,n)) = \int_{|x_1|+\cdots+|x_n| \leq x} dx_1 dx_2 ... dx_n =$$

$$= \int_{|x_n| \leq x} \int_{|x_1|+\cdots+|x_{n-1}| \leq x - |x_n|} dx_1 ... dx_{n-1}$$

$$= \int_{|x_n| \leq x} \mathrm{Vol}(O(x - |x_n|, n-1)) dx_n = \mathrm{Vol}(O(1, n-1)) \cdot \int_{|x_n| \leq x} (x - |x_n|)^{n-1} dx_n$$

$$= \mathrm{Vol}(O(1, n-1)) \cdot \frac{2x^n}{n}.$$

An immediate induction shows that $\mathrm{Vol}(O(x,n)) = \frac{(2x)^n}{n!}$. Now, the problem asks to prove that there exists some nonzero integer vector X such that AX lies in the figure $O(\sqrt[n]{n! \cdot |\det A|}, n)$. Stated otherwise, we need to prove that the image of this figure by the linear application determined by A^{-1} contains a nonzero lattice point. But the volume of this centrally symetric convex body is 2^n (just replace x by $\sqrt[n]{n! \cdot |\det A|}$). Unfortunately, we cannot directly apply Minkowski's theorem, because this volume is not strictly greater than 2^n. However, we can imitate the argument used after the solution of the previous exercise in order to obtain the desired result: for all $\epsilon > 0$ we know that the octahedron $O(\sqrt[n]{n! \cdot |\det A|} + \epsilon, n)$ contains some AX_ϵ (where X_ϵ is a nonzero integer vector). One shows that these vectors X_ϵ are uniformly bounded, then extracts a constant family of vectors and takes the limit. We leave to the reader the details.

The highlight of the IMO 1997, the very beautiful problem 6 also has a magnificent solution using geometry of numbers. Actually, we will prove much more than the result asked in the contest, which shows that, for large values of n, one of the bounds asked by the IMO problem is very weak:

Example 8. For each positive integer n, let $f(n)$ denote the number of ways of representing n as a sum of powers of two with nonnegative integer coefficients. Representations that differ only in the ordering of their summands are considered to be the same. For instance, $f(4) = 4$. Prove that there are two constants a, b such that

$$2^{\frac{n^2}{2} - n\log_2(n) - an} < f(2^n) < 2^{\frac{n^2}{2} - n\log_2(n) - bn}$$

for all sufficiently large n.

Adapted after IMO 1997

Solution. It is clear that $f(2^n)$ is just the number of nonnegative integer solutions of the equation $a_0 + 2a_1 + \cdots + 2^n a_n = 2^n$, which is the same as the number of solutions in nonnegative integers of the inequation $2a_1 + 4a_2 + \cdots + 2^n a_n \leq 2^n$. For any such solution different from $(0, 0, ..., 0, 2^n)$ we have $a_n = 0$ and we will consider the hypercube $H(a_1, a_2, ..., a_{n-1}) = [a_1, a_1 + 1) \times [a_2, a_2 + 1) \times ... \times [a_{n-1}, a_{n-1} + 1)$. It is clear that these hypercubes are pairwise disjoint for distinct solutions $(a_1, a_2, ..., a_{n-1})$. So the number of solutions of the inequation is the total volume of these hypercubes. Now, observe that any such hypercube is included in the set of points $(x_1, x_2, ..., x_{n-1})$ with $x_i \geq 0$ and $\sum\limits_{i=1}^{n-1} 2^i(x_i - 1) < 2^n$. Also, the union of these cubes covers the region consisting of those points $(x_1, x_2, ..., x_{n-1})$ with $x_i \geq 0$ and $\sum\limits_{i=1}^{n-1} 2^i x_i \leq 2^n$. Indeed, take a point $(x_1, x_2, ..., x_{n-1})$ in this region. Then $([x_1], [x_2], ..., [x_{n-1}], 0)$ is a solution of the inequation and the point belongs to the corresponding hypercube. Now, let us consider more generally the region $R(a_1, a_2, ..., a_n, A)$ defined by the inequations $x_i \geq 0$ and $a_1 x_1 + a_2 x_2 + \cdots + a_n x_n \leq A$. Its volume is

$$\mathrm{Vol}(R(a_1, ..., a_n, A)) = \int_{x_i \geq 0, a_1 x_1 + \cdots + a_n x_n \leq A} dx_1 dx_2 ... dx_n =$$

$$\int_{0 \leq x_n \leq \frac{A}{a_n}} \int_{x_1, ..., x_{n-1} \geq 0, a_1 x_1 + \cdots + a_{n-1} x_{n-1} \leq A - a_n x_n} dx_1 ... dx_{n-1}$$

$$= \int_0^{\frac{A}{a_n}} \mathrm{Vol}(R(a_1, ..., a_{n-1}, A - a_n x_n)) dx_n =$$

$$= \mathrm{Vol}(R(a_1, ..., a_{n-1}, 1)) \cdot \int_0^{\frac{A}{a_n}} (A - a_n x_n)^{n-1} dx_n =$$

$$= \frac{A^n}{n a_n} \cdot \mathrm{Vol}(R(a_1, ..., a_{n-1}, 1)).$$

This relation easily implies by induction that $\mathrm{Vol}(R(a_1, a_2, ..., a_n)) = \frac{A^n}{n! \cdot a_1 a_2 ... a_n}$. Thus, because the sum of the volumes of the hypercubes is between the volume of $R(2, 4, ..., 2^{n-1}, 2^n)$ and $R(2, 4, ..., 2^{n-1}, 2 + 2^2 + \cdots + 2^{n-1} + 2^n) =$

$R(2, 4, ..., 2^{n-1}, 2^{n+1} - 2)$, counting the solution $(0, 0, ..., 0, 2^n)$, we deduce that the number of solutions satisfies the inequalities

$$1 + \frac{2^{\frac{n^2-n}{2}}}{(n-1)!} \leq f(2^n) \leq 1 + \frac{(2^{n+1}-2)^{n-1}}{2^{\frac{n^2-n}{2}} \cdot (n-1)!}.$$

Now, note that

$$\frac{2^{\frac{n^2-n}{2}}}{(n-1)!} = 2^{\frac{n^2}{2} + O(n) - \log_2((n-1)!)}$$

and that

$$\log_2((n-1)!) = \frac{1}{\ln 2}((n-1)\ln(n-1) + O(n)) = n\log_2(n) + O(n)$$

by Stirling's formula. Similarly,

$$\frac{(2^{n+1}-2)^{n-1}}{2^{\frac{n^2-n}{2}} \cdot (n-1)!} = \frac{n^2}{2} - n\log_2(n) + O(n).$$

The existence of the two constants is now obvious.

We end this chapter with some difficult problems concerning representations of solutions of some Diophantine equations. We will show, using Minkowski's theorem, that if $n \leq 4$ and A is a symmetric and positive matrix (that is, ${}^t x A x \geq 0$ for all vectors $x \in \mathbb{R}^n$) in $SL_n(\mathbb{Z})$ (the set of integer $n \times n$ matrices with determinant 1), then there exists a matrix B with integer coefficients such that $A = B \cdot B^t$ (a result which actually holds for $n \leq 7$, for $n = 8$ being false). This will have some nice applications in the study of some Diophantine equations. Let us start with some notations and easy observations. A bases of $\mathbb{Z}^n$ will be a family $B = (v_1, v_2, ..., v_p)$ of vectors in $\mathbb{Z}^n$, such that any vector $x \in \mathbb{Z}^n$ can be uniquely expressed as $k_1v_1 + k_2v_2 + \cdots + k_pv_p$ for some integers $k_1, k_2, ..., k_p$. For instance, it is clear that the canonical bases $(e_1, e_2, ..., e_n)$ of $\mathbb{R}^n$ is a basis of $\mathbb{Z}^n$, where e_i is the vector which has 1 on position i and 0 otherwise. But there are many other bases of $\mathbb{Z}^n$. Actually, in the chapter **A Little Introduction to Algebraic Number Theory** we proved that any integer vector whose coordinates are relatively prime can be completed to a basis of $\mathbb{Z}^n$. We can easily prove that any two bases $B_1 = (v_1, v_2, ..., v_p)$ and $B_2 = (w_1, w_2, ..., w_q)$ have the same number of elements. Indeed, let us write $v_i = a_{i1}w_1 + a_{i2}w_2 + \cdots + a_{iq}w_q$ and $w_i = b_{i1}v_1 + b_{i2}v_2 + \cdots + b_{ip}v_p$ for some integers a_{ij}, b_{ij}. Then if A and B are the matrices with entries a_{ij} and b_{ij}, we have $AB = I_p$ (just replace in $v_i = a_{i1}w_1 + a_{i2}w_2 + \cdots + a_{iq}w_q$ each w_i by $b_{i1}v_1 + b_{i2}v_2 + \cdots + b_{ip}v_p$ and then use the linear independence of v_i). Thus $p = \text{rank}(AB) \leq \text{rank}(A) \leq q$ and by symmetry we also have $q \leq p$, so $q = p$. (Now one can see that, due to the existence of the canonical basis mentioned above, any basis of $\mathbb{Z}^n$ has n elements.) Now, for an $n \times n$ matrix A with integer coefficients we can define a bilinear form $g_A(x, y) = \sum\limits_{1 \leq i \leq j \leq n} a_{ij}x_i y_j = x^t Ay$, where $x = x_1e_1 + x_2e_2 + \cdots + x_ne_n$ and $y = y_1e_1 + y_2e_2 + \cdots + y_ne_n$. Let f_A be the quadratic form associated with this bilinear form, that is $f_A(x) = g_A(x, x)$. Now, take $B = (v_1, v_2, ..., v_n)$ a basis in $\mathbb{Z}^n$ and suppose that $v_i = v_{1i}e_1 + v_{2i}e_2 + \cdots + v_{ni}e_n$. By the previous argument (showing that two basis have the same cardinality) we know that $V = (v_{ij})$ is invertible. For an integer vector whose coordinates are x_i in the canonical basis and x'_i in B, we have $x = Vx'$, and a short computation shows that $g_A(x, y) = x'^t(V^tAV)y'$. On the other hand, a direct computation shows that $g_A(x, y) = x'^tGy'$ where $G = (g_A(v_i, v_j))$, and this shows that $G = V^tAV$.

Example 9. If $A \in SL_n(\mathbb{Z})$, $n \leq 4$, is a symmetric positive matrix, then there is a matrix B with integer entries such that $A = B^tB$.

Solution. We will keep the notations used in the introduction to this example. Let us start with a very modest result, but one which, as we will see immediately, is the key idea for solving the problem.

Lemma 13.3. *There exists a vector $v_1 \in \mathbb{Z}^n$ such that $f_A(v_1) = 1$.*

Proof. The proof is a direct consequence of Minkowski's theorem. Indeed, we have seen that the volume of the ellipsoid defined by $f_A(x) < 2$ is equal to

$\frac{(\sqrt{2\pi})^n}{\Gamma(1+\frac{n}{2})}$. For $n \leq 4$, using the fact that $\Gamma(n) = (n-1)!$ and $\Gamma(n+\frac{1}{2}) = \frac{1\cdot 3\cdot \ldots \cdot (2n-1)}{2^n} \cdot \sqrt{\pi}$, you can easily verify that $\left(\frac{\pi}{2}\right)^n > (\Gamma\left(1+\frac{n}{2}\right))^2$. This shows that the volume of the ellipsoid is greater than 2^n and thus it contains a nontrivial lattice point, which we call v_1. Because $f_A(v_1) > 0$ (A is invertible and positive) and $f_A(v_1) \in \mathbb{Z}$, it is clear that v_1 is a good choice.

□

Now, we will extend this vector v_1 to a basis $B = (v_1, v_2, \ldots, v_n)$ so as to have the first line and column of G constructed:

Lemma 13.4. *Let v_1 be a vector as found in lemma 13.1. Then there exist integer vectors $v_2, v_3, \ldots, v_n$ such that $B = (v_1, v_2, \ldots, v_n)$ is a basis of $\mathbb{Z}^n$ and $g_A(v_1, v_i) = 0$ for all $i \geq 2$.*

Proof. The proof is very beautiful. Consider $H = \{x \in \mathbb{Z}^n | g_A(v_1, x) = 0\}$. Clearly, H is a submodule of $\mathbb{Z}^n$, thus it is of the form $\mathbb{Z}v_2 + \cdots + \mathbb{Z}v_r$ for some linearly independent integer vectors $v_2, v_3, \ldots, v_r$. We claim that $B = (v_1, v_2, \ldots, v_r)$ is a basis of $\mathbb{Z}^n$. Indeed, take $x \in \mathbb{Z}^n$. We need to study the equation $x = k_1 v_1 + v$, where $v \in H$. All we need is $g_A(v_1, x - k_1 v_1) = 0$, which is the same as $k_1 f_A(v_1) = g_A(v_1, x)$, thus $k_1 = g_A(v_1, x)$. Thus k_1 exists and is uniquely determined. This means (because $v_2, \ldots, v_r$ are linearly independent) that there exist unique integers $k_1, k_2, \ldots, k_r$ such that $x = k_1 v_1 + k_2 v_2 + \cdots + k_r v_r$. Thus B is a basis of $\mathbb{Z}^n$, and consequently we also have $r = n$. This finishes the proof of lemma 2.

□

Now, we can proceed to an inductive proof. We will prove that the assertion holds for $n \geq 1$ by induction. Of course, the case $n = 1$ is trivial, so assume that the result holds for $n-1$. Using lemma 1 and lemma 2, we know that for some matrix S with integer coefficients we have $A = S^t \cdot \begin{pmatrix} 1 & 0 \\ 0 & A' \end{pmatrix} \cdot S$, where clearly A' is a symmetric positive matrix in $SL_{n-1}(\mathbb{Z})$. Applying the inductive hypothesis, we can write $A' = B'^t B'$ for some matrix B' with integer entries.

Therefore $A = B^t B$ where $B = \begin{pmatrix} 1 & 0 \\ 0 & B' \end{pmatrix} \cdot S$.

Let us now discuss two beautiful applications of this result. The first is quite classical; it was among the results obtained by Fermat. However, it appeared as an old proposal for the IMO, as well as in the Iranian Olympiad in 2001.

Example 10. Let x, y, z be positive integers such that $xy = z^2 + 1$. Prove that there exist integers a, b, c, d such that $x = a^2 + b^2, y = c^2 + d^2$ and $z = ac + bd$.

Solution. Let us consider the matrix $A = \begin{pmatrix} x & z \\ z & y \end{pmatrix}$. Then $A \in SL_2(\mathbb{Z})$ because $xy = z^2 + 1$. Also, $\operatorname{tr}(A) = x + y > 2$, thus A has positive eigenvalues, so A is symmetric and positive. (This could have been established directly, too, by showing that

$$xu^2 + 2zuv + yv^2 = \frac{(xu + zv)^2 + v^2}{x} > 0$$

for all u, v not both equal to 0.) By the previous result, A can be written as $B \cdot B^t$ for some matrix $B = \begin{pmatrix} a & b \\ c & d \end{pmatrix}$. By identifying entries in the equality $A = B \cdot B^t$, we deduce the desired representation. Note that the last example implies a famous theorem of Fermat: each prime number of the form $4k + 1$ is a sum of two squares. Indeed, we saw that for such a prime p there is always an n such that $p | n^2 + 1$. However, the last theorem shows that any divisor of a number of the form $n^2 + 1$ is a sum of two squares.

Next, let us see a very difficult Diophantine equation, whose solution follows in a few lines from the important result proved above.

Example 11. Find all integers a, b, c, x, y, z such that $ax^2 + by^2 + cz^2 = abc + 2xyz - 1$, $ab + bc + ca \geq x^2 + y^2 + z^2$, and $a, b, c > 0$.

[Gabriel Dospinescu] Mathematical Reflections

Solution. Let us consider the matrix $M = \begin{pmatrix} a & z & y \\ z & b & x \\ y & x & c \end{pmatrix}$. Clearly, M is symmetric. The condition

$$ax^2 + by^2 + cz^2 = abc + 2xyz - 1$$

implies that $M \in SL_3(\mathbb{Z})$. Now, let us prove that A is positive. Because M is symmetric and invertible, it is enough to prove that its eigenvalues are positive. Let these eigenvalues be u, v, w. Then we know that u, v, w are real numbers (because M is symmetric), that $uvw = 1$ and $u+v+w = \operatorname{tr}(M) = a+b+c > 0$. On the other hand, it is not difficult to see that

$$uv + vw + wu = ab - z^2 + bc - x^2 + ac - y^2 \geq 0,$$

the sum of the principal second-order minors. Thus u, v, w are zeros of a polynomial of the form $X^3 - UX^2 + VX - 1$ for some nonnegative U, V. Clearly, such a polynomial can have only nonnegative zeros, thus $u, v, w \geq 0$. Because $\det(M) = 1$, it follows that M satisfies all conditions of the previous theorem, so M is of the form ${}^tNN$ for some integer matrix N. If we write $N = \begin{pmatrix} a_1 & a_2 & a_3 \\ b_1 & b_2 & b_3 \\ c_1 & c_2 & c_3 \end{pmatrix}$, we deduce that $a = ||A||^2$, $b = ||B||^2$, $c = ||C||^2$, $z = \langle A, B\rangle$, $y = \langle A, C\rangle$ and finally $x = \langle B, C\rangle$ for some integer vectors A, B, C (here $||\cdot||$ and $\langle\cdot\rangle$ are the Euclidean norm and inner product respectively) that form a basis of $\mathbb{Z}^3$ (they are the rows of the matrix N). All these triples found are actually solutions. Indeed, if A, B, C form a basis in $\mathbb{Z}^3$, then the matrix N whose rows are A, B, C is in $GL_3(\mathbb{Z})$, that is its determinant is -1 or 1, so $\det({}^tNN) = (\det(N))^2 = 1$. Thus $\det(A) = 1$ and

$$ax^2 + by^2 + cz^2 = abc + 2xyz - 1.$$

Also,

$$x^2 + y^2 + z^2 \leq ab + bc + ca$$

is a consequence of the Cauchy-Schwarz inequality, because

$$x^2 = \langle B, C\rangle^2 \leq ||B||^2 \cdot ||C||^2.$$

Thus these are the solutions of the Diophantine equation.

And last but not least, let us prove a beautiful theorem which, although not related to Minkowski's theorems, has strong connections with the geometry of numbers. You will notice, if you know the three-squares theorem, that the problem is trivial. But if not, what would you do? Without such an advanced result, the problem is not easy at all, but as we will see, a good geometric argument is the key of a very elementary solution:

Example 12. Prove that any integer which can be written as the sum of the squares of three rational numbers can also be written as the sum of the squares of three integers.

[Davenport-Cassels]

Solution. Let us suppose by contradiction that the property does not hold. We will use a geometric argument combined with the extremal principle. Let S be the sphere of radius $\sqrt{n}$ in $\mathbb{R}^3$ and suppose that $a \in S$ has all coordinates rational numbers. There exists an integer vector $v \in \mathbb{Z}^3$ and an integer $d > 1$ such that $a = \frac{v}{d}$. Choose the pair (a, v) for which d is minimal. We claim that there exists a vector $b \in \mathbb{Z}^3$ such that $||a - b|| < 1$, where $||x||$ is the Euclidean norm of the vector x. Indeed, it is enough to write $a = (x, y, z)$ and to consider $b = (X, Y, Z)$, where integers X, Y, Z are such that

$$|X - x| \le \frac{1}{2}, \qquad |Y - y| \le \frac{1}{2}, \qquad |Z - z| \le \frac{1}{2}.$$

Now, since a is assumed to have at least one non-integer coordinate, $a \neq b$. Consider the line ab. It will cut the sphere S in a and another point c. Let us determine precisely this point. Writing $c = b + \lambda \cdot (a - b)$ and imposing the condition $||c||^2 = n$ yields a quadratic equation in λ, with an obvious solution $\lambda = 1$. Using Viéte's formula for this equation, we deduce that another solution is $\lambda = \frac{||b||^2 - n}{||a-b||^2}$. On the other hand, the identity

$$||a - b||^2 = n + ||b||^2 - \frac{2}{d} \langle b, v \rangle$$

and the fact that $0 < ||a - b|| < 1$ show that $||a - b||^2 = \frac{A}{d}$ for a certain positive integer A smaller than d. Therefore, $\lambda = \frac{d}{A}(||b||^2 - n)$ and $c = b + \frac{||b||^2 - n}{A}(v - db) = \frac{w}{A}$ for an integer vector w. This shows that the pair (c, w) contradicts the minimality of (a, v) and proves the result.

13.2 Practice problems

1. Suppose that a and b are rational numbers such that the equation $ax^2 + by^2 = 1$ has at least one rational solution. Then it has infinitely many rational solutions.

Kurschak Competition

2. Is there a sphere in $\mathbb{R}^3$ which has exactly one point with all coordinates rational numbers?

Tournament of the Towns

3. Two sequences of integers $a_1, a_2, a_3, \dots$ and $b_1, b_2, b_3, \dots$, satisfy the equation

$$(a_n - a_{n-1})(a_n - a_{n-2}) + (b_n - b_{n-1})(b_n - b_{n-2}) = 0$$

for each integer n greater than 2. Prove that there is a positive integer k such that $a_k = a_{k+2008}$.

USA TST 2008

4. In the plane consider a polygon of area greater than n. Prove that it contains $n+1$ points $A_i(x_i, y_i)$ such that $x_i - x_j, y_i - y_j \in \mathbb{Z}$ for all $1 \le i, j \le n+1$.

Chinese TST 1988

5. Suppose that a, b, c are positive integers such that $ac = b^2 + 1$. Prove that the equation $ax^2 + 2bxy + cy^2 = 1$ is solvable in integers.

Komal

6. Let $A = (a_{ij})$ be a symmetric matrix with rational entries such that

$$\sum_{1 \le i,j \le n} a_{ij} x_i x_j > 0$$

for all $x = (x_1, \dots, x_n) \in \mathbb{R}^n \setminus \{0\}$. Prove that there are integers $x_1, \dots, x_n$ (not all zero) such that

$$\sum_{1 \le i,j \le n} a_{ij} x_i x_j < n \sqrt[n]{\det A}.$$

Minkowski

7. Prove that if $A = (a_{ij})$ is an $n \times n$ invertible matrix with real entries, then there exist integers $x_1, x_2, \dots, x_n$, not all zero, such that

$$\prod_{i=1}^{n} \left| \sum_{j=1}^{n} a_{ij} x_j \right| \le \frac{n!}{n^n} \cdot |\det A|.$$

Product theorem for Homogeneous Linear Forms

8. Let $f(X) = (X - x_1)(X - x_2) \cdots (X - x_n)$ be an irreducible polynomial over the field of rational numbers, with integer coefficients and real zeros. Prove that

$$\prod_{1 \le i < j \le n} |x_i - x_j| \ge \frac{n^n}{n!}.$$

Siegel

9. Prove that there is no position in which an n by n square can cover more than $(n+1)^2$ integral lattice points.

D.J.Newman, AMM E 1954

10. Let a, b, c, d be positive integers such that there are 2004 pairs (x, y) with $x, y \in [0, 1]$ for which $ax + by$, $cx + dy \in \mathbb{Z}$. If $\gcd(a, c) = 6$, find $\gcd(b, d)$.

Nikolai Nikolov, Oleg Mushkarov, Bulgaria 2005

11. There are $k > 0$ lattice points in the interior of a polygon P with at least four vertices. Prove that

$$|P \cap \mathbb{Z}^2| \leq 3k + 6.$$

Scott's theorem

12. For each positive integer n, let $f(n)$ be the number of ways to make $n!$ cents using an unordered collection of coins, each worth $k!$ cents for some k, $1 \leq k \leq n$. Prove that for some constant C, independent of n,

$$n^{n^2/2-Cn}e^{-n^2/4} \leq f(n) \leq n^{n^2/2+Cn}e^{-n^2/4}.$$

Gabriel Dospinescu, Titu Andreescu, Putnam 2007

13. Consider the graph G whose vertices are the points with rational coordinates in $\mathbb{R}^n$, two vertices being connected if the distance between the corresponding points is 1. Prove that G is connected if and only if $n \geq 5$.

Iran 1998

14. Prove that for a positive integer n the following assertions are equivalent:

    (a) n is the sum of three squares of integers;

    (b) the set of points with all coordinates rational on the sphere centered at the origin and having radius $\sqrt{n}$ is dense in this sphere.

15. Consider a disc of radius R. At each lattice point of this disc, except for the origin, one plants a circular tree of radius r. Suppose that r is optimal with respect to the following property: if you look from the origin, you can see at least one point situated at the exterior of the disc. Prove that
$$\frac{1}{\sqrt{R^2+1}} \leq r < \frac{1}{R}.$$

George Polya, AMM

16. Suppose that $x_1, x_2, \ldots, x_n$ are algebraic integers such that for each $1 \leq i \leq n$ there is at least one conjugate of x_i which is not among $x_1, x_2, \ldots, x_n$. Prove that the set of n-tuples $(f(x_1), f(x_2), \ldots, f(x_n))$ with $f \in \mathbb{Z}[X]$ is dense in $\mathbb{R}^n$.

17. Let $n \geq 2$ be an integer. A convex polygon inside the square $[0, n] \times [0, n]$ has area greater than n. Prove that there exists at least one lattice point inside or on the edges of the polygon.

18. A 2004×2004 array of points is drawn. Find the largest positive integer n with the following property: one can draw a convex n-gon with vertices on the points of the array.

Ricky Liu, USA TST 2004

19. Prove that there exist $n+2$ points in $\mathbb{R}^n$ such that the distance between any two of them is an odd integer if and only if $16|n+2$.

Graham, Rothschild, Straus theorem

20. An r-dimensional polytope (ie convex hull of finitely many points) $P \subset \mathbb{R}^n$ has vertices in lattice points. Prove that there exists a polynomial f of degree r such that for all positive integers m we have $f(m) = |\mathbb{Z}^n \cap mP|$.

Ehrhart's theorem

21. Let n, k be positive integers. Prove that there exists a constant $B(k, n)$ such that for any n-dimensional polytope P with vertices in lattice points and having exactly k interior lattice points we have $\text{Vol}(P) \leq B(k, n)$.

Hensley's theorem

22. Let k, n be positive integers. Prove that there are only finitely many equivalence classes of n-dimensional polytopes with vertices at lattice points and having exactly k interior lattice points.

Lagarias-Ziegler's theorem

Chapter 14

14.1 Theory and examples

Quite often, a collection of simple ideas can make very difficult problems look easy. We have seen or will see a few such examples in our journey through the world of numbers: congruences that readily solve Diophantine equations, properties of the primes of the form $4k+3$, or even facts about complex numbers, analysis or higher algebra, cleverly applied.
In this unit, we will discuss a fundamental concept in number theory, the order of an element. It may seem contradictory for us to talk about simple ideas and then say "a fundamental concept". Well, what we are going to talk about is the bridge between simplicity and complexity. The reason for which we say it is a simple idea can be guessed easily from the definition: given an integer $n > 1$ and an integer a such that $\gcd(a, n) = 1$, the least positive integer d for which $n | a^d - 1$ is called the order of a modulo n. The definition is correct, since from Euler's theorem we have $n | a^{\varphi(n)} - 1$, so such numbers d exist. The complexity of this concept will be illustrated in the examples that follow.

We will denote by $o_n(a)$ the order of a modulo n. A simple property of $o_n(a)$ has important consequences: if k is a positive integer such that $n | a^k - 1$ and $d = o_n(a)$, then $d | k$. Indeed, because $n | a^k - 1$ and $n | a^d - 1$, it follows that $n | a^{\gcd(k,d)} - 1$. But from the definition of d we have $d \leq \gcd(k, d)$, which cannot hold unless $d | k$. Nice and easy. But could such a simple idea be of any use? The answer is yes, and the solutions of the problems to come will vouch for it. But before that we note a first application of this simple observation: $o_n(a) | \varphi(n)$. This is a consequence of the above property and of Euler's theorem.
Now an old and nice problem, which may seem really trivial after this introduction. It appeared in Saint Petersburg Mathematical Olympiad and also in Gazeta Matematică.

Example 1. Prove that $n | \varphi(a^n - 1)$ for all positive integers a and n.

Solution. What is $o_{a^n-1}(a)$? It may seem a silly question, since of course $o_{a^n-1}(a) = n$. (because if $a^k - 1$ is a multiple of $a^n - 1$, then $a^k - 1 \geq a^n - 1$

and so $k \geq n$-we assumed $a > 1$, otherwise the conclusion is clear; thus the order is at least n and on the other hand it obviously divides n) Using the observation in the introduction, we obtain exactly $n|\varphi(a^n - 1)$.

Here is another beautiful application of the order of an element. It is the first case of Dirichlet's theorem that we intend to discuss, a classical property.

Example 2. Prove that any prime factor of the n-th Fermat number $2^{2^n}+1$ is congruent to 1 modulo 2^{n+1}. Then show that there are infinitely many prime numbers of the form $2^n k + 1$ for any fixed n.

Solution. Let us consider a prime p such that $p|2^{2^n} + 1$. Then p divides $(2^{2^n} + 1)(2^{2^n} - 1) = 2^{2^{n+1}} - 1$ and consequently $o_p(2)|2^{n+1}$. This ensures the existence of a positive integer $k \leq n + 1$ such that $o_p(2) = 2^k$. We will prove that in fact $k = n + 1$. Indeed, if this is not the case, then $o_p(2)|2^n$, and so $p|2^{o_p(2)} - 1|2^{2^n} - 1$. But this is impossible, since $p|2^{2^n} + 1$ and p is odd. Hence we found that $o_p(2) = 2^{n+1}$ and we need to prove that $o_p(2)|p - 1$ to finish the first part of the question. But this follows from the introduction of this chapter. The second part is a direct consequence of the first. Indeed, it is enough to prove that there exists an infinite set of pairwise relatively prime Fermat's numbers $(2^{2^{n_k}} + 1)_{n_k > n}$. Then we could take a prime factor of each such number and apply the first part to obtain that each such prime is of the form $2^n k + 1$. But not only is it easy to find such a sequence of pairwise relatively prime numbers, but in fact, any two different Fermat numbers are relatively prime. Indeed, suppose that $d|\gcd(2^{2^n} + 1, 2^{2^{n+k}} + 1)$. Then $d|2^{2^{n+1}} - 1$ and so $d|2^{2^{n+k}} - 1$. Combining this with $d|2^{2^{n+k}} + 1$, we obtain a contradiction. Hence both parts of the problem are solved.

We continue with another special case of the well-known and difficult theorem of Dirichlet on arithmetical sequences. Though classical, the following problem is not straightforward, and this probably explains its presence on a Korean TST in 2003.

Example 3. For a prime p, let $f_p(x) = x^{p-1} + x^{p-2} + \cdots + x + 1$.

a) If $p|m$, prove that any prime factor of $f_p(m)$ is relatively prime to $m(m-1)$.

b) Prove that there are infinitely many positive integers n such that $pn + 1$ is prime.

Solution. a) Take a prime divisor q of $f_p(m)$. Because $q|1 + m + \cdots + m^{p-1}$, it is clear that $\gcd(q, m) = 1$. Moreover, if $\gcd(q, m-1) \neq 1$, then $q|m-1$ and because $q|1 + m + \cdots + m^{p-1}$, it follows that $q|p$. But $p|m$ and we find that $q|m$, which is clearly impossible.
More difficult is b). We are tempted to use a) and explore the properties of $f_p(m)$, just like in the previous problem. So, let us take a prime $q|f_p(m)$ for a certain positive integer m that is divisible by p. Then we have $q|m^p - 1$. But this implies that $o_q(m)|p$ and consequently $o_q(m) \in \{1, p\}$. If $o_q(m) = p$, then $q \equiv 1 \pmod{p}$. Otherwise, $q|m-1$, and because $q|f_p(m)$, we deduce that $q|p$. Hence $q = p$. But, while solving a), we have seen that this is not possible, so the only choice is $p|q-1$. Now, we need to find a sequence $(m_k)_{k\geq 1}$ of multiples of p such that $f_p(m_k)$ are pairwise relatively prime. This is not as easy as in the first example. Anyway, just by trial and error, it is not too difficult to find such a sequence. There are many other approaches, but we like the following one: take $m_1 = p$ and $m_k = pf_p(m_1)f_p(m_2)\cdots f_p(m_{k-1})$. Let us prove that $f_p(m_k)$ is relatively prime to $f_p(m_1), f_p(m_2), \ldots, f_p(m_{k-1})$. But this is easy, since $f_p(m_1)f_p(m_2)\cdots f_p(m_{k-1})|f_p(m_k) - f_p(0) = f_p(m_k) - 1$.

Let us use this special case of Dirichlet's theorem to prove the following non-trivial result:

Example 4. Let $k \geq 2$ be an integer. Prove that there are infinitely many composite numbers n with the property that $n|a^{n-k} - 1$ for all integers a relatively prime to n.

[A.Makowski]

Solution. Let us choose these numbers of the form $n = kp$ for some suitable prime number p. We need $p|a^{n-k} - 1$, so it is enough to have $p - 1|n - k$, which is clearly true. Next, we need $k|a^{n-k}-1$, which (by Euler's theorem) is true if $n-k$ is divisible by $\varphi(k)$. So, it would be enough to have $\varphi(k)|p-1$ and to be sure that $p>k$ so that $\gcd(p,k)=1$. But from the previous problem there are infinitely many prime numbers $p \equiv 1 \pmod{\varphi(k)}$ and those numbers greater than k furnish infinitely many good numbers n.

The following problem has become a classic, and variants of it appeared in mathematics competitions. It seems to be a favorite Olympiad problem, since it uses only elementary facts and the method is nothing less than beautiful.

Example 5. Find the least n such that $2^{2005}|17^n-1$.

Solution. The problem actually asks for $o_{2^{2005}}(17)$. We know that

$$o_{2^{2005}}(17)|\varphi(2^{2005}) = 2^{2004},$$

so $o_{2^{2005}}(17)=2^k$, for some $k \in \{1,2,\dots,2004\}$. The order of an element has done its job. Now, it is time to work with exponents. We have $2^{2005}|17^{2^k}-1$. Using the factorization

$$17^{2^k}-1=(17-1)(17+1)(17^2+1)\dots(17^{2^{k-1}}+1),$$

we proceed by finding the exponent of 2 in each factor of this product. But this is not difficult, because for all $i \geq 0$ the number $17^{2^i}+1$ is a multiple of 2, but not a multiple of 4. Hence $v_2(17^{2^k}-1)=4+k$ and the order is found by solving the equation $k+4=2005$. Thus $o_{2^{2005}}(17)=2^{2001}$.

Another simple, but not straightforward, application of the order of an element is the following divisibility problem. Here, we also need some properties of the prime numbers.

Example 6. Find all prime numbers p and q such that $p^2+1|2003^q+1$ and $q^2+1|2003^p+1$.

[Gabriel Dospinescu]

Solution. Without loss of generality, we may assume that $p \leq q$. We discuss first the trivial case $p = 2$. In this case, $5|2003^q + 1$ and it is easy to deduce that q is even, hence $q = 2$, which is a solution to the problem. Now, suppose that $p > 2$ and let r be a prime factor of $p^2 + 1$. Because $r|2003^{2q} - 1$, it follows that $o_r(2003)|2q$. Suppose that $\gcd(q, o_r(2003)) = 1$. Then $o_r(2003)|2$ and $r|2003^2 - 1 = 2^3 \cdot 3 \cdot 7 \cdot 11 \cdot 13 \cdot 167$. It seems that this is a dead end, since there are too many possible values for r. Another simple observation narrows the number of possible cases: because $r|p^2+1$, r must be of the form $4k+1$ or equal to 2, and now we do not have many possibilities: $r \in \{2, 13\}$. The case $r = 13$ is also impossible, because $2003^q + 1 \equiv 2 \pmod{13}$ and $r|2003^q + 1$. So, we have found that for any prime factor r of $p^2 + 1$, we have either $r = 2$ or $q|o_r(2003)$, which in turn implies $q|r - 1$. Because $p^2 + 1$ is even but not divisible by 4, and because any odd prime factor of it is congruent to 1 modulo q, we must have $p^2 + 1 \equiv 2 \pmod{q}$. This implies that $q|(p-1)(p+1)$. Combining this with the assumption that $p \leq q$ yields $q|p+1$ and in fact $q = p+1$. It follows that $p = 2$, contradicting the assumption $p > 2$. Therefore the only solution is $p = q = 2$.

A bit more difficult is the following 2003 USA TST problem.

Example 7. Find all ordered triples of primes (p, q, r) such that

$$p|q^r + 1, q|r^p + 1, r|p^q + 1.$$

[Reid Barton] USA TST 2003

Solution. It is quite clear that p, q, r are distinct. Indeed, if for example $p = q$, then the relation $p|q^r + 1$ is impossible. We will prove that we cannot have $p, q, r > 2$. Suppose this is the case. The first condition $p \mid q^r + 1$ implies $p \mid q^{2r} - 1$ and so $o_p(q) \mid 2r$. If $o_p(q)$ is odd, it follows that $p|q^r - 1$, which combined with $p|q^r + 1$ yields $p = 2$, which is impossible. Thus, $o_p(q)$ is either 2 or $2r$. Could we have $o_p(q) = 2r$? No, since this would imply that $2r|p - 1$ and so $0 \equiv p^q + 1 \pmod{r} \equiv 2 \pmod{r}$, that is $r = 2$, false. Therefore, the only possibility is $o_p(q) = 2$ and so $p|q^2 - 1$. We cannot have $p|q - 1$, because

$p|q^r+1$ and $p \neq 2$. Thus, $p|q+1$ and in fact $2p|q+1$. In the same way, we find that $2q|r+1$ and $2r|p+1$. This is clearly impossible, just by looking at the greatest among p, q, r. So, our assumption is wrong, and one of the three primes must equal 2. Suppose without loss of generality that $p = 2$. Then q is odd, $q|r^2+1$ and $r|2^q+1$. Similarly, $o_r(2)|2q$. If $q|o_r(2)$, then $q|r-1$ and so $q|r^2+1-(r^2-1) = 2$, which contradicts the already established result that q is odd. Thus, $o_r(2)|2$ and $r|3$. As a matter of fact, this implies that $r = 3$ and $q = 5$, yielding the triple $(2, 5, 3)$. It is immediate to verify that this triple satisfies all conditions of the problem. Moreover, all solutions are given by cyclic permutations of this triple.

Can you find the least prime factor of the number $2^{2^5}+1$? Yes, with a large amount of work, you will probably find it. But what about the number $12^{2^{15}}+1$? It has more than 30000 digits, so you will probably be bored before finding its least prime factor. But here is a beautiful and short solution, which does not need a single division.

Example 8. Find the least prime factor of the number $12^{2^{15}}+1$.

__Solution.__ Let p be this prime number. Because p divides $\left(12^{2^{15}}+1\right) \cdot \left(12^{2^{15}}-1\right) = 12^{2^{16}}-1$, we find that $o_p(12)|2^{16}$. Exactly as in the solution of the first example, we find that $o_p(12) = 2^{16}$ and so $2^{16}|p-1$. Therefore $p \geq 1+2^{16}$. But it is well-known that $2^{16}+1$ is a prime (and if you do not believe it, you can check it!). So, we might try to see if this number divides $12^{2^{15}}+1$. Let $q = 2^{16}+1$. Then $12^{2^{15}}+1 = 2^{q-1} \cdot 3^{\frac{q-1}{2}}+1 \equiv 3^{\frac{q-1}{2}}+1 \pmod{q}$. It remains to see whether $\left(\frac{3}{q}\right) = -1$. But this is done in the chapter **Quadratic reciprocity** and the answer is positive, so indeed $3^{\frac{q-1}{2}}+1 \equiv 0 \pmod{q}$ and $2^{16}+1$ is the least prime factor of the number $12^{2^{15}}+1$.

OK, you must be already tired of this old fashioned idea that any prime factor of $2^{2^n}+1$ is congruent to 1 modulo 2^{n+1}. Yet, you might find the energy to devote attention to the following interesting problems.

Example 9. Prove that for any $n > 2$ the greatest prime factor of $2^{2^n} + 1$ is greater than or equal to $n \cdot 2^{n+2} + 1$.

Chinese TST 2005

Solution. You will not imagine how simple this problem really is. If the start is right... Indeed, let us write $2^{2^n} + 1 = p_1^{k_1} p_2^{k_2} \cdots p_r^{k_r}$ where $p_1 < \cdots < p_r$ are prime numbers. We know that we can find odd positive integers q_i such that $p_i = 1 + 2^{n+1} q_i$. Now, reduce the relation $2^{2^n} + 1 = p_1^{k_1} p_2^{k_2} \cdots p_r^{k_r}$ modulo 2^{2n+2}. It follows that $1 \equiv 1 + 2^{n+1} \sum_{i=1}^{r} k_i q_i \pmod{2^{2n+2}}$ and so $\sum_{i=1}^{r} k_i q_i \geq 2^{n+1}$. But then $q_r \sum_{i=1}^{r} k_i \geq 2^{n+1}$. Now everything becomes clear, since

$$2^{2^n} + 1 > (1 + 2^{n+1})^{k_1 + k_2 + \cdots + k_r} > 2^{(n+1)(k_1 + k_2 + \cdots + k_r)}$$

and so $k_1 + k_2 + \cdots + k_r \leq \dfrac{2^n}{n+1}$. Then $q_r \geq 2(n+1)$ and we are done.

Example 10. It is not known whether there are infinitely many primes of the form $2^{2^n} + 1$. Yet, prove that the sum of the reciprocals of the proper divisors of $2^{2^n} + 1$, converges to 0.

[Paul Erdős] AMM 4590

Solution. Note that the sum of the reciprocals of all the divisors of n is $\frac{\sigma(n)}{n}$, where $\sigma(n)$ is the sum of all the divisors of n. It suffices to prove that $\frac{\sigma(2^{2^n}+1)}{2^{2^n}+1}$ converges to 1. Let $p_1^{k_1} \cdots p_r^{k_r}$ be the prime factorization of $2^{2^n} + 1$ and observe that $1 < \frac{\sigma(2^{2^n}+1)}{2^{2^n}+1} < \frac{1}{\prod_{i=1}^{r} (1 - \frac{1}{p_i})} < \frac{1}{(1 - \frac{1}{2^n})^r}$. Because $2^{2^n} + 1 > 2^{n(k_1 + \cdots + k_r)} \geq 2^{rn}$, $r = O\left(\frac{2^n}{n}\right)$ and so $\frac{1}{(1 - \frac{1}{2^n})^r}$ converges to 1 for $n \to \infty$. From the above inequality, $\frac{\sigma(2^{2^n}+1)}{2^{2^n}+1}$ converges to 1 and the conclusion follows.

We have seen that the order of a modulo n is a divisor of $\varphi(n)$. Therefore a natural question appears: given a positive integer n, can we always find an integer a whose order modulo n is exactly $\varphi(n)$? We call such a number a a primitive root modulo n. The answer to this question turns out to be negative, but in some cases primitive roots exist. We will prove here that primitive roots mod p^n exist whenever $p > 2$ is a prime number and n is a positive integer. The proof is quite long and complicated, but breaking it into smaller pieces will make it easier to understand. So, let us start with a lemma due to Gauss:

Lemma 14.1. *For each integer* $n > 1$, $\sum\limits_{d|n} \varphi(d) = n$.

Proof. One of the (many) proofs goes like this: imagine that you are trying to reduce the fractions $\frac{1}{n}, \frac{2}{n}, \dots, \frac{n}{n}$ in lowest terms. The denominator of any new fraction will be a divisor of n and it is clear that for any divisor d of n we obtain $\varphi(d)$ fractions with denominator d. By counting in two different ways the total number of fractions obtained, we can conclude. □

Take now $p > 2$ a prime number and observe that any element of $\mathbb{Z}/p\mathbb{Z}$ has an order which divides $p-1$. Consider d a divisor of $p-1$ and define $f(d)$ to be the number of elements in $\mathbb{Z}/p\mathbb{Z}$ that have order d. Suppose that x is an element of order d. Then $1, x, ..., x^{d-1}$ are distinct solutions of the equation $u^d = 1$, an equation which has at most d solutions in the field $\mathbb{Z}/p\mathbb{Z}$. Therefore $1, x, ..., x^{d-1}$ are all solutions of this equation and any element of order d is among these elements. Clearly, x^i has order d if and only if $\gcd(i, d) = 1$. Thus at most $\varphi(d)$ elements have order d, which means that $f(d) \leq \varphi(d)$ for all d. But since any nonzero elements has an order which divides $p-1$, we deduce that

$$\sum_{d|p-1} f(d) = p - 1 = \sum_{d|p-1} \varphi(d)$$

(we used in the last equality the lemma above). This identity combined with the previous inequality shows that $f(d) = \varphi(d)$ for all $d|p-1$. We have thus proved the following:

Theorem 14.2. *For any divisor d of $p-1$ there are exactly $\varphi(d)$ elements of order d in $\mathbb{Z}/p\mathbb{Z}$.*

The above theorem implies the existence of primitive roots modulo any prime p (the case $p = 2$ being obvious). If g is a primitive root mod p, then the p elements $0, 1, g, g^2, ..., g^{p-2}$ are distinct and so they represent a permutation of $\mathbb{Z}/p\mathbb{Z}$. Let us fix now a prime number $p > 2$ and a positive integer k and show the existence of a primitive root mod p^k. First of all, let us observe that for any $j \geq 2$ and any integer x we have $(1+xp)^{p^{j-2}} \equiv 1 + xp^{j-1} \pmod{p^j}$. Establishing this property is immediate by induction on j and the binomial formula. With this preparatory result, we will prove now the following:

Theorem 14.3. *If p is an odd prime, then for any positive integer k there exists a primitive root mod p^k.*

Proof. Indeed, take g a primitive root mod p. Clearly, $g+p$ is also a primitive root mod p. Using again the binomial formula, it is easy to prove that one of the two elements g and $g+p$ is not a root of $X^{p-1} - 1$ mod p^2. This shows that there exists y a primitive root mod p for which $y^{p-1} \not\equiv 1 \pmod{p^2}$. Let $y^{p-1} = 1 + xp$. Then by using the previous observation we can write $y^{p^{k-2}(p-1)} \equiv (1+xp)^{p^{k-2}} \equiv 1 + xp^{k-1} \pmod{p^k}$ and so p^k does not divide $y^{p^{k-2}(p-1)} - 1$. Thus the order of y mod p^k is a multiple of $p-1$ (because y is a primitive root mod p) which divides $p^{k-1}(p-1)$ but does not divide $p^{k-2}(p-1)$. So, y is a primitive root mod p^k.

□

In order to finish this (long) theoretical part, let us present a very efficient criterion for primitive roots modulo p^k:

Theorem 14.4. *Each primitive root mod p and p^2 is a primitive root modulo any power of p.*

Proof. Let us prove first that if g is a primitive root mod p and p^2 then it is also a primitive root mod p^3. Let k be the order of g mod p^3. Then k is a divisor of $p^2(p-1)$. Because p^2 divides g^k-1, k must be a multiple of $p(p-1)$. It remains to prove that k is not $p(p-1)$. Supposing the contrary, let $g^{p-1}=1+rp$, then we know that $p^3|(1+rp)^p-1$. Using again the binomial formula, we deduce that p divides r and so p^2 divides $g^{p-1}-1$, which contradicts the fact that g is a primitive root mod p^2. Now, we use induction. Suppose that $n \geq 4$ and that g is a primitive root mod p^{n-1}. Let k be the order of g mod p^n. Because p^{n-1} divides g^k-1, k must be a multiple of $p^{n-2}(p-1)$. Also, k is a divisor of $p^{n-1}(p-1)=\varphi(p^n)$. So, all we have to do is to prove that k is not $p^{n-2}(p-1)$. Otherwise, by Euler's theorem we can write $g^{p^{n-3}(p-1)}=1+rp^{n-2}$ and from the binomial formula it follows that r is a multiple of p and so p^{n-1} divides $g^{p^{n-3}(p-1)}-1$, contradicting the fact that g has order $p^{n-2}(p-1)$ modulo p^{n-1}. The theorem is thus proved. □

It is important to note that the previous results allow us to find all positive integers that have primitive roots. First off all, observe that such a number n cannot be written in the form $n=n_1n_2$ with $\gcd(n_1,n_2)=1$ and $n_1,n_2>2$. Indeed, if $\gcd(g,n)=1$ then $g^{\frac{\varphi(n)}{2}} \equiv (g^{\varphi(n_1)})^{\frac{\varphi(n_2)}{2}} \equiv 1 \pmod{n_1}$ and similarly $g^{\frac{\varphi(n)}{2}} \equiv 1 \pmod{n_2}$. Thus n divides $g^{\frac{\varphi(n)}{2}}-1$ and g cannot have order $\varphi(n)$. Also, the fact that $g^{2^{k-2}} \equiv 1 \pmod{2^k}$ for any odd integer g and any $k \geq 3$ (whose the proof is immediate by induction) shows that there are no primitive roots mod 2^k, for $k \geq 3$. This shows that the only candidates are $2, 4, p^k$ and $2p^k$ for an odd prime number p. And these numbers have primitive roots. For 2 and 4 it is obvious, while for powers of odd primes it has been proved above. For $2p^k$ observe that $\varphi(2p^k)=\varphi(p^k)$, so the odd number among $g, g+p^k$ (where g is a primitive root mod p^k) is a primitive root mod $2p^k$.

Now, let us solve some problems. However, make sure you correctly remember Fermat's little theorem before attempting to solve the following problem.

Example 11. Find all positive integers n such that $n|a^{n+1}-a$ for all $a \in Z$.

Solution. Consider such an integer $n > 1$ and observe first that is must be squarefree. Indeed, if p is a prime divisor of n, just choose $a = p$. Next, write $n = p_1 p_2 ... p_k$ for some pairwise distinct prime numbers $p_1, p_2, ..., p_k$. Fix some $1 \leq i \leq k$ and choose a a primitive root modulo p_i. Then clearly the condition $n | a^{n+1} - a$ implies that n is a multiple of $p_i - 1$. Now, it is very easy to determine all such numbers n. Assume that $p_1 < p_2 < \cdots < p_k$ and observe that $p_1 = 2$ (because $p_1 - 1$ divides n), then $p_2 - 1 | 2$ (the same argument), thus $p_2 = 3$. Continue in this manner to obtain $p_3 = 7, p_4 = 43$. And things change after this, because we would find that $p_5 - 1$ divides 1806 and it is easy to see that this is not possible, because the only divisors d of 1806 such that $d + 1$ is a prime are $1, 2, 6, 42$, which is not a prime number. Therefore $k \leq 4$ and such numbers are $1, 2, 6, 42, 1806$.
A very beautiful and difficult problem comes now. We will see that using the previous results on primitive roots we can obtain a quick and elegant solution.

Example 12. Find all positive integers n such that $n^2 | 2^n + 1$.

[Laurenţiu Panaitopol] IMO 1990

Solution. It is clear that any solution must be odd and that 1 and 3 are solutions, so assume that $n \geq 5$. Because 2 is a primitive root mod 3 and mod 9 (as you can immediately check), it follows from the above results that 2 is a primitive root mod 3^k for all k. In particular, if $3^k | 2^n + 1$ then $3^k | 2^{2n} - 1$ and because the order of 2 mod 3^k is $2 \cdot 3^{k-1}$, we deduce that $3^{k-1} | n$. This shows that $v_3(2^n + 1) \leq v_3(n) + 1$ for all n. In particular, for any solution n of the problem we have $2v_3(n) = v_3(n^2) \leq v_3(2^n + 1) \leq 1 + v_3(n)$, so $v_3(n) \leq 1$. Let us prove that we actually have $v_3(n) = 1$, if $n > 1$. Let p be the smallest prime divisor of n. Then $p | 2^{2n} - 1$, so $o_p(2) | 2n$ and $o_p(2) | p - 1$. By the definition of p we have $\gcd(2n, p - 1) = 2$, so $p | 3$ and thus $p = 3$ and $3 | n$. This shows that we can write $n = 3a$ where $\gcd(3, a) = 1$. Now, we would like to prove that $a = 1$ (therefore, the only solution of the problem which is greater than 1 is $n = 3$). Assuming the contrary, let q be its smallest prime divisor. Then $q | 2^n + 1$ and $q | 2^{6a} - 1$. As above, we deduce that $o_q(2)$ is a divisor of $6a$ and $q - 1$, and because $\gcd(a, q - 1) = 1$, it follows that $o_q(2) | 6$ and so $q | 63$. Because

$\gcd(a,3) = 1$, the only possibility is $q = 7$. But then $7|2^n + 1 = 8^a + 1$, which is clearly impossible. This shows that $a = 1$ and $n = 3$, a contradiction with $n \geq 5$. Hence 1 and 3 are the only solutions of the problem.

Finally, a chestnut from the celebrated contest Miklos Schweitzer, which uses the previous theoretical results as well as a large dose of creativity.

Example 13. Let $p \equiv 3 \pmod 4$ be a prime number. Prove that

$$\prod_{1\leq x<y\leq \frac{p-1}{2}} (x^2+y^2) \equiv (-1)^{\left\lfloor \frac{p+1}{8} \right\rfloor} \pmod p.$$

[J.Suranyi] Miklos Schweitzer Competition

Solution. Let $p = 4k+3$ and take g to be a primitive root modulo p and $x = g^2$. Then the squares of the residues mod p are exactly $1, x, x^2, ..., x^{2k}$, so the product we need to evaluate is $\prod\limits_{0\leq i<j\leq 2k} (x^i + x^j) \pmod p$. Therefore, if P is the desired product, we have

$$P \cdot \prod_{0\leq i<j\leq 2k} (x^i - x^j) \equiv \prod_{0\leq i<j\leq 2k} (x^{2i} - x^{2j}) \pmod p.$$

Observe that each of the two products is actually a Vandermonde determinant and because $x^{2k+1} = 1$, the generators of the second determinant are exactly $1, x^2, ..., x^{2k}, x, x^3, ..., x^{2k-1}$. Hence the second determinant is obtained from the first one by $k + (k-1) + \cdots + 2 + 1$ transpositions of the lines and so $P \equiv (-1)^{\frac{k(k+1)}{2}} \pmod p$. An easy case examination shows that $\frac{k(k+1)}{2} - \left\lfloor \frac{p+1}{8} \right\rfloor$ is even and the conclusion follows.

14.2 Practice problems

1. Find all positive integers m, n for which $n \mid m^{2\cdot 3^n} + m^{3^n} + 1$.

Bulgaria 1997

2. Let q be a prime such that q^2 divides at least one Mersenne number $2^p - 1$ with p a prime number. Prove that $q > 3 \cdot 10^9$. You may take for granted that the only primes q such that $q^2 \mid 2^{q-1} - 1$ and which are smaller than $3 \cdot 10^9$ are 1093 and 3511.

3. Prove that there exists a function f with integer values such that $2^n \mid 19^{f(n)} - 97$ for any positive integer n.

Vietnamese TST 1997

4. Prove that for any prime number $p > 3$ we have $\binom{2p}{p} \equiv 2 \pmod{p^3}$.

5. Let $m > 1$ be an odd number. Find the least n such that $2^{1989} \mid m^n - 1$.

IMO 1989 Shortlist

6. Let m, n be two positive integers. Prove that the remainders of the numbers $1^n, 2^n, \ldots, m^n$ modulo m are pairwise distinct if and only if m is square-free and n is relatively prime to $\varphi(m)$.

7. Prove that the equation

$$\frac{x^7 - 1}{x - 1} = y^5 - 1$$

has no solution in integers.

IMO 2006 Shortlist

8. Let a be an integer greater than 1. Prove that the function

$$f : \{2, 3, 5, 7, 11, \dots\} \to \mathbb{N}, f(p) = \frac{p-1}{o_p(a)}$$

is unbounded. Here $o_p(a)$ is the order of a modulo p.

Jon Froemke, Jerrold W Grossman, AMM E 3216

9. Let $f(n)$ be the greatest common divisor of $2^n - 2, 3^n - 3, 4^n - 4, \dots$. Determine $f(n)$ and prove that $f(2n) = 2$.

AMM

10. Let f be a polynomial with integer coefficients such that for some prime number p we have $f(i) \equiv 0 \pmod{p}$ or $f(i) \equiv 1 \pmod{p}$ for any integer i. If $f(0) = 0$ and $f(1) = 1$, prove that $\deg(f) \geq p - 1$.

IMO 1997 Shortlist

11. A Carmichael number n satisfies $n \mid a^n - a$ for all integers a. Find all Carmichael numbers of the form $3pq$ with p, q prime numbers. Using the existence of Carmichael composite numbers, prove that there are infinitely many pseudo-primes (composite n such that $n \mid 2^n - 2$).

12. Find all prime numbers p, q such that $pq \mid 5^p + 5^q$.

China 2009

13. Find the sum of the m-th powers of the primitive roots mod p for a given prime p and a positive integer m.

14. Find all positive n such that n and $2^n + 1$ have the same prime factors.

Gabriel Dospinescu

15. Let p be a prime number and m, n be integers greater than 1 such that $n \mid m^{p(n-1)} - 1$. Prove that $\gcd(m^{n-1} - 1, n) > 1$.

MOSP 2001

16. Prove that there are infinitely many pairs of distinct prime numbers (p, q) such that p divides $2^{q-1} - 1$ and q divides $2^{p-1} - 1$.

Romania TST 2009

17. Let n be a positive integer, and let A_n be the the set of all a such that $n \mid (a^n + 1)$, $1 \leq a \leq n$ and $a \in \mathbb{Z}$.

    (a) Find all n such that $A_n \neq \emptyset$.
    (b) Find all n such that $|A_n|$ is even and non-zero.
    (c) Is there n such that $|A_n| = 130$?

Italian TST 2006

18. The sequence $\{x_n\}$ is defined by $x_1 = 2, x_2 = 12$ and $x_{n+2} = 6x_{n+1} - x_n$. Let p be an odd prime and let $q > 3$ be a prime divisor of x_p. Prove that $q \geq 2p - 1$.

Chinese TST 2008

19. Let A be a finite set of prime numbers and let a be an integer greater than 1. Prove that there are only finitely many positive integers n such that all prime factors of $a^n - 1$ are in A.

Iranian Olympiad

20. Find all positive integers n such that n divides $2^n + 3^n + \cdots + (n-1)^n$.

IMAR Contest 2004

21. Let $n \geq 3$ be a positive integer. Compute the greatest common divisor of the numbers $2^n - 2, 3^n - 3, \ldots, n^n - n$.

Dorin Andrica, Mihai Piticari, Romania TST 2008

22. Find all positive integers n with the following property: there is a unique a such that $0 \leq a < n!$ and $n! \mid a^n + 1$.

IMO Shortlist 2007

23. Let x, y be two integers with $2 \leq x, y \leq 100$. Prove that $x^{2^n} + y^{2^n}$ is not a prime for some positive integer n.

Russia 2009

24. Is there a positive integer n such that every nonzero digit appears the same number of times in each of the numbers $n, 2n, \ldots, 2000n$?

Komal

25. Prove that for any prime p there is a prime q that does not divide any of the numbers $n^p - p$, with $n \geq 1$.

IMO 2003

26. Let a be an integer greater than 1. Prove that for infinitely many n the largest prime factor of $a^n - 1$ is greater than $n \log_a n$.

Gabriel Dospinescu

27. Let $\varepsilon > 0$. Prove the existence of a constant c such that for all odd primes p there exists a primitive root mod p less than $cp^{\frac{1}{2}+\varepsilon}$.

Vinogradov

28. Let a and b be integers greater than 1. Prove that there exists a prime p such that the order of a mod p is b, unless $b = 2$ and $a + 1$ is a power of 2 or $a = 2$ and $b = 6$.

Zsigmondy's theorem

Chapter 15

15.1 Theory and examples

Recall that the sequence $(\{na\})_{n\geq 1}$ is dense in [0,1] if a is an irrational number, a classical theorem of Kronecker. Various applications of this nice result have appeared in different contests and will probably make the object of many more Olympiad problems. Yet, there are some examples in which this result is inefficient. A simple one is as follows: using Kronecker's theorem, one can prove that for any positive integer a that is not a power of 10 there exists a positive integer n such that a^n begins with 2008. The natural question – what fraction of numbers between 1 and n have this property (speaking here about large values of n) – is much more difficult, and to answer it we need some stronger tools. This is the reason we now discuss some classical approximation theorems, particularly the very effective Weil criterion and its consequences. The proofs of these results are nontrivial and require some heavy duty analysis. Yet, the consequences that will be discussed here are almost elementary. Of course, one cannot start a topic about approximation theorems without talking first about Kronecker's theorem. We skip the proof, not only because it is very well-known, but because we will prove a much stronger result about the sequence $(\{na\})_{n\geq 1}$. Instead, we will discuss two beautiful problems, corollaries of this theorem.

Example 1. Prove that the sequence $(\lfloor n\sqrt{2003}\rfloor)_{n\geq 1}$ contains arbitrarily long geometric progressions with arbitrarily large ratio.

[Radu Gologan] Romanian TST 2003

Solution. Let p be any positive integer. We will prove that there are arbitrarily long geometric sequences with ratio p. Given $n \geq 3$, we will find a positive integer m such that $\lfloor p^k m\sqrt{2003}\rfloor = p^k \lfloor m\sqrt{2003}\rfloor$ for all $1 \leq k \leq n$. If the existence of such a number is proved, then the conclusion is immediate. Observe that $\lfloor p^k m\sqrt{2003}\rfloor = p^k \lfloor m\sqrt{2003}\rfloor$ is equivalent to $\lfloor p^k\{m\sqrt{2003}\}\rfloor = 0$, or to $\{m\sqrt{2003}\} < \dfrac{1}{p^n}$. The existence of a positive integer m with the last property is ensured by Kronecker's theorem.

Here is a problem that is apparently very difficult, but which is again a simple consequence of Kronecker's theorem.

Example 2. Consider a positive integer k and a real number a such that $\log a$ is irrational. For each $n \geq 1$ let x_n be the number formed by the first k digits of $\lfloor a^n \rfloor$. Prove that the sequence $(x_n)_{n \geq 1}$ is not eventually periodical.

[Gabriel Dospinescu] Mathlinks Contest

Solution. First of all, the number formed with the first k digits of a number m is $\left\lfloor 10^{k-1+\{\log m\}} \right\rfloor$. The proof of this claim is not difficult. Indeed, let us write $m = \overline{a_1 a_2 \dots a_p}$, with $p \geq k$. Then $m = \overline{a_1 \dots a_k} \cdot 10^{p-k} + \overline{a_{k+1} \dots a_p}$, hence $\overline{a_1 \dots a_k} \cdot 10^{p-k} \leq m < (\overline{a_1 \dots a_k} + 1) \cdot 10^{p-k}$. It follows that $\overline{a_1 \dots a_k} = \left\lfloor \frac{m}{10^{p-k}} \right\rfloor$ and, since $p = 1 + \lfloor \log m \rfloor$, the claim is proved.
Now consider the claim false: thus there is some T for which $x_{n+T} = x_n$ for any large enough n. Another observation is the following: there is a positive integer r such that $x_{rT} > 10^{k-1}$. Indeed, assuming the contrary, we find that for all $r > 0$ we have $x_{rT} = 10^{k-1}$. Using the first observation, it follows that $k - 1 + \{\log \lfloor a^{rT} \rfloor\} < \log(1 + 10^{k-1})$ for all r. Thus

$$\log\left(1 + \frac{1}{10^{k-1}}\right) > \log \lfloor a^{rT} \rfloor - \lfloor \log \lfloor a^{rT} \rfloor \rfloor > \log(a^{rT} - 1) - \lfloor \log a^{rT} \rfloor$$

$$= \{rT \log a\} - \log \frac{a^{rT}}{a^{rT} - 1}.$$

It suffices now to consider a sequence of positive integers (r_n) such that $1 - \frac{1}{n} < \{r_n T \log a\}$ (the existence is a direct consequence of Kronecker's theorem) and we deduce that

$$\log\left(1 + \frac{1}{10^{k-1}}\right) + \frac{1}{n} + \log \frac{a^{r_n T}}{a^{r_n T} - 1} > 1 \text{ for all } n.$$

The last inequality is clearly impossible.
Finally, assume the existence of such an r. It follows that for $n > r$ we have $x_{nT} = x_{rT}$, thus

$$\{\log \lfloor a^{nT} \rfloor\} \geq \log\left(1 + \frac{1}{10^{k-1}}\right).$$

This shows that

$$\log\left(1 + \frac{1}{10^{k-1}}\right) \leq \log \lfloor a^{nT} \rfloor - \lfloor \log \lfloor a^{nT} \rfloor \rfloor \leq nT \log a - \lfloor \log a^{nT} \rfloor$$

$$= \{nT \log a\} \text{ for all } n > r.$$

In the last inequality, we used the fact that $\lfloor \log \lfloor x \rfloor \rfloor = \lfloor \log x \rfloor$, which is not difficult to establish: indeed, if $\lfloor \log x \rfloor = k$, then $10^k \leq x < 10^{k+1}$, and thus $10^k \leq \lfloor x \rfloor < 10^{k+1}$, which means that $\lfloor \log \lfloor x \rfloor \rfloor = k$. Finally, note that the relation $\log\left(1 + \frac{1}{10^{k-1}}\right) \leq \{nT \log a\}$ contradicts Kronecker's theorem. This finishes the proof.

We continue with two subtle results, based on Kronecker's lemma.

Example 3. For a pair (a, b) of real numbers let $F(a, b)$ denote the sequence of general term $c_n = \lfloor an + b \rfloor$. Find all pairs (a, b) such that $F(x, y) = F(a, b)$ implies $(x, y) = (a, b)$.

[Roy Streit] AMM E 2726

Solution. Let us see what happens when $F(x, y) = F(a, b)$. We must have $\lfloor an + b \rfloor = \lfloor xn + y \rfloor$ for all positive integers n. Dividing this equality by n and taking the limit, we infer that $a = x$. Now, if a is rational, the sequence of fractional parts of $an + b$ takes only a finite number of values, so if r is chosen sufficiently small (but positive) we will have $F(a, b + r) = F(a, b)$, so no pair (a, b) can be a solution of the problem. On the other hand, we claim that any irrational number a is a solution for any real number b. Indeed, take $x_1 < x_2$ and a positive integer n such that $na + x_1 < m < na + x_2$

for a certain integer m. The existence of such an n follows immediately from Kronecker's theorem. But the last inequality shows that $F(a, x_1) \neq F(a, x_2)$ and so a is a solution. Therefore the answer is: all pairs (a, b) with a irrational.

Finally, an equivalent condition for the irrationality of a real number:

Example 4. Let r be a real number in $(0, 1)$ and let $S(r)$ be the set of positive integers n for which the interval $(nr, nr + r)$ contains exactly one integer. Prove that r is irrational if and only if for all integers M there exists a complete system of residues modulo M, contained in $S(r)$.

[Klark Kimberling]

Solution. One part of the solution is very easy: if r is rational, let M be its denominator. Then clearly if n is a multiple of M there is no integer k in the desired interval. Now, suppose that r is irrational and take integers m, M such that $0 \leq m < M$. By Kronecker's theorem, the integer multiples of $\frac{1}{r}$ form a dense set modulo M. So, there exists an integer k such that the image of $\frac{k}{r}$ is in $(m, m+1)$, that is for a certain integer s we have $sM+m < \frac{k}{r} < sM+m+1$. It is then clear that if we take $n = sM + m$ we have $n \equiv m \pmod{M}$ and $nr < k < nr + r$. This finishes the solution.

Before getting into the quantitative results stated at the beginning of this chapter, we must talk about a surprising result, which turns out to be very useful when dealing with real numbers and their properties. Sometimes, it will help us reduce a complicated problem concerning real numbers to integers, as we will see in one of the examples. But first, let us state and prove this result.

Example 5. Let $x_1, x_2, \ldots, x_k$ be real numbers and let $\varepsilon > 0$. There exists a positive integer n and integers $p_1, p_2, \ldots, p_k$ such that $|nx_i - p_i| < \varepsilon$ for all i.

[Dirichlet]

Solution. We need to prove that if we have a finite set of real numbers, we can multiply all its elements by a suitable integer such that the elements of the new set are as close to integers as we want.

Let us choose an integer $N > \frac{1}{\varepsilon}$ and partition the interval $[0,1)$ into N intervals,

$$[0,1) = \bigcup_{s=1}^{N} J_s, \quad J_s = \left[\frac{s-1}{N}, \frac{s}{N}\right).$$

Now, choose $n = N^k + 1$ and assign to each q in the set $\{1, 2, \ldots, n\}$ a sequence of k positive integers $\alpha_1, \alpha_2, \ldots, \alpha_k$, where $\alpha_i = s$ if and only if $\{qx_i\} \in J_s$. We obtain at most N^k sequences corresponding to these numbers, and so by the pigeonhole principle we can find $1 \leq u < v \leq n$ such that the same sequence is assigned to u and v. This means that for all $1 \leq i \leq k$ we have

$$|\{ux_i\} - \{vx_i\}| < \frac{1}{N} \leq \varepsilon \tag{15.1}$$

It suffices to pick $n = v - u$, $p_i = \lfloor vx_i \rfloor - \lfloor ux_i \rfloor$.

And here is how we can use this result in problems where it is more comfortable to work with integers. But don't kid yourself, there are not many such problems. The one we are going to discuss next has meandered between world's Olympiads: proposed at the 1949 Moscow Olympiad, it appeared next at the W.L. Putnam Competition in 1973 and later on in an IMO Shortlist, proposed by Mongolia.

Example 6. Let $x_1, x_2, \ldots, x_{2n+1}$ be real numbers with the property: for any $1 \leq i \leq 2n+1$ one can make two groups of n numbers by using all the x_j, $j \neq i$, such that the sum of the numbers in each group is the same. Prove that all the numbers must be equal.

Solution. For integers the solution is well-known and not difficult: it suffices to note that in this case all numbers x_i have the same parity, and the use of infinite descent solves the problem (either they are all even and in this case we divide each number by 2 and obtain a new set with smaller sum of magnitudes and the same properties; otherwise, we subtract 1 from each number and then divide by 2). Now, assume that they are real numbers, which is definitely a more subtle case. First of all, if they are all rational, it suffices to multiply by their common denominator and apply the first case. Suppose at least one of the numbers is irrational. Consider $\varepsilon > 0$, a positive integer m, and some integers $p_1, p_2, \dots, p_{2n+1}$ such that $|mx_i - p_i| < \varepsilon$ for all i. We claim that if $\varepsilon > 0$ is small enough, then $p_1, p_2, \dots, p_{2n+1}$ have the same property as $x_1, x_2, \dots, x_{2n+1}$. Indeed, take some i and write the given condition as

$$\sum_{j \neq i} a_{ij} m x_j = 0 \text{ or } \sum_{j \neq i} a_{ij}(mx_j - p_j) = -\sum_{j \neq i} a_{ij} p_j$$

(where $a_{ij} \in \{-1, 1\}$). Then

$$\left| \sum_{j \neq i} a_{ij} p_j \right| = \left| \sum_{j \neq i} a_{ij}(mx_j - p_j) \right| \leq 2n\varepsilon.$$

Thus if we choose $\varepsilon < \frac{1}{2m}$, then $\sum_{j \neq i} a_{ij} p_j = 0$ and so $p_1, p_2, \dots, p_{2n+1}$ have the same property. Because they are all integers, $p_1, p_2, \dots, p_{2n+1}$ must be all equal (again, because of the first case). Hence we have proved that for any $N > 2m$ there are integers n_N, p_N such that $|n_N x_i - p_N| \leq \frac{1}{N}$.
Because at least one of the numbers $x_1, x_2, \dots, x_{2n+1}$ is irrational, it is not difficult to prove that the sequence $(n_N)_{N>2m}$ is unbounded. But $\frac{2}{N} > |n_N| \max_{i,j} |x_i - x_j|$, hence $\max_{i,j} |x_i - x_j| = 0$ and the problem is solved.

If you thought the last problem was too classical, here is another one, a little bit less known, but with the same flavor:

Example 7. Let $a_1, a_2, ..., a_{2007}$ be real numbers with the following property: no matter how we choose 13 numbers among them, there exist 8 numbers among the 2007 which have the same arithmetic mean as the 13 chosen ones. Prove that they are all equal.

Solution. Note (again) that the problem is quite easy for integers. Indeed, the assumption implies that the sum of any 13 numbers is a multiple of 13. Let a_i, a_j be among the 2007 numbers and let $x_1, x_2, ..., x_{12}$ be some a_k with $k \neq i$ and $k \neq j$. Then $a_i + x_1 + x_2 + \cdots + x_{12}$ and $a_j + x_1 + x_2 + \cdots + x_{12}$ are multiples of 13, so $a_i \equiv a_j \pmod{13}$. Thus all numbers give the same remainder r modulo 13. It suffices to subtract r from all a_i, to divide by 13 in order to obtain a new collection of 2007 integers, with smaller absolute values and still satisfying the property given in the statement of the problem. Repeating this procedure, we will finally obtain a collection of zeros, which means that the initial numbers were all equal.
Now, let us pass to the case when all numbers are known only to be real. The idea is the same as in the previous example: we will approximate, using Dirichlet's theorem, all numbers by rational numbers with a common denominator. Explicitly, take some $\epsilon > 0$ and n and p_i some integers (with $n > 0$) such that $|na_i - p_i| < \epsilon$ for all i. Take some indices $i_1, i_2, ..., i_{13}$. We know that for some indices $j_1, j_2, ..., j_8$ we have

$$\frac{na_{i_1} + na_{i_2} + \cdots + na_{i_{13}}}{13} = \frac{na_{j_1} + na_{j_2} + \cdots + na_{j_8}}{8}.$$

If $x_i = na_i - p_i$, it follows that

$$\left| \frac{p_{i_1} + p_{i_2} + \cdots + p_{i_{13}}}{13} - \frac{p_{j_1} + p_{j_2} + \cdots + p_{j_8}}{8} \right| < 2\epsilon,$$

because $|x_i| < \epsilon$. Now, observe that if

$$\left| \frac{p_{i_1} + p_{i_2} + \cdots + p_{i_{13}}}{13} - \frac{p_{j_1} + p_{j_2} + \cdots + p_{j_8}}{8} \right|$$

is nonzero, it is at least equal to $\frac{1}{8 \cdot 13}$. Thus, if we take $\epsilon < \frac{1}{16 \cdot 13}$, we know that the corresponding p_i have the same property as a_i. By the first case, we must therefore have $p_1 = p_2 = \cdots = p_{2007}$. Thus $2\epsilon > n|a_i - a_j|$ for all i, j and all $\epsilon < \frac{1}{16\cdot 13}$. Clearly, this implies $a_1 = a_2 = \cdots = a_{2007}$ and finishes the proof.

Now, let us turn to more quantitative results about the set of fractional parts of natural multiples of different real numbers. The following criterion, due to Weyl, deserves to be discussed because of its beauty and apparent simplicity.

Theorem 15.1 (Weyl's theorem). *Let $(a_n)_{n\geq 1}$ be a sequence of real numbers from the interval [0,1]. Then the following statements are equivalent:*
a) For any real numbers $0 \leq a \leq b \leq 1$,

$$\lim_{n\to\infty} \frac{|\{i|\ 1 \leq i \leq n,\ a_i \in [a, b]\}|}{n} = b - a;$$

b) For any continuous function $f : [0, 1] \to \mathbb{R}$,

$$\lim_{n\to\infty} \frac{1}{n}\sum_{k=1}^{n} f(a_k) = \int_0^1 f(x)dx;$$

c) For any positive integer $r \geq 1$,

$$\lim_{n\to\infty} \frac{1}{n}\sum_{k=1}^{n} e^{2i\pi r a_k} = 0.$$

In this case we will say that the sequence is equidistributed.

Proof. We will present just a sketch of the solution, but containing all the necessary ingredients. First, we observe that a) says precisely that b) is true for the characteristic function of any subinterval of [0,1]. By linearity, this remains true for any piecewise constant function. Now, there is a well-known and easy to verify property of continuous functions: they can be uniformly approximated with piecewise constant functions. That is, given $\varepsilon > 0$, we can find a piecewise constant function g such that $|g(x) - f(x)| < \varepsilon$ for all $x \in [0, 1]$. But then if we write

$$\left|\frac{1}{n}\sum_{k=1}^{n} f(a_k) - \int_0^1 f(x)dx\right| \leq \frac{1}{n}\sum_{k=1}^{n} |f(a_k) - g(a_k)| + \int_0^1 |f(x) - g(x)|dx$$

$$+\left|\frac{1}{n}\sum_{k=1}^{n} g(a_k) - \int_0^1 g(x)dx\right|$$

and apply the result in b) for the function g, we easily deduce that b) is true for any continuous function. The fact that b) implies c) is immediate. More subtle is that b) implies a). Let us consider the subinterval $I = [a, b]$ with $0 < a < b < 1$. Next, consider two sequences of continuous functions f_k, g_k such that f_k is zero on $[0, a]$, $[b, 1]$ and 1 on $\left[a + \frac{1}{k}, b - \frac{1}{k}\right]$ (being affine otherwise), while g_k has "the same" properties but is greater than or equal to λ_I (the characteristic function of $I = [a, b]$). Therefore

$$\frac{1}{n}\sum_{j=1}^{n} f_k(a_j) \le \frac{|\{i|\ 1 \le i \le n,\ a_i \in [a, b]\}|}{n} \le \frac{1}{n}\sum_{j=1}^{n} g_k(a_j).$$

But from the hypothesis,

$$\frac{1}{n}\sum_{j=1}^{n} f_k(a_j) \to \int_0^1 f_k(x)dx = b - a - \frac{1}{k}$$

and

$$\frac{1}{n}\sum_{j=1}^{n} g_k(a_j) \to \int_0^1 g_k(x)dx = b - a + \frac{1}{k}.$$

Now, let us take $\varepsilon > 0$ and k sufficiently large. The above inequalities show that actually for all sufficiently large positive integers n

$$\left|\frac{|\{i|\ 1 \le i \le n,\ a_i \in [a, b]\}|}{n} - b + a\right| \le 2\varepsilon$$

and the conclusion follows. You have already seen how to adapt this proof for the case $a = 0$ or $b = 1$. Finally, let us prove that c) implies b). Of course, a linearity argument allows us to assume that b) is true for any trigonometric polynomial. Because any continuous function $f : [0, 1] \to \mathbb{R}$ satisfying $f(0) = f(1)$ can be uniformly approximated by trigonometric polynomials (this is a really nontrivial result due to Weierstrass), we deduce that b) is true for continuous functions f for which $f(0) = f(1)$. Now, given a continuous $f : [0,1] \to \mathbb{R}$, it is immediate that for any $\varepsilon > 0$ we can find two continuous functions g, h, both having equal values at 0 and 1 and such that

$$|f(x) - g(x)| \leq h(x) \text{ and } \int_0^1 h(x)dx \leq \varepsilon.$$

Using the same arguments as those used to prove that b) implies a), one can easily see that b) is true for any continuous function. □

Before presenting the next problem, we need another definition: we say that the sequence $(a_n)_{n\geq 1}$ is uniformly distributed mod 1 if the sequence of fractional parts of a_n is equidistributed. We invite the reader to find an elementary proof for the following problem in order to appreciate the power of Weyl's criterion. So, here is the classical example.

Example 8. Let a be an irrational number. Then the sequence $(na)_{n\geq 1}$ is uniformly distributed mod 1.

Solution. Well, after so much work, you deserve a reward: this is a simple consequence of Weyl's criterion. Indeed, it suffices to prove that c) is true, which reduces to proving that

$$\lim_{n\to\infty} \frac{1}{n} \sum_{k=1}^{n} e^{2i\pi pka} = 0 \tag{15.2}$$

for all integers $p \geq 1$. But this is just a geometric series!!! A one-line computation shows that (15.1) is satisfied and thus we obtain the desired result.

It is probably time to solve the problem mentioned at the very beginning of this note: how to compute the density of those numbers n for which 2^n begins with (for example) 2006. Well, again a reward: this is going to be equally easy (of course, you need some rest before looking at some deeper results).

Example 9. What is the density of the set of positive integers n for which 2^n begins with 2006?

Solution. 2^n begins with 2006 if and only if there is a $p \geq 1$ and some digits $a_1, a_2, \ldots, a_p \in \{0, 1, \ldots, 9\}$ such that $2^n = \overline{2006a_1a_2\ldots a_p}$, which is clearly equivalent to the existence of $p \geq 1$ such that

$$2007 \cdot 10^p > 2^n \geq 2006 \cdot 10^p.$$

This can be rewritten as

$$\log 2007 + p > n \log 2 \geq \log 2006 + p,$$

implying $\lfloor n \log 2 \rfloor = p + 3$. Hence $\log \frac{2007}{1000} > \{n \log 2\} > \log \frac{2006}{1000}$ and the density of our set is the density of the set of positive integers n satisfying

$$\log \frac{2007}{1000} > \{n \log 2\} > \log \frac{2006}{1000}.$$

From Example 8, the last set has density $\log \frac{2007}{2006}$ and this is the answer to our problem.

We saw a beautiful proof of the fact that if a is irrational, then $(na)_{n\geq 1}$ is uniformly distributed mod 1. Actually, much more is true, but this is also much more difficult to prove. The following two examples are important theorems. The first is due to Van der Corput and shows how a brilliant combination of algebraic manipulations and Weyl's criterion can yield difficult and important results.

Example 10. Let (x_n) be a sequence of real numbers such that the sequences $(x_{n+p} - x_n)_{n\geq 1}$ are equidistributed for all $p \geq 1$. Then (x_n) is also equidistributed.

[Van der Corput]

Solution. This is not an Olympiad problem!!! But mathematics is not just about Olympiads and from time to time (in fact, from a certain time on) one should try to discover what is behind such great results. This is the reason we present a proof of this theorem based on a technical lemma of Van der Corput, which turned out to be fundamental in studying exponential sums.

Lemma 15.2 (Van der Corput). *For any complex numbers $z_1, z_2, \ldots, z_n$ and any $h \in \{1, 2, \ldots, n\}$, the following inequality is true (with the convention that $z_i = 0$ for any integer i not in $\{1, 2, \ldots, n\}$):*

$$h^2\left|\sum_{i=1}^{n} z_i\right|^2 \leq (n+h-1)\left[2\sum_{r=1}^{h-1}(h-r)\mathrm{Re}\left(\sum_{i=1}^{n-r} z_i\overline{z_{i+r}}\right) + h\sum_{i=1}^{n}|z_i|^2\right].$$

Proof. The simple observation that

$$h\sum_{i=1}^{n} z_i = \sum_{i=1}^{n+h-1}\sum_{j=0}^{h-1} z_{i-j}$$

allows us to write (via Cauchy Schwarz's inequality):

$$h^2\left|\sum_{i=1}^{n} z_i\right|^2 \leq (n+h-1)\sum_{i=1}^{n+h-1}\left|\sum_{j=0}^{h-1} z_{i-j}\right|^2.$$

And next? Well, we expand $\sum_{i=1}^{n+h-1}\left|\sum_{j=0}^{h-1} z_{i-j}\right|^2$ and see that it is nothing other than

$$2\sum_{r=1}^{h-1}(h-r)\mathrm{Re}\left(\sum_{i=1}^{n-r} z_i\overline{z_{i+r}}\right) + h\sum_{i=1}^{n}|z_i|^2.$$

We will now prove Van der Corput's theorem, by using this lemma and Weyl's criterion.

Of course, the idea is to show that

$$\lim_{n\to\infty} \frac{1}{n}\sum_{k=1}^{n} e^{2i\pi p x_k} = 0$$

for all $p \geq 1$. Fix such a p and take for the moment a positive real number h and $\varepsilon \in (0,1)$ (h may depend on ε). Setting $z_j = e^{2i\pi p x_j}$, we have

$$\left|\frac{1}{n}\sum_{j=1}^{n} z_j\right|^2 \leq \frac{1}{n^2}\cdot\frac{n+h-1}{h^2}\left[hn + 2\sum_{i=1}^{h-1}(h-i)\mathrm{Re}\left(\sum_{j=1}^{n-i} z_j\cdot\overline{z_{i+j}}\right)\right].$$

Now, observe that

$$\mathrm{Re}\left(\sum_{j=1}^{n-i} z_j\cdot\overline{z_{i+j}}\right) = \mathrm{Re}\left(\sum_{j=1}^{n-i} e^{2i\pi p(x_j - x_{i+j})}\right) \leq \left|\sum_{j=1}^{n-i} e^{2i\pi p(x_j - x_{i+j})}\right|.$$

Using Weyl's criterion for the sequences $(x_{n+i} - x_n)_{n\geq 1}$ for $i = 1, 2, \ldots, h-1$, we deduce that for all sufficiently large n we have

$$\left|\sum_{j=1}^{n-i} e^{2i\pi p(x_j - x_{i+j})}\right| \leq \varepsilon n.$$

Therefore

$$\left|\frac{1}{n}\sum_{j=1}^{n} z_j\right|^2 \leq \frac{1}{n^2}\cdot\frac{n+h-1}{h^2}\left[hn + 2\varepsilon n\sum_{i=1}^{h-1}(h-i)\right]$$

$$< \frac{n+h-1}{nh}(1+\varepsilon) < \frac{2(1+\varepsilon)}{h} < \varepsilon^2$$

for n large enough. Now, by choosing $h > \dfrac{2(1+\varepsilon)}{\varepsilon^2}$, we deduce that for all sufficiently large n we have

$$\left|\frac{1}{n}\sum_{j=1}^{n} z_j\right| \leq \varepsilon.$$

Hence Weyl's criterion is satisfied and thus $(x_n)_{n\geq 1}$ is equidistributed. □

This was surely the most difficult result of this chapter, but why not take one more step once we are already here? Let us prove the following weaker (but as the reader will probably agree, absolutely nontrivial) version of a famous theorem of Weyl. It is related to the equidistribution of the sequence $(f(n))_{n\geq 1}$ where f is a real polynomial having at least one irrational coefficient other than the constant term. We will not prove this here, but focus on the following result.

Example 11. If f is a polynomial with real coefficients and irrational leading coefficient, then the sequence $(f(n))_{n\geq 1}$ is equidistributed.

[Weyl]

Solution. You have probably noticed that this is an immediate consequence of Van der Corput's theorem (but just imagine the amount of work done to arrive at this conclusion!!!): the proof by induction is immediate. Indeed, if f has degree 1, then the conclusion is clear (see example 5). Now, if the result holds for polynomials of degree at most k, it suffices (by Van der Corput's theorem) to prove that for all positive integers p, the sequence $(f(n+p) - f(n))_{n\geq 1}$ is equidistributed. But this is exactly the induction hypothesis applied to the polynomial $f(X+p) - f(X)$ (whose leading coefficient is clearly irrational). The proof by induction finishes here.

The solution of the following problem, which is a consequence of Weyl's criterion, is due to Marian Tetiva:

Example 12. If α is an irrational number and P is a nonconstant polynomial with integer coefficients, then there are infinitely many pairs (m, n) of integers such that $P(m) = \lfloor n\alpha \rfloor$.

[H. A. ShahAli]

Solution. Of course, we can assume that $\alpha > 0$. If $\alpha < 1$, no heavy machinery is required: all we need is to note that for all integers m the interval

$\left(\frac{P(m)}{\alpha}, \frac{P(m)+1}{\alpha}\right)$ has length greater than 1, thus it contains an integer n_m. It is clear that $\lfloor n_m \cdot \alpha \rfloor = P(m)$, so we have infinitely many solutions (at least one for each m). The difficult part is when $\alpha > 1$. Let us consider $\beta = \frac{\alpha}{\alpha-1}$. By a well-known result of Beaty, the sets $A = \{\lfloor n\alpha \rfloor \mid n \geq 1\}$ and $B = \{\lfloor n\beta \rfloor \mid n \geq 1\}$ give a partition of the set of positive integers. A second of observation shows that it is enough to prove the statement for polynomials P whose leading coefficient is positive. Thus starting from a certain point m_0, $P(m)$ is a positive integer, thus belonging to A or to B. Suppose that the equation $P(m) = \lfloor n\alpha \rfloor$ has finitely many solutions, that is for all sufficiently large m, $P(m) \in B$. Hence for some N we have the existence of a sequence of positive integers $(n_m)_{m>N}$ such that $P(m) = \lfloor n_m\beta \rfloor$. This clearly implies $\left\lfloor \frac{P(m)}{\beta} \right\rfloor = n_m - 1$, that is the fractional part of $\frac{P(m)}{\beta}$ is in $\left(1 - \frac{1}{\beta}, 1\right)$ for all sufficiently large m. Or, $\frac{1}{\beta}P$ clearly satisfies the conditions of Weyl's criterion, so the sequence of fractional parts of $\frac{P(m)}{\beta}$ is dense in $[0, 1]$, which is impossible, because all but finitely many terms are in $\left(1 - \frac{1}{\beta}, 1\right)$. This finishes the proof of the case $\alpha > 1$ and ends the solution.

15.2 Practice problems

1. Let $z_1, z_2, \ldots, z_n$ be arbitrary complex numbers. Prove that for any $\varepsilon > 0$ there are infinitely many positive integers n such that

$$\varepsilon + \sqrt[k]{|z_1^k + z_2^k + \cdots + z_n^k|} > \max\{|z_1|, |z_2|, \ldots, |z_n|\}.$$

2. (a) Prove that for any real number x and any positive integer N one can find integers p and q such that $0 < q \le N$ and $|qx-p| \le \dfrac{1}{N+1}$.

   (b) Suppose that a is a divisor of a number of the form n^2+1. Prove that a is a sum of two squares of integers.

3. Compute $\sup\limits_{n\ge 1}\left(\min\limits_{\substack{p,q\in\mathbb{N}\\ p+q=n}} |p - q\sqrt{3}|\right)$.

Putnam Competition

4. Is it true that for any irrational number x the sequence $(\{10^n x\})_n$ is equidistributed in $[0,1]$?

5. Prove that by using different terms of the sequence $\lfloor n^2\sqrt{2006}\rfloor$ one can construct geometric sequences of any length.

6. Does the sequence $\sin(n^2) + \sin(n^3)$ converge?

Gabriel Dospinescu

7. A flea moves in the positive direction of an axis, starting from the origin. It can only jump over distances equal to $\sqrt{2}$ and $\sqrt{2005}$. Prove that there exists an n_0 such that the flea will be able to arrive in any interval $[n, n+1]$ for each $n \ge n_0$.

IMAR Contest 2005

8. Let $x > 1$ be a real number and $a_n = \lfloor x^n \rfloor$. Can the number $S = 0.a_1a_2a_3...$ be rational? The expansion is formed by writing down the decimal digits of $a_1, a_2, ...$ in turn.

Mo Song-Qing, AMM 6540

9. Let $a \neq 0$ be a rational number and b an irrational number, then the sequence $nb \lfloor na \rfloor$ is uniformly distributed mod 1. What if a is irrational?

L.Kuipers

10. Prove that the sequence of the first digit of $2^n + 3^n$ is not periodical.

Tuymaada Olympiad

11. Let a, b be positive real numbers such that $\{na\} + \{nb\} < 1$ for all n. Prove that at least one of them is an integer.

12. Let a, b, c be positive real numbers. Prove that the sets $A = \{\lfloor na \rfloor \mid n \geq 1\}$, $B = \{\lfloor nb \rfloor \mid n \geq 1\}$, $C = \{\lfloor nc \rfloor \mid n \geq 1\}$ cannot form a partition of the set of positive integers.

Putnam Competition

13. Suppose that f is a real, continuous, and periodical function such that the sequence $\left(\sum_{k=1}^{n} \frac{|f(k)|}{k}\right)_{n\geq 1}$ is bounded. Prove that $f(k) = 0$ for all positive integers k. Give a necessary and sufficient condition ensuring the existence of a constant $c > 0$ such that $\sum_{k=1}^{n} \frac{|f(k)|}{k} > c \ln n$ for all n.

Gabriel Dospinescu

14. (a) Let f be a polynomial with integral coefficients and let a be a positive irrational number. Can we have

$$f(\mathbb{N}) \subset \{[na] | n \geq 1\}?$$

(b) Is it true that any set of positive integers with positive density contains an infinite arithmetical sequence?

15. Let $x_1, x_2, \ldots$ be a sequence of numbers in $[0, 1)$ such that at least one of its sequential limit points is irrational. For $0 \leq a < b \leq 1$, let $N_n(a, b)$ be the number of n-tuples $(a_1, a_2, \ldots, a_n) \in \{\pm 1\}^n$ such that $\{a_1x_1 + a_2x_2 + \cdots + a_nx_n\} \in [a, b)$. Prove that $\frac{N_n(a,b)}{2^n}$ converges to $b - a$.

Andrew Odlyzko, AMM 6542

16. Let $n \in N$ and let $0 \leq a_1 \leq a_2 \leq \cdots \leq a_n \leq \pi$ and $b_1, b_2, \ldots, b_n$ be nonnegative real numbers such that

$$|\sum_{i=1}^{n} b_i \cos(ka_i)| < \frac{1}{k}$$

for all positive integers k. Prove that $b_1 = b_2 = \cdots = b_n = 0$

Bulgarian TST

17. Let n be a positive integer. Prove that there exists $\varepsilon > 0$ such that for any positive real numbers $a_1, a_2, \ldots, a_n$ there exists $t > 0$ such that

$$\{ta_1\}, \{ta_2\}, \ldots, \{ta_n\} \in \left(\varepsilon, \frac{1}{2}\right).$$

St. Petersburg 1998

18. Let x be a real number. Prove the existence of a constant $c > 0$ with the following property: for any $n \geq 1$ there exists a natural number $k \leq n$ such that

$$d(k^2 x, \mathbb{Z}) \leq c \cdot \frac{\log n}{\sqrt[3]{n}},$$

where $d(a, \mathbb{Z}) = \min_{n \in \mathbb{Z}} |a - n|$.

19. For a real number x, prove the equivalence of the following statements:

    (a) For any $\varepsilon > 0$ there exists a sequence $(a_n)_n$ such that $|a_n - n| \leq \varepsilon$ for all n and such that $(xa_n)_n$ is equidistributed modulo 1.

    (b) x is transcendental.

Yves Meyer

Chapter 16

16.1 Theory and examples

Problems about the sum of the digits of a positive integer often occur in mathematical contests because of their difficulty and lack of standard ways to tackle them. This is why a synthesis of the most frequent techniques used in such problems is useful. We have selected several representative problems to illustrate how the main results and techniques work and why they are so important.

We will only work in base 10 and will denote the decimal sum of the digits of the positive integer x by $s(x)$. The following "formula" can be easily checked:

$$s(n) = n - 9\sum_{k\geq 1} \left\lfloor \frac{n}{10^k} \right\rfloor \tag{16.1}$$

From (16.1) we can deduce immediately some well-known results about $s(n)$, such as $s(n) \equiv n \pmod 9$ and $s(m+n) \leq s(m) + s(n)$. Unfortunately, (16.1) is a clumsy formula, which can hardly be used in applications. On the other hand, there are several more or less known results about the sum of the digits, results which may offer simple ways to attack harder problems.

The easiest of these techniques is probably just the careful analysis of the structure of the numbers and their digits. This can work surprisingly well, as we will see in the following examples.

Example 1. Prove that among any 79 consecutive numbers, we can choose at least one whose sum of digits is a multiple of 13.

Baltic Contest 1997

Solution. [Adrian Zahariuc] Note that among the first 40 numbers, there are exactly four multiples of 10. Also, it is clear that the next to last digit of one of them is at least 6. Let x be this number. Clearly, x, $x+1$,..., $x+39$ are among our numbers, so $s(x), s(x)+1, ..., s(x)+12$ occur as sums of digits in some of our numbers. One of these sums is a multiple of 13 and we are done.

We will continue with two harder problems, which still do not require any special result or technique.

Example 2. Find the greatest N for which there are N consecutive positive integers such that the sum of digits of the k-th number is divisible by k, for $k = 1, 2, ..., N$.

Tournament of Towns 2000

Solution. [Adrian Zahariuc] The answer, which is not trivial at all, is 21. The main idea is that among $s(n+2)$, $s(n+12)$ and $s(n+22)$ there are two consecutive numbers, which is impossible since all of them should be even. Indeed, we carry over at $a+10$ only when the next to last digit of a is 9, but this situation can occur at most once in our case. So, for $N > 21$, we have no solution. For $N = 21$, we can choose $N+1$, $N+2$,..., $N+21$, where $N = 291 \cdot 10^{11!} - 12$. For $i = 1$ we have nothing to prove. For $2 \le i \le 11$, $s(N+i) = 2+9+0+9(11!-1)+i-2 = i+9 \cdot 11!$ while for $12 \le i \le 21$, $s(N+i) = 2+9+1+(i-12) = i$, so our numbers have the desired property.

Example 3. How many positive integers $n \le 10^{2005}$ can be written as the sum of two positive integers with the same sum of digits?

[Adrian Zahariuc]

Solution. Answer: $10^{2005} - 9023$. At first glance, it is seemingly impossible to find the exact number of positive integers with this property. In fact, the following is true: a positive integer cannot be written as the sum of two numbers with the same sum of digits if and only if all of its digits except for the first are 9 and the sum of its digits is odd.

Let n be such a number. Suppose there are positive integers a and b such that $n = a+b$ and $s(a) = s(b)$. The main fact is that when we add $a+b = n$, there are no carry overs. This is clear enough. It follows that $s(n) = s(a) + s(b) = 2s(a)$, which is impossible since $s(n)$ is odd.
Now we will prove that any number n which is not one of the numbers above, can be written as the sum of two positive integers with the same sum of digits. We will start with the following:

Lemma 16.1. *There is $a \leq n$ such that $s(a) \equiv s(n-a) \pmod 2$.*

Proof. If $s(n)$ is even, take $a = 0$. If $s(n)$ is odd, then n must have a digit which is not the first one and is not equal to 9, otherwise it would have one of the forbidden forms mentioned in the beginning of the solution. Let c be this digit and let p be its position (from right to left). Choose $a = 10^{p-1}(c+1)$. In the addition $a + (n-a) = n$ there is exactly one carry over, so

$$s(a) + s(n-a) = 9 + s(n) \equiv 0 \bmod 2 \Rightarrow s(a) \equiv s(n-a) \bmod 2$$

which proves our claim. □

Back to the original problem. All we have to do now is take one-by-one a "unit" from a number and give it to the other until the two numbers have the same sum of digits. This will happen because they have the same parity. So, let us do this rigorously. Set

$$a = \overline{a_1 a_2 \ldots a_k}, \qquad n - a = \overline{b_1 b_2 \cdots b_k}.$$

Let I be the set of those $1 \leq i \leq k$ for which $a_i + b_i$ is odd. The lemma shows that the number of elements of I is even, so it can be divided into two sets with the same number of elements, say I_1 and I_2. For $i = 1, 2, ..., k$ define $A_i = \frac{a_i+b_i}{2}$ if $i \notin I$, $\frac{a_i+b_i+1}{2}$ if $i \in I_1$ or $\frac{a_i+b_i-1}{2}$ if $i \in I_2$ and $B_i = a_i + b_i - A_i$. It is clear that the numbers

$$A = \overline{A_1 A_2 ... A_k}, \qquad B = \overline{B_1 B_2 ... B_k}$$

have the properties $s(A) = s(B)$ and $A + B = n$. The proof is complete.

We have previously seen that $s(n) \equiv n \pmod 9$. This is probably the most well-known property of the function s and it has a series of remarkable applications. Sometimes it is combined with simple inequalities such as $s(n) \leq 9(\lfloor \log n \rfloor + 1)$. Some immediate applications are the following:

Example 4. Find all n for which one can find a and b such that

$$s(a) = s(b) = s(a + b) = n.$$

[Vasile Zidaru, Mircea Lascu]

Solution. We have $a \equiv b \equiv a + b \equiv n \pmod 9$, so 9 divides n. If $n = 9k$, we can take $a = b = 10^k - 1$ and we are done, since $s(10^k - 1) = s(2 \cdot 10^k - 2) = 9k$.

Example 5. Find all the possible values of the sum of the digits of a perfect square.

Iberoamerican Olympiad 1995

Solution. What does the sum of the digits have to do with perfect squares? Apparently, nothing, but perfect squares do have something to do with remainders mod 9. In fact, it is easy to prove that the only possible values of a perfect square mod 9 are 0, 1, 4 and 7. So, we deduce that the sum of the digits of a perfect square must be congruent to 0, 1, 4, or 7 mod 9. To prove that all such numbers work, we will use a small and very common (but worth remembering!) trick: use numbers that consist almost only of 9-s. We have the following identities:

$$\underbrace{99...99}_{n}{}^2 = \underbrace{99...99}_{n-1}8\underbrace{00...00}_{n-1}1 \Rightarrow s(\underbrace{99...99}_{n}{}^2) = 9n$$

$$\underbrace{99...99}_{n-1}1^2 = \underbrace{99...99}_{n-2}82\underbrace{00...00}_{n-2}81 \Rightarrow s(\underbrace{99..99}_{n-1}1^2) = 9n+1$$

$$\underbrace{99...99}_{n-1}2^2 = \underbrace{99...99}_{n-2}84\underbrace{00...00}_{n-2}64 \Rightarrow s(\underbrace{99..99}_{n-1}2^2) = 9n+4$$

$$\underbrace{99...99}_{n-1}4^2 = \underbrace{99...99}_{n-2}88\underbrace{00...00}_{n-2}36 \Rightarrow s(\underbrace{99..99}_{n-1}4^2) = 9n+7$$

and since $s(0) = 0$, $s(1) = 1$, $s(4) = 4$ and $s(16) = 7$ the proof is complete.

Example 6. Compute $s(s(s(4444^{4444})))$.

IMO 1975

Solution. Using the inequality $s(n) \le 9(\lfloor \log n \rfloor + 1)$ several times we have

$$s(4444^{4444}) \le 9(\lfloor \log 4444^{4444} \rfloor + 1) < 9 \cdot 20000 = 180000;$$

$$s(s(4444^{4444})) \le 9(\lfloor \log s(4444^{4444}) \rfloor + 1) \le 9(\log 180000 + 1) \le 63,$$

so $s(s(s(4444^{4444}))) \le 14$ (indeed, among the numbers from 1 to 63, the maximum value of the sum of digits is 14). On the other hand, $s(s(s(n))) \equiv s(s(n)) \equiv s(n) \equiv n \pmod 9$ and since

$$4444^{4444} \equiv 7^{4444} = 7 \cdot 7^{3 \cdot 1481} \equiv 7 (\mathrm{mod} 9),$$

the only possible answer is 7.
Finally, we present two beautiful problems which appeared in the Russian Olympiad and, later, in Kvant.

Example 7. Prove that for any N there is an $n \ge N$ such that $s(3^n) \ge s(3^{n+1})$.

Solution. Suppose by way of contradiction that there is an $N \geq 2$ such that $s(3^{n+1}) - s(3^n) > 0$ for all $n \geq N$. But, for $n \geq 2$, $s(3^{n+1}) - s(3^n) \equiv 0 \pmod 9$, so $s(3^{n+1}) - s(3^n) \geq 9$ for all $n \geq N$. It follows that

$$\sum_{k=N+1}^{n} \left(s\left(3^{k+1}\right) - s\left(3^k\right)\right) \geq 9(n-N) \Rightarrow s(3^{n+1}) \geq 9(n-N)$$

for all $n \geq N+1$. But $s(3^{n+1}) \leq 9(\lfloor \log 3^{n+1} \rfloor + 1)$, so $9n - 9N \leq 9 + 9(n+1)\log 3$, for all $n \geq N+1$. This is obviously a contradiction.

Example 8. Find all positive integers k for which there exists a positive constant c_k such that $\frac{s(kN)}{s(N)} \geq c_k$ for all positive integers N. For any such k, find the best c_k.

[I. N. Bernstein]

Solution. It is not difficult to observe that any k of the form $2^r \cdot 5^q$ is a solution of the problem. Indeed, in that case we have (by using the properties presented in the beginning of the chapter):

$$s(N) = s(10^{r+q}N) \leq s(2^q \cdot 5^r)s(kN) = \frac{1}{c_k}s(kN)$$

where clearly $c_k = \frac{1}{s(2^q \cdot 5^r)}$ is the best constant (we have equality for $N = 2^q \cdot 5^r$).

Now, assume that $k = 2^r \cdot 5^q \cdot Q$ with $Q > 1$ relatively prime to 10. Let $m = \varphi(Q)$ and write $10^m - 1 = QR$ for some integer R. If $R_n = R(1 + 10^m + \cdots + 10^{m(n-1)})$ then $10^{mn} - 1 = QR_n$ and so $s(Q(R_n+1)) = s(10^{mn} + Q - 1) = s(Q)$ and $s(R_n + 1) \geq (n-1)s(R)$ (note than the condition $Q > 1$, which is the same as $R < 10^m - 1$, is essential for this last inequality, because it guarantees that $R + 1$ has at most m digits and thus when adding $R + 1$ and $10^m \cdot R$, we obtain the digits of R followed by the digits of $R + 1$; if we proceed in the same manner for each addition, we see that $R_n + 1$ has among its digits at least $n-1$ copies of the sequence of digits of R). By taking n sufficiently large, we conclude that for any $\epsilon > 0$ there exists $N = R_n + 1$ such that

$$\frac{s(kN)}{s(N)} \leq \frac{s(2^r \cdot 5^q)s(Q)}{(n-1)s(R)} < \epsilon.$$

This shows that the numbers found in the first part of the solution are the only solutions of the problem.

If so far we have studied some remarkable properties of the function s, which were quite well-known, it is time to present some problems and results which are less familiar, but interesting and hard. The first result is the following:

Lemma 16.2. *If* $1 \leq x \leq 10^n$, *then* $s(x(10^n - 1)) = 9n$.

Proof. The idea is very simple: all we have to do is write $x = \overline{a_1 a_2 ... a_j}$ with $a_j \neq 0$ (we can ignore the trailing 0's of x) and note that

$$x(10^n - 1) = \overline{a_1 a_2 ... a_{j-1}(a_j - 1)\underbrace{99...99}_{n-j}(9 - a_1)...(9 - a_{j-1})(10 - a_j)},$$

which obviously has the sum of digits equal to $9n$. □

The previous result is by no means hard, but we will see that it can be the key in many situations. A first application is:

Example 9. Evaluate $s(9 \cdot 99 \cdot 9999 \cdot ... \cdot \underbrace{99...99}_{2^n})$.

USAMO 1992

Solution. The problem is trivial if we know the previous result. We have

$$N = 9 \cdot 99 \cdot 9999 \cdot \ldots \cdot \underbrace{99...99}_{2^{n-1}} < 10^{1+2+\cdots+2^{n-1}} < 10^{2^n} - 1$$

so $s(\underbrace{99...99}_{2^n} N) = 9 \cdot 2^n$.

However, there are very hard applications of this apparently unimportant result, such as the following problem.

Example 10. Prove that for each n there is a positive integer with n nonzero digits, that is divisible by the sum of its digits.

IMO 1998 Shortlist

Solution. Just to assure our readers that this problem did not appear on the IMO Shortlist out of nowhere, such numbers are called Niven numbers and they are an important research source in number theory. Now, let us solve it. We will see that constructing such a number is difficult. First, we will dispose of the case $n = 3^k$, when we can take the number $\underbrace{11...11}_{3^k}$ (it can be easily proved by induction that $3^{k+2} | 10^{3^k} - 1$). From the idea that we should search numbers with many equal digits and the last result, we decide that the required number p should be of the form $\underbrace{aa...aa}_{s} b \cdot (10^t - 1)$, with $\underbrace{aa...aa}_{s} b \leq 10^t - 1$. This number has $s + t + 1$ digits and its sum of digits is $9t$. Therefore, we require $s + t = n - 1$ and $9t | \underbrace{aa...aa}_{s} b \cdot (10^t - 1)$. We now use the fact that if t is a power of 3, then $9t | 10^t - 1$. So, let us take $t = 3^k$ where k is chosen such that $3^k < n < 3^{k+1}$. If we also take into account the condition $\underbrace{aa...aa}_{s} b \leq 10^t - 1$, it is natural to pick $p = \underbrace{11...11}_{n-3^k-1} 2(10^{3^k} - 1)$ when $n \leq 2 \cdot 3^k$ and $p = \underbrace{22...22}_{2 \cdot 3^k}(10^{2 \cdot 3^k} - 1)$ otherwise.

We continue our investigations of finding suitable techniques for problems involving sum of digits with a very beautiful result, which has several interesting and difficult consequences.

Lemma 16.3. *Any multiple of* $\underbrace{99...99}_{k}$ *has sum of its digits at least* $9k$.

Proof. We will use the extremal principle. Suppose by way of contradiction that the statement is false, and take M to be the smallest multiple of a such that $s(M) < 9k$, where $a = \underbrace{99...99}_{k}$. Clearly, $M > 10^k$, hence $M = \overline{a_p a_{p-1} ... a_0}$, with $p \geq k$ and $a_p \neq 0$. Take $N = M - 10^{p-k}a$, which is a multiple less than M of a. We will prove that $s(N) < 9k$. Observe that

$$N = M - 10^p + 10^{p-k} = (a_p - 1) \cdot 10^p + a_{p-1}10^{p-1} + \cdots + (a_{p-k} + 1)10^{p-k} + \cdots + a_0,$$

so that we can write

$$s(N) \leq a_p - 1 + a_{p-1} + \cdots + (a_{p-k} + 1) + \cdots + a_0 = s(M) < 9k.$$

In this way, we contradict the minimality of M and the proof is completed. □

We will show three applications of this fact, which might seem simple, but seemingly unsolvable without it. But before that, let us insist a little bit on a very similar (yet more difficult) problem proposed by Radu Todor for the 1993 IMO: if $b > 1$ and a is a multiple of $b^n - 1$, then a has at least n nonzero digits when expressed in base b. The solution uses the same idea, but the details are not obvious, so we will present a full solution. Arguing by contradiction, assume that there exists A, a multiple of $b^n - 1$ with less than n nonzero digits in base b, and among all these numbers consider that number A with minimal number of nonzero digits in base b and with minimal sum of digits in base b. Suppose that a has exactly s nonzero digits (everything is in base b) and let $A = a_1 b^{n_1} + a_2 b^{n_2} + \cdots + a_s b^{n_s}$ with $n_1 > n_2 > \cdots > n_s$. We claim that $s = n$. First of all, we will prove that any two numbers among $n_1, n_2, ..., n_s$ are not congruent mod n. It will follow that $s \leq n$. Indeed, if $n_i \equiv n_j \pmod{n}$ let $0 \leq r \leq n-1$ be the common value of n_i and n_j modulo n. The number $B = A - a_i b^{n_i} - a_j b^{n_j} + (a_i + a_j) b^{nn_1 + r}$ is clearly a multiple of $b^n - 1$. If $a_i + a_j < b$ then B has $s-1$ nonzero digits, which contradicts the minimality of s. So $b \leq a_i + a_j < 2b$. If $q = a_i + a_j - b$, then

$$B = b^{nn_1+r+1} + qb^{nn_1+r} + a_1 b^{n_1} + \cdots + a_{i-1} b^{n_{i-1}}$$

$$+ a_{i+1}^{n_{i+1}} + \cdots + a_{j-1} b^{n_{j-1}} + a_{j+1} b^{n_{j+1}} + \cdots + a_s b^{n_s}.$$

Therefore the sum of digits of B in base b is $a_1 + a_2 + \cdots + a_s + 1 + q - (a_i + a_j) < a_1 + a_2 + \cdots + a_s$. This contradiction shows that $n_1, n_2, \ldots, n_s$ give distinct remainders $r_1, r_2, \ldots, r_s$ when divided by n. Finally, suppose that $s < n$ and consider the number $C = a_1 b^{r_1} + \cdots + a_s b^{r_s}$. Clearly, C is a multiple of $b^n - 1$. But $C < b^n - 1$! This shows that $s = n$ and finishes the solution.

Example 11. Prove that for every k, we have

$$\lim_{n \to \infty} \frac{s(n!)}{(\ln(\ln(n)))^k} = \infty.$$

Solution. Due to the simple fact that $10^{\lfloor \log n \rfloor} - 1 \leq n \Rightarrow 10^{\lfloor \log n \rfloor} - 1 | n!$, we have $s(n!) \geq \lfloor \log n \rfloor$, from which our conclusion follows.

Example 12. Let S be the set of positive integers whose decimal representation contains at most 1988 ones and the rest zeros. Prove that there is a positive integer which does not divide any element of S.

Tournament of Towns 1988

Solution. Again, the solution follows directly from our result. We can choose the number $10^{1989} - 1$, whose multiples have sum of digits greater than 1988.

Example 13. Prove that for each $k > 0$, there is an infinite arithmetical progression with a common difference relatively prime to 10, such that all its terms have the sum of digits greater than k.

IMO 1999 Shortlist

Solution. Let us remind you that this is the last problem from IMO 1999 Shortlist, so it is one of the hardest. The official solution seems to confirm this. But, due to the above lemma we can chose the sequence $a_n = n(10^m - 1)$, where $m > k$ and we are done.

Now, as a final proof of the utility of these two results, we will present a hard problem from the USAMO.

Example 14. Let n be a fixed positive integer. Denote by $f(n)$ the smallest k for which one can find a set X of n positive integers with the property

$$s\left(\sum_{x \in Y} x\right) = k$$

for all nonempty subsets Y of X. Prove that

$$C_1 \log n < f(n) < C_2 \log n$$

for some constants C_1 and C_2.

[Titu Andreescu, Gabriel Dospinescu] USAMO 2005

Solution. We will prove that

$$\lfloor \log(n+1) \rfloor \leq f(n) \leq 9 \log \left\lceil \frac{n(n+1)}{2} + 1 \right\rceil,$$

which is enough to establish our claim. Let l be the smallest integer such that

$$10^l - 1 \geq \frac{n(n+1)}{2}.$$

Consider the set $X = \{j(10^l - 1) : 1 \le j \le n\}$. By the previous inequality and our first lemma, it follows that

$$s\left(\sum_{x\in Y} x\right) = 9l$$

for all nonempty subsets Y of X, so $f(n) \le 9l$, and the upper bound is proved. Now, let m be the greatest integer such that $n \ge 10^m - 1$. We will use the following well-known lemma:

Lemma 16.4. *Any set $M = \{a_1, a_2, ..., a_m\}$ has a nonempty subset whose element sum is divisible by m.*

Proof. Consider the sums $a_1, a_1+a_2, \ldots, a_1+a_2+\cdots+a_m$. If one of then is a multiple of m, them we are done. Otherwise, there are two of them congruent mod m, say the i-th and the j-th. Then, $m|a_{i+1} + a_{i+2} + \cdots + a_j$ and we are done. □

From the lemma, it follows that any n-element set X has a subset Y whose element sum is divisible by $10^m - 1$. By our second lemma, it follows that

$$s\left(\sum_{x\in Y} x\right) \ge m \Rightarrow f(n) \ge m,$$

and the proof is complete.

The last solved problem is one we consider to be very hard, and which uses different techniques than the ones we have mentioned so far.

Example 15. Let a and b be positive integers such that $s(an) = s(bn)$ for all n. Prove that $\log \frac{a}{b}$ is an integer.

[Adrian Zahariuc, Gabriel Dospinescu]

Solution. We start with an observation. If $\gcd(\max\{a,b\},10)=1$, then the problem becomes trivial. Indeed, suppose that $a=\max\{a,b\}$. Then, by Euler's theorem, $a|10^{\varphi(a)}-1$, so there is an n such that $an=10^{\varphi(a)}-1$, and since numbers consisting only of 9-s have a digit sum greater than all previous numbers, it follows that $an=bn$, so $a=b$. Let us now solve the harder problem. For any $k\geq 1$ there is an n_k such that $10^k \leq an_k \leq 10^k+a-1$. It follows that $s(an_k)$ is bounded, so $s(bn_k)$ is bounded. On the other hand,

$$10^k\frac{b}{a} \leq bn_k < 10^k\frac{b}{a}+b,$$

so, for sufficiently large k, the first p nonzero digits of $\frac{b}{a}$ are exactly the same as the first p digits of bn_k. This means that the sum of the first p digits of $\frac{b}{a}$ is bounded, which could only happen when this fraction has finitely many decimals. Analogously, we can prove the same result about $\frac{a}{b}$. Let $a=2^x5^ym$ and $b=2^z5^tm'$, where $\gcd(m,10)=\gcd(m',10)=1$. It follows that $m|m'$ and $m'|m$, so $m=m'$. Now, we can write the hypothesis as

$$s(2^z5^umn2^{c-x}5^{c-y}) = s(2^x5^ymn2^{c-x}5^{c-y}) = s(mn)$$

for all $c\geq\max\{x,y\}$. Now, if $p=max\{z+c-x,u+c-y\}-min\{z+c-x,u+c-y\}$, we find that there is a $k\in\{2,5\}$ such that $s(mn)=s(mk^pn)$ for all positive integer n. It follows that

$$s(mn)=s(k^pmn)=s(k^{2p}mn)=s(k^{3p}mn)=\cdots$$

Let $t=a^p$, so $\log t\in\mathbb{R}-\mathbb{Q}$ unless $p=0$. Now, we will use the following:

Lemma 16.5. *If $\log t\in\mathbb{R}-\mathbb{Q}$, then for any sequence of digits, there is a positive integer n such that t^nm starts with the selected sequence of digits.*

Proof. If we prove that $\{\{\log t^nm\}|n\in\mathbb{Z}^+\}$ is dense in $(0,1)$, then we are done. But $\log t^nm = n\log t+m$ and by Kronecker's theorem $\{\{n\log t\}|n\in\mathbb{Z}^+\}$ is dense in $(0,1)$, so the proof of the lemma is complete. □

The lemma implies the very important result that $s(t^nm)$ is unbounded for $p\neq 0$, which is a contradiction. Hence $p=0$ and $z+c-x=u+c-y$, so

$a = 10^{x-z}b$. The main proof is complete. This problem can be nicely extended to any base. The proof of the general case is quite similar, although there are some very important differences.

The aforementioned methods are just a starting point in solving such problems since the spectrum of problems involving the sum of the digits is very large. The techniques are even more useful when they are applied creatively.

16.2 Practice problems

1. We start with an even perfect number (which is equal to the sum of its divisors, except itself) different from 6 and calculate its sum of digits. Then, we calculate the sum of digits of the new number and so on. Prove that we will eventually get 1.

2. Prove that for any positive integer n there are infinitely many numbers m not containing any zero, such that $s(m) = s(mn)$.

   Russian Olympiad 1970

3. Prove that among any 39 consecutive positive integers there is one whose digit sum is divisible by 11.

   Russian Olympiad 1961

4. Prove that
$$\sum_{n\geq 1} \frac{s(n)}{n(n+1)} = \frac{10}{9} \ln 10.$$

   O. Shallit, AMM

5. Are there positive integers n such that $s(n) = 1000$ and $s(n^2) = 1000000$?

   Russian Olympiad 1985

6. Prove that there are infinitely many positive integers n such that n is relatively prime to 10 and
$$s(n) + s(n^2) = s(n^3).$$

   Gabriel Dospinescu

7. If $s(n) = 100$ and $s(44n) = 800$, find $s(3n)$.

Rusia 1999

8. Let a and b be positive real numbers. Prove that the sequence $s(\lfloor an+b \rfloor)$ contains a constant subsequence.

Laurenţiu Panaitopol, Romanian TST 2002

9. Are there arbitrarily long arithmetic sequences whose terms have the same digit sum? What about infinite arithmetic sequences?

10. Let a be a positive integer such that $s(a^n + n) = 1 + s(n)$ for any sufficiently large n. Prove that a is a power of 10.

Gabriel Dospinescu

11. Are there polynomials $p \in \mathbb{Z}[X]$ such that

$$\lim_{n \to \infty} s(p(n)) = \infty?$$

12. Let a, b, c, d be prime numbers such that $2 < a \leq c$ and $a \neq b$. Suppose that for sufficiently large n, the numbers $an+b$ and $cn+d$ have the same digit sum in any base between 2 and $a-1$. Prove that $a = c$ and $b = d$.

Gabriel Dospinescu

13. Let S be a set of positive integers such that for any $\alpha \in \mathbb{R} - \mathbb{Q}$, there is a positive integer n such that $\lfloor \alpha^n \rfloor \in S$. Prove that S contains numbers with arbitrarily large digit sum.

Gabriel Dospinescu

14. Prove that the sequence $\frac{s(n)}{s(n^2)}$ is unbounded.

15. Prove that there is a constant $c > 0$ such that $s(2^n) \geq c \ln n$ for all n.

Schinzel

16. Prove that there are arbitrarily long sequences of consecutive numbers which do not contain any Niven numbers.

Mathlinks Contest

17. Let a, b, c, d be prime numbers such that $5 < a \leq c$ and $a \neq b$. If $s(an + b) = s(cn + d)$ for all sufficiently large n, then $a = c$ and $b = d$.

Richard Stong

18. Let k be a positive integer. Prove that there is a positive integer m such that the equation $n + s(n) = m$ has exactly k solutions.

Mihai Manea, Romanian TST 2003

19. Let x_n be a strictly increasing sequence of positive integers such that $v_2(x_n) - v_5(x_n)$ has the limit ∞ or $-\infty$. Prove that $s(x_n)$ tends to ∞.

Bruno Langlois

20. Is there an infinite arithmetic sequence which contains no Niven number?

Gabriel Dospinescu

21. Prove that the sum of digits of 9^n is greater than 9 for $n > 2$.

Mark Sapir, AMM

22. Is there an increasing arithmetic sequence with 10000 terms such that the digit sum of its terms forms again an increasing arithmetic sequence?

Tournament of the Towns

23. A number m is called *special* if there is no k such that $k + s(k) = m$. Prove that there are infinitely many special numbers of the form $10^n + b$ if and only if $b - 1$ is special.

Christopher D. Long

24. Prove that there exists a constant C such that for all N, the number of Niven numbers smaller than N is at most $C\frac{x}{(\ln x)^{2/3}}$.

Chapter 17

17.1 Theory and examples

"Olympiad problems can be solved without using concepts from analysis (or linear algebra)" is a sentence often heard when talking about problems given at various mathematics competitions. This is true, but the essence of some of these problems lies in analysis, and this is the reason that such problems are always the highlight of a contest. Their elementary solutions are very tricky and sometimes extremely difficult to design, while when using analysis they can fall apart rather quickly. Well, of course, "quickly" only if you see the right sequence (or function) that hides behind each such problem. Practically, in this chapter our aim is to exhibit convergent integer sequences. Clearly, these sequences must eventually become constant, and from here the problem becomes much easier. The difficulty lies in finding those sequences. Sometimes this is not so challenging, but most of the time it turns out to be a very difficult task. We develop skills in "hunting" for these sequences first by solving some easier questions, and after that we tackle the real problem.

As usual, we begin with a classical and beautiful problem, which has many applications and extensions.

Example 1. Let $f, g \in \mathbb{Z}[X]$ be two nonconstant polynomials such that $f(n)|g(n)$ for infinitely many n. Prove that f divides g in $\mathbb{Q}[X]$.

Solution. Indeed, we need to look at the remainder of g when divided by f in $\mathbb{Q}[X]$. Let us write $g = f \cdot q + r$, were q, r are polynomials in $\mathbb{Q}[X]$ with $\deg r < \deg f$. Now, multiplying by the common denominator of all coefficients of the polynomials q and r, the hypothesis becomes: there exist two infinite integer sequences $(a_n)_{n\geq 1}$, $(b_n)_{n\geq 1}$ and a positive integer N such that $b_n = N\dfrac{r(a_n)}{f(a_n)}$ (we could have some problems with the zeros of f, but they are only finitely many, so for n large enough, a_n is not a zero of f). Because $\deg r < \deg f$, it follows that $\dfrac{r(a_n)}{f(a_n)} \to 0$, thus $(b_n)_{n\geq 1}$ is a sequence of integers that converges to 0. This implies that this sequence will eventually become the zero sequence. Well, this is the same as $r(a_n) = 0$ from a certain point n_0, which is practically the same as $r = 0$ (do not forget that any nonzero polynomial has only finitely many zeros). The problem is solved.

The next problem is a special case of a much more general and classical result: if f is a polynomial with integer coefficients, k is an integer greater than 1, and $\sqrt[k]{f(n)} \in \mathbb{Q}$ for all n, then there exists a polynomial $g \in \mathbb{Q}[X]$ such that $f(x) = g^k(x)$. We will not discuss here this general result (the reader will find a proof in the chapter **Arithmetic Properties of Polynomials**).

Example 2. Let a, b, c be integers with $a \neq 0$ such that $an^2 + bn + c$ is a perfect square for any positive integer n. Prove that there exist integers x and y such that $a = x^2$, $b = 2xy$, $c = y^2$.

Solution. Let us begin by writing $an^2 + bn + c = x_n^2$ for a certain sequence $(x_n)_{n\geq 1}$ of nonnegative integers. We would expect that $x_n - n\sqrt{a}$ converges. And yes, it does, but it is not a sequence of integers, so its convergence is more or less useless. In fact, we need another sequence. The easiest way is to work with $(x_{n+1} - x_n)_{n\geq 1}$, since this sequence certainly converges to $\sqrt{a}$ (you have already noticed why it was not useless to find that $x_n - n\sqrt{a}$ is convergent; we used this to establish the convergence of $(x_{n+1} - x_n)_{n\geq 1}$). This time, the sequence consists of integers, so it is eventually constant. Hence we can find a positive integer M such that $x_{n+1} = x_n + \sqrt{a}$ for all $n \geq M$. Thus a must be a perfect square, that is $a = x^2$ for some integer x. A simple induction shows that $x_n = x_M + (n - M)x$ for $n \geq M$ and so $(x_M - Mx + nx)^2 = x^2n^2 + bn + c$ for all $n \geq M$. Identifying the coefficients finishes the solution, since we can take $y = x_M - Mx$.

Even this very particular case is interesting. Indeed, here is a very nice application of the previous problem:

Example 3. Prove that there cannot exist three polynomials P, Q, R with integer coefficients, of degree 2, and such that for all integers x, y there exists an integer z such that $P(x) + Q(y) = R(z)$.

Tuymaada Olympiad

Solution. Using the above result, the problem becomes straightforward. Indeed, suppose that $P(X) = aX^2 + bX + c, Q(X) = dX^2 + eX + f$ and $R(X) = mX^2 + nX + p$ are such polynomials. Fix two integers x, y. Then the equation $mz^2 + nz + p - P(x) - Q(y) = 0$ has an integer solution, so the discriminant is a perfect square. It means that $m(4P(x)+4Q(y)-4p)+n^2$ is a perfect square and this for all integers x, y. Now, for a fixed y, the polynomial of second degree $4mP(X) + m(4Q(y) - 4p) + n^2$ transforms all integers into perfect squares. By the previous problem, it is the square of a polynomial of first degree. In particular, its discriminant is zero. Because y is arbitrary, it follows that Q is constant, which is not possible because $\deg(Q) = 2$.

Another easy example is the following problem, in which finding the right convergent sequence of integers in not difficult at all. But, attention must be paid to details!

Example 4. Let $a_1, a_2, \ldots, a_k$ be positive real numbers such that at least one of them is not an integer. Prove that there exit infinitely many positive integers n such that n and $\lfloor a_1 n \rfloor + \lfloor a_2 n \rfloor + \cdots + \lfloor a_k n \rfloor$ are relatively prime.

[Gabriel Dospinescu]

Solution. The solution to such a problem needs to be indirect. So, let us assume that there exists a number M such that n and $\lfloor a_1 n \rfloor + \lfloor a_2 n \rfloor + \cdots + \lfloor a_k n \rfloor$ are not relatively prime for all $n \geq M$. Now, what are the most efficient numbers n to be used? They are the prime numbers, since if n is prime and it is not relatively prime with $\lfloor a_1 n \rfloor + \lfloor a_2 n \rfloor + \cdots + \lfloor a_k n \rfloor$, then it must divide

$\lfloor a_1 n \rfloor + \lfloor a_2 n \rfloor + \cdots + \lfloor a_k n \rfloor$. This suggests considering the sequence of prime numbers $(p_n)_{n \geq 1}$. Since this sequence is infinite, there is N such that $p_n \geq M$ for all $n \geq N$. According to our assumption, this implies that for all $n \geq N$ there exist a positive integer x_n such that $\lfloor a_1 p_n \rfloor + \lfloor a_2 p_n \rfloor + \cdots + \lfloor a_k p_n \rfloor = x_n p_n$. And now, you have already guessed what is the convergent sequence! Yes, it is $(x_n)_{n \geq N}$. This is clear, since $\frac{\lfloor a_1 p_n \rfloor + \lfloor a_2 p_n \rfloor + \cdots + \lfloor a_k p_n \rfloor}{p_n}$ converges to $a_1 + a_2 + \cdots + a_k$. Thus we can find P such that $x_n = a_1 + a_2 + \cdots + a_k$ for all $n \geq P$. But this is the same as $\{a_1 p_n\} + \{a_2 p_n\} + \cdots + \{a_k p_n\} = 0$. This says that $a_i p_n$ are integers for all $i = 1, 2, ..., k$ and $n \geq P$ and so a_i are integers for all i, contradicting the hypothesis.

Step by step, we start to build some experience in "guessing" the sequences. It is then time to solve some more difficult problems. The next one may seem obvious after reading its solution. In fact, it is just that type of problem whose solution is very short, but difficult to find.

Example 5. Let a and b be integers such that $a \cdot 2^n + b$ is a perfect square for all positive integers n. Prove that $a = 0$.

Polish TST

Solution. Suppose that $a \neq 0$. Then $a > 0$, otherwise for large values of n the number $a \cdot 2^n + b$ is negative. From the hypothesis, there exists a sequence of positive integers $(x_n)_{n \geq 1}$ such that $x_n = \sqrt{a \cdot 2^n + b}$ for all n. A direct computation shows that $\lim_{n \to \infty} (2x_n - x_{n+2}) = 0$. This implies the existence of a positive integer N such that $2x_n = x_{n+2}$ for all $n \geq P$. But $2x_n = x_{n+2}$ is equivalent to $b = 0$. Then a and $2a$ are both perfect squares, which is impossible for $a \neq 0$. This shows that our assumption is wrong, and so $a = 0$.

Schur proved that if f is a non constant polynomial with integer coefficients, then the set of primes dividing at least one of the numbers $f(1), f(2), \ldots$ is infinite. The following problem is an extension of this result.

Example 6. Suppose that f is a polynomial with integer coefficients and that (a_n) is a strictly increasing sequence of positive integers such that $a_n \leq f(n)$ for all n. Then the set of prime numbers dividing at least one term of the sequence (a_n) is infinite.

Solution. The idea is very nice: for any finite set of prime numbers $p_1, \ldots, p_r$ and any $k > 0$, we have

$$\sum_{\alpha_1, \alpha_2, \ldots, \alpha_N \geq 0} \frac{1}{p_1^{k\alpha_1} \ldots p_N^{k\alpha_N}} < \infty.$$

Indeed, it suffices to observe that we actually have

$$\sum_{\alpha_1, \alpha_2, \ldots, \alpha_N \geq 0} \frac{1}{p_1^{k\alpha_1} \ldots p_N^{k\alpha_N}} = \prod_{j=1}^{N} \sum_{i \geq 0} \frac{1}{p_j^{ki}} = \prod_{j=1}^{n} \frac{p_j^k}{p_j^k - 1}.$$

On the other hand, by taking $k = \dfrac{1}{2\deg(f)}$ we have

$$\sum_{n \geq 1} \frac{1}{(f(n))^k} = \infty.$$

Thus, if the conclusion of the problem is not true, we can find $p_1, \ldots, p_r$ such that any term of the sequence is of the form $p_1^{k\alpha_1} \ldots p_N^{k\alpha_N}$ and thus

$$\sum_{n \geq 1} \frac{1}{a_n^k} \leq \sum_{\alpha_1, \alpha_2, \ldots, \alpha_N \geq 0} \frac{1}{p_1^{k\alpha_1} \ldots p_N^{k\alpha_N}} < \infty.$$

On the other hand,

$$\sum_{n \geq 1} \frac{1}{a_n^k} \geq \sum_{n \geq 1} \frac{1}{(f(n))^k} = \infty,$$

a contradiction.

The same idea is used in the following problem.

Example 7. Let a and b be integers greater than 1. Prove that there is a multiple of a which contains all digits $0, 1, \ldots, b-1$ when written in base b.

Adapted after a Putnam Competition problem

Solution. Let us suppose the contrary. Then any multiple of a misses at least one digit when written in base b. Since the sum of inverses of all multiples of a diverges (because $1 + \frac{1}{2} + \frac{1}{3} + \cdots = \infty$), it suffices to show that the sum of inverses of all positive integers missing at least one digit in base b is convergent, and we will reach a contradiction. But of course, it suffices to prove it for a fixed (but arbitrary) digit j. For any $n \geq 1$, there are at most $(b-1)^n$ numbers which have n digits in base b, all different from j. Thus, since each one of them is at least equal to b^{n-1}, the sum of inverses of numbers that miss the digit j when written in base b is at most equal to $\sum_{n\geq 1} b\left(\frac{b-1}{b}\right)^n$, which converges. The conclusion follows.

The following example generalizes an old Kvant problem.

Example 8. Find all polynomials f with real coefficients such that if n is a positive integer which is written in base 10 only with ones, then $f(n)$ has the same property.

[Titu Andreescu, Gabriel Dospinescu] Putnam 2007

Solution. Let f be such a polynomial and observe that from the hypothesis it follows that there exists a sequence $(a_n)_{n\geq 1}$ of positive integers such that $f\left(\frac{10^n-1}{9}\right) = \frac{10^{a_n}-1}{9}$. But this sequence $(a_n)_{n\geq 1}$ cannot be really arbitrary: actually we can find precious information from an asymptotic study. Indeed, suppose that $\deg(f) = d \geq 1$. Then there exists a nonzero number A such that $f(x) \approx Ax^d$ for large values of x. Therefore $f\left(\frac{10^n-1}{9}\right) \approx \frac{A}{9^d} \cdot 10^{nd}$. Thus $10^{a_n} \approx \frac{A}{9^{d-1}} \cdot 10^{nd}$. This shows that the sequence $(a_n - nd)_{n\geq 1}$ converges to a limit l such that $A = 9^{d-1} \cdot 10^l$. Because this sequence consists of integers, it becomes eventually equal to the constant sequence l. Thus from a certain point we have $f\left(\frac{10^n-1}{9}\right) = \frac{10^{nd+l}-1}{9}$. If $x_n = \frac{10^n-1}{9}$, we deduce that the equation $f(x) = \frac{(9x+1)^d \cdot 10^l - 1}{9}$ has infinitely many solutions, so $f(X) = \frac{(9X+1)^d \cdot 10^l - 1}{9}$. Thus this is the general term of these polynomials (not including here obvious constant solutions), being clear that all such polynomials satisfy the conditions of the problem.

We return to classical mathematics and discuss a beautiful problem that appeared in the Tournament of the Towns in 1982, in a Russian Team Selection Test in 1997, and also in the Bulgarian Olympiad in 2003. Its beauty explains why the problem was so popular among the exam writers.

Example 9. Let f be a monic polynomial with integer coefficients such that for any positive integer n the equation $f(x) = 2^n$ has at least one positive integer solution. Prove that $\deg(f) = 1$.

Solution. The problem states that there exists a sequence of positive integers $(x_n)_{n\geq 1}$ such that $f(x_n) = 2^n$. Let us suppose that $\deg(f) = k > 1$. Then, for large values of x, $f(x)$ behaves like x^k. So, trying to find the right convergent sequence, we could try first to "think big": we have $x_n^k \cong 2^n$, that is for large n, x_n behaves like $2^{\frac{n}{k}}$. Then, a good possible convergent sequence could be $x_{n+k} - 2x_n$. Now, the hard part: proving that this sequence is indeed convergent. First, we will show that $\dfrac{x_{n+k}}{x_n}$ converges to 2. This is easy, since the relation $f(x_{n+k}) = 2^k f(x_n)$ implies

$$\frac{f(x_{n+k})}{x_{n+k}^k}\left(\frac{x_{n+k}}{x_n}\right)^k = 2^k \cdot \frac{f(x_n)}{x_n^k}$$

and since $\lim\limits_{x\to\infty} \frac{f(x)}{x^k} = 1$ and $\lim\limits_{n\to\infty} x_n = \infty$, we find that indeed $\lim\limits_{n\to\infty} \frac{x_{n+k}}{x_n} = 2$. We see that this will help us a lot. Indeed, write $f(x) = x^k + \sum_{i=0}^{k-1} a_i x^i$.

Then $f(x_{n+k}) = 2^k f(x_n)$ can be also written as

$$x_{n+k} - 2x_n = \frac{\sum_{i=0}^{k-1} a_i (2^k x_n^i - x_{n+k}^i)}{\sum_{i=0}^{k-1} (2x_n)^i x_{n+k}^{k-i-1}}$$

But from the fact that $\lim_{n\to\infty} \frac{x_{n+k}}{x_n} = 2$, it follows that the right-hand side of the above relation is also convergent. Hence $(x_{n+k} - 2x_n)_{n\geq 1}$ converges and so there exist M, N such that for all $n \geq M$ we have $x_{n+k} = 2x_n + N$. But now the solution is almost over, since the last result combined with $f(x_{n+k}) = 2^k f(x_n)$ yields $f(2x_n + N) = 2^k f(x_n)$ for $n \geq M$, that is $f(2x + N) = 2^k f(x)$. So, an arithmetical property of the polynomial turned into an algebraic one by using analysis. This algebraic property helps us to finish the solution. Indeed, we see that if z is a complex zero of f, then $2z + N, 4z + 3N, 8z + 7N, \ldots$ are all zeros of f. Since f is nonzero, this sequence must be finite and this can happen only for $z = -N$. Because $-N$ is the only zero of f, we deduce that $f(x) = (x + N)^k$. But since the equation $f(x) = 2^{2k+1}$ has positive integer roots, we find that $2^{\frac{1}{k}} \in \mathbb{Z}$, which implies $k = 1$, a contradiction. Thus, our assumption was wrong and $\deg(f) = 1$.

The following problem generalizes the problem above.

Example 10. Find all complex polynomials f with the following property: there exists an integer a greater than 1 such that for all sufficiently large positive integer n, the equation $f(x) = a^{n^2}$ has at least one solution in the set of positive integers.

[Gabriel Dospinescu] Mathlinks Contest

Solution. From the beginning we exclude the constant polynomials, so let f be a solution of degree $d \geq 1$. Let $(x_n)_{n\geq n_0}$ be a sequence of positive integers such that $f(x_n) = a^{n^2}$ for some integer a greater than 1. Now, observe that we can choose A such that the polynomial $g(X) = f(x + A)$ has no term of degree $d - 1$. Define $y_n = x_n - A$ and observe that $g(y_n) = a^{n^2}$. Now, what really interests us is the asymptotic behavior of the sequence y_n. This boils down to finding the behavior of the solution of the equation $g(y) = z$ when z is very large. In order to do this, put $g(y) = By^d + Cy^e + \cdots$ with $B > 0$ (the fact that $B > 0$ is obvious because $g(x)$ remains positive for arbitrarily large values of x). Now, suppose that $C \neq 0$. The choice of A ensures that $e \leq d - 2$. Therefore, if we define $z = u^d$ and $By^d = v^d$, $E = \frac{C}{B^{\frac{e}{d}}}$ and finally $m = d - e$, then we have $u^d = v^d(1 + Ev^{-m} + o(v^{-m}))$. Thus

$$u = v(1 + Ev^{-m} + o(v^{-m}))^{\frac{1}{d}} = v\left(1 + \frac{E}{d}v^{-m} + o(v^{-m})\right)$$
$$= v + \frac{E}{d}v^{1-m} + o(v^{1-m}).$$

This shows that $u \approx v$, and combining this observation with the previous result gives $v = u - \frac{E}{d}u^{1-m} + o(u^{1-m})$. Coming back to our notations, we infer that $B^{\frac{1}{d}}y = z^{\frac{1}{d}} - \frac{E}{d}z^{-\frac{p}{d}} + o(z^{-\frac{p}{d}})$ where $p = m - 1$. Finally, this can be written in the form $y = Fz^{\frac{1}{d}} + Gz^{-\alpha} + o(z^{-\alpha})$ (the definitions of F, G and α are obvious from the last formula). Coming back to the relation $g(y_n) = a^{n^2}$ we deduce that $y_n = Fa^{\frac{n^2}{d}} + Ga^{-\alpha n^2} + o(a^{-\alpha n^2})$. Therefore

$$y_{d+n} = Fa^{\frac{n^2}{d}}a^{2n+d} + o(a^{2n+d-\alpha n^2}).$$

This shows that if we define $z_n = y_{n+d} - a^{2n+d}y_n$ then $z_n = o(1)$. On the other hand, by definition of y_n we obtain that $\alpha_{n+1} = z_n + A(1 - a^{2n+2+d})$ is an integer. Therefore, the relation

$$z_{n+1} - a^2 z_n = \alpha_{n+1} - a^2\alpha_n + A(a^2 - 1)$$

and the fact that $z_n = o(1)$ shows that $\alpha_{n+1} - a^2\alpha_n$ is eventually constant, equal to $A(1 - a^2)$. Thus for sufficiently large n we have $z_{n+1} = a^2 z_n$, so we have proved the existence of a constant K such that $z_n = Ka^{2n}$ for sufficiently large n. Because $a > 1$ and $z_n = o(1)$, it follows that $K = 0$ and thus $z_n = 0$ for sufficiently large n. But the assumption $C \neq 0$ implies that $G \neq 0$ and by one of the previous relations we also have $z_n \approx -Ga^{2n+d-\alpha n^2}$, which is not true if $z_n = 0$ from a certain point on. This contradiction shows that $f(X) = B(X-A)^d$ for some rational numbers A, B (because f takes rational values for infinitely many rational values of the variable, it is equal to its Lagrange interpolation polynomial, thus it has integer coefficients). Let $B = \frac{p}{q}$ and $A = \frac{r}{s}$. Then $p(sx_n - r)^d = qs^d a^{n^2}$. By taking n a multiple of d greater than n_0 we obtain the existence of integers p_1, q_1 such that $p = p_1^d, q = q_1^d$. Thus $p_1(sx_n - r) = \pm q_1 s a^{\frac{n^2}{d}}$, which shows that a^{n^2} is a d-th power for all sufficiently large n. This implies the existence of an integer b such that $a = b^d$. Now, by taking $p_1, q_1, s > 0$ (we can do that, without loss of generality), we deduce that for some n_1 (which we will identify with n_0 from now on, by eventually enlarging n_0) we have $sx_n = r + \frac{q_1 s b^{n^2}}{p_1}$. Let $\alpha = \gcd(s, p_1)$ and write $s = \alpha u, p_1 = \alpha v$ with $\gcd(u, v) = 1$. Then $\alpha u x_n = r + \frac{q_1 u b^{n^2}}{v}$ thus $v | q_1 b^{n^2}$ and so for all $n \geq n_0$, $\alpha u x_n = r + \frac{q_1 b^{n_0^2}}{v} u b^{n^2 - n_0^2}$. By taking $n = n_0$ we deduce that $u | r$. Because $u | s$, it follows that $u = 1$ and so $sx_n = r + \frac{q_1 b^{n^2}}{v}$. Note that $\gcd(v, q_1) = 1$ because $v | p_1$, so $v | b^{n_0^2}$. Let $b^{n_0^2} = mv$. Thus $sx_n = r + mq_1 b^{n^2 - n_0^2}$. By taking again $n = n_0$, we obtain that $mq_1 \equiv -r \pmod{s}$, so $r(1 - b^{n^2 - n_0^2}) \equiv 0 \pmod{s}$ and so $b^{n^2 - n_0^2} \equiv 1 \pmod{s}$ for all $n > n_0$. Applying this relation to $n+1$ and making the division in the group of invertible residues mod s, we infer that $b^{2n+1} \equiv 1 \pmod{s}$ for all sufficiently large n. Repeating this procedure, we deduce that $b^2 \equiv 1 \pmod{s}$ and so $b \equiv 1 \pmod{s}$. This implies $mv = b^{n_0^2} \equiv 1 \pmod{s}$ and since $r \equiv -mq_1 \pmod{s}$ and $\gcd(s, v) = 1$, we finally obtain the necessary condition $rv \equiv -q_1 \pmod{s}$. Now, let us show that the conditions $\gcd(p_1, q_1) = \gcd(r, s) = 1$ and $p_1 = sv, \gcd(s, v) = 1, rv \equiv -q_1 \pmod{s}$ are sufficient for the polynomial $f(X) = \left(\frac{p_1}{q_1}(X - \frac{r}{s})\right)^d$ to be a solution of the problem. Indeed, using the Chinese Remainder Theorem, we can choose b such that $b \equiv 0 \pmod{v}$ and $b \equiv 1 \pmod{s}$. Thus $v | rv + q_1 b^{n^2}$ and also $s | rv + q_1 b^{n^2}$. Because $\gcd(s, v) = 1$ it follows that there exists a sequence x_n of positive integers such that $rv + q_1 b^{n^2} = svx_n$. Thus $f(x_n) = b^{dn^2}$ and the problem is finally solved.

The idea behind the following problem is so beautiful that any reader who attempts to solve it will feel generously rewarded by discovering this mathematical gem either by herself or himself, or in the solution provided.

Example 11. Let $\pi(n)$ be the number of prime numbers not exceeding n. Prove that there exist infinitely many n such that $\pi(n)|n$.

[S. Golomb] AMM

Solution. Let us prove the following result, which is the key to the problem.

Lemma 17.1. *For any increasing sequence of positive integers* $(a_n)_{n\geq 1}$ *such that* $\lim\limits_{n\to\infty} \frac{a_n}{n} = 0$, *the sequence* $\left(\frac{n}{a_n}\right)_{n\geq 1}$ *contains all positive integers. In particular,* a_n *divides* n *for infinitely many* n.

Proof. Even if it seems unbelievable, this is true. Moreover, the proof is extremely short. Let m be a positive integer. Consider the set

$$A = \left\{n \geq 1 | \ \frac{a_{mn}}{mn} \geq \frac{1}{m}\right\}.$$

This set contains 1 and it is bounded, since $\lim\limits_{n\to\infty} \frac{a_{mn}}{mn} = 0$. Thus it has a maximal element k. If $\frac{a_{mk}}{mk} = \frac{1}{m}$, then m is in the sequence $\left(\frac{n}{a_n}\right)_{n\geq 1}$. Otherwise, we have $a_{m(k+1)} \geq a_{mk} \geq k+1$, which shows that $k+1$ is also in the set, in contradiction with the maximality of k. The lemma is proved. $\square$

Thus, all we need to show now is that $\lim\limits_{n\to\infty} \frac{\pi(n)}{n} = 0$. Fortunately, this is well known and not difficult to prove. There are easier proofs than the following one, but we prefer to deduce it from a famous and beautiful result of Erdős: $\prod\limits_{p\leq n} p < 4^{n-1}$. This was proved in chapter **Look at the Exponent**, but really we expect you to know how to prove it (it is one of those marvelous proofs that cannot be forgotten). Now, the fact that $\lim_{n\to\infty} \frac{\pi(n)}{n} = 0$ follows easily. Indeed, fix $k \geq 1$. We have for all large n the inequality

$$(n-1)\log 4 \geq \sum_{k\leq p\leq n} \log p \geq (\pi(n) - \pi(k))\log k,$$

which shows that

$$\pi(n) \leq \pi(k) + \frac{(n-1)\log 4}{\log k}.$$

This proves that $\lim_{n\to\infty} \frac{\pi(n)}{n} = 0$. The problem is finally solved.

A somewhat tricky, but less technical problem follows now. A special case of it was proposed by the USA for IMO 1990:

Example 12. Let f be a polynomial with rational coefficients, of degree at least 2, and let $(a_n)_{n\geq 1}$ be a sequence of rational numbers such that $f(a_{n+1}) = a_n$ for all n. Prove that this sequence is periodic.

[Bjorn Poonen] AMM 10369

Solution. First of all, it is clear that the sequence is bounded. Indeed, because $\deg(f) \geq 2$ there exists M such that $|f(x)| \geq |x|$ if $|x| \geq M$. By taking M sufficiently large one can also assume that $M > |a_1|$. Then an immediate induction shows that $|a_n| \leq M$ for all n. We will now prove that for some positive integer N we have $Na_n \in \mathbb{Z}$ for all n. Indeed, let $a_1 = \frac{p}{q}$ for some integers p, q and let k be a positive integer such that $kf = f_s X^s + \cdots + f_1 X + f_0 \in \mathbb{Z}[X]$. Define $N = qf_s$. Then $Na_1 = pf_s \in \mathbb{Z}$, and clearly if $Na_n \in \mathbb{Z}$ then Na_{n+1} is a rational zero of the monic polynomial with integer coefficients $\frac{kN^s}{f_s}\left(f\left(\frac{X}{N}\right) - a_n\right)$, so it is an integer. This shows that $(Na_n)_{n\geq 1}$ is a bounded sequence of integers, therefore it takes only a finite number of values. Suppose that the sequence $(a_n)_{n\geq 1}$ takes at most m different values. Consider $(m+1)$-tuples

$(a_i, a_{i+1}, ..., a_{i+m})$ for positive integers i. There are at most m^{m+1} such $(m+1)$-tuples that can be formed and in each such $(m+1)$-tuple there exists a value taken at least twice. Therefore there exists a pattern that is repeated infinitely many times, which means that there exists k such that for all positive integers n there exists $j > n$ for which $a_j = a_{j+k}$. But applying f^r to this last relation and taking into account that $f^r(a_{n+r}) = a_n$ shows that $a_n = a_{n+k}$ for all n. That is, the sequence is periodical.

A fine concoction of number theory and analysis is used in the solution of the next (very) difficult problem. We will see one of the thousands of unexpected applications of Pell's equation:

Example 13. Find all polynomials p and q with integer coefficients such that $p(X)^2 = (X^2 + 6X + 10)q(X)^2 - 1$.

Vietnamese TST 2002

Solution. One easy step is to notice that $X^2 + 6X + 10 = (X + 3)^2 + 1$, so by taking $f(X) = p(X - 3)$ and $g(X) = q(X - 3)$ the problem "reduces" to solving the equation $(X^2 + 1)f(X)^2 = g(X)^2 + 1$ in polynomials with integer coefficients. Of course, we may assume that the leading coefficients of f and g are positive and also that both polynomials are nonconstant. Therefore there exists an M such that $f(n) > 2, g(n) > 2$ for all $n > M$. As it is well known, the solutions in positive integers to the Pell equation $x^2 + 1 = 2y^2$ are (x_n, y_n) where

$$x_n = \frac{(1+\sqrt{2})^{2n-1} + (1-\sqrt{2})^{2n-1}}{2}; \qquad y_n = \frac{(1+\sqrt{2})^{2n-1} - (1-\sqrt{2})^{2n-1}}{2}.$$

Observe that $g^2(x_n) + 1 = 2(y_n f(x_n))^2$. There exist two sequences of positive integers $(a_n)_{n>M}$ and $(b_n)_{n>M}$ such that $g(x_n) = x_{a_n}$ and $y_n f(x_n) = y_{b_n}$. Let $k = \deg(g)$ and $m = \deg(f)$. Because the sequence

$$2\frac{g(x_n)}{x_n^k} \cdot \left(\frac{x_n}{(1+\sqrt{2})^{2n-1}}\right)^k$$

clearly converges to a nonzero limit, so does the sequence $\frac{2x_{a_n}}{(1+\sqrt{2})^{k(2n-1)}}$ and therefore the sequence $(1+\sqrt{2})^{2a_n-1-k(2n-1)}$ converges to a nonzero limit. This sequence having integer terms, it becomes constant from a certain point. Hence there exists $n_0 > M$ and an integer u such that $2a_n - 1 - k(2n-1) = u$ for all $n > n_0$. Thus $g\left(\frac{x-\frac{1}{x}}{2}\right) = \frac{x^k(1+\sqrt{2})^u + (-\frac{1}{x})^k(1-\sqrt{2})^u}{2}$ holds for all x of the form $(1+\sqrt{2})^{2n-1}$. Because this equality between two rational functions holds for infinitely many values of the argument, it follows that it is actually true for all x. By looking at the leading coefficient in both sides of the equality (after multiplication by X^k) we deduce that $(1+\sqrt{2})^u$ is rational, which cannot hold unless $u = 0$. Thus

$$g(X) = \frac{(X+\sqrt{X^2+1})^k + (X-\sqrt{X^2+1})^k}{2}.$$

The expression in the right-hand side of the last equality is a polynomial with integer coefficients only for odd values of k. This also gives the expression of f:

$$f(X) = \frac{(X+\sqrt{X^2+1})^k + (-X+\sqrt{X^2+1})^k}{2\sqrt{X^2+1}}.$$

The solutions of the original problem are easily deduced from f and g by a translation.

The previous example deserves a little digression. Actually, one can find all polynomials with *real* coefficients that satisfy $(X^2+1)f(X)^2 = g(X)^2 + 1$. Indeed, it is clear that f and g are relatively prime. By differentiation, the last relation can be written as $(X^2+1)f(X)f'(X) + Xf^2(X) = g(X)g'(X)$. Thus f divides gg', and by Gauss's lemma we deduce that $f|g'$. The relation $(X^2+1)f^2(X) = g(X)^2 + 1$ also shows that $\deg(f) = \deg(g')$ and so there exists a constant k such that $f(X) = kg'(X)$. Therefore $k^2(X^2+1)g'(X)^2 = g(X)^2 + 1$. By identifying the leading coefficient of g in the two sides, we immediately find that $k^2 = n^2$. This shows that $\frac{g'(X)^2}{1+g(X)^2} = \frac{n^2}{1+X^2}$. By changing g and $-g$ we may assume that $g'(X) > 0$ for sufficiently large x and thus for such values of the variable we have $\frac{g'(x)}{\sqrt{g(x)^2+1}} = \frac{n}{\sqrt{x^2+1}}$. This shows that the function $\ln\left(\frac{g(x)+\sqrt{g(x)^2+1}}{(x+\sqrt{x^2+1})^n}\right)$ is constant in a neighborhood of infinity. This allows us to find g in such a neighborhood and thus to find g on the whole real line.

It is time now for the last problem, which is, as usual, very hard. We do not exaggerate when we say that the following problem is exceptionally difficult.

Example 14. Let a and b be integers greater than 1 such that $a^n - 1 | b^n - 1$ for every positive integer n. Prove that b is a natural power of a.

[Marius Cavachi] AMM

Solution. This time we will be able to find the right convergent sequence only after examining a few recursive sequences. Let us see. So, initially we are given that there exists a sequence of positive integers $(x_n^{(1)})_{n\geq 1}$ such that $x_n^{(1)} = \frac{b^n - 1}{a^n - 1}$. Then, $x_n^{(1)} \cong \left(\frac{b}{a}\right)^n$ for large values of n. So, we could expect that the sequence $(x_n^{(2)})_{n\geq 1}$, $x_n^{(2)} = bx_n^{(1)} - ax_{n+1}^{(1)}$, to be convergent. Unfortunately,

$$x_n^{(2)} = \frac{b^{n+1}(a-1) - a^{n+1}(b-1) + a - b}{(a^n - 1)(a^{n+1} - 1)},$$

which is not necessarily convergent. But... if we look again at this sequence, we see that for large values of n it grows like $\left(\frac{b}{a^2}\right)^n$, so much slower. And this is the good idea: repeat this procedure until the final sequence behaves like $\left(\frac{b}{a^{k+1}}\right)^n$, where k is chosen such that $a^k \leq b < a^{k+1}$. Thus the final sequence will converge to 0. Again, the hard part has just begun, since we have to prove that if we define $x_n^{(i+1)} = bx_n^{(i)} - a^i x_{n+1}^{(i)}$ then $\lim\limits_{n\to\infty} x_n^{(k+1)} = 0$. This is not easy at all. The idea is to compute $x_n^{(3)}$ and after that to prove the following statement: for any $i \geq 1$ the sequence $(x_n^{(i)})_{n\geq 1}$ has the form

$$\frac{c_i b^n + c_{i-1}a^{(i-1)n} + \cdots + c_1 a^n + c_0}{(a^{n+i-1}-1)(a^{n+i-2}-1)\dots(a^n-1)}$$

for some constants $c_0, c_1, \dots, c_i$. Proving this is not so hard, the hard part was to think of it. How can we prove the statement other than by induction? And induction turns out to be quite easy. Supposing that the statement is true for i, then the corresponding statement for $i+1$ follows from $x_n^{(i+1)} = bx_n^{(i)} - a^i x_{n+1}^{(i)}$ directly (note that in order to compute the difference, we just have to multiply the numerator $c_i b^n + c_{i-1}a^{(i-1)n} + \cdots + c_1 a^n + c_0$ by b and $a^{n+i}-1$. Then, we proceed in the same way with the second fraction and the term $b^{n+1}a^{n+i}$ will vanish). So, we have found a formula which shows that as soon as $a^i > b$ we have $\lim_{n\to\infty} x_n^{(i)} = 0$. So, $\lim_{n\to\infty} x_n^{(k+1)} = 0$. Another step of the solution is to take the minimal index j such that $\lim_{n\to\infty} x_n^{(j)} = 0$. Clearly, $j > 1$ and the recursive relation $x_n^{(i+1)} = bx_n^{(i)} - a^i x_{n+1}^{(i)}$ shows that $x_n^{(i)} \in \mathbb{Z}$ for all n and i. Thus, there exists an M such that whenever $n \geq M$ we have $x_n^{(j)} = 0$. This is the same as $bx_n^{(j-1)} = a^j x_{n+1}^{(j-1)}$ for all $n \geq M$, which implies $x_n^{(j-1)} = \left(\frac{b}{a^j}\right)^{n-M} x_M^{(j-1)}$ for all $n \geq M$. Let us suppose that b is not a multiple of a. Because $\left(\frac{b}{a^j}\right)^{n-M} x_M^{(j-1)} \in \mathbb{Z}$ for all $n \geq M$, we must have $x_M^{(j-1)} = 0$ and so $x_n^{(j-1)} = 0$ for $n \geq M$, which means $\lim_{n\to\infty} x_n^{(j-1)} = 0$. But this contradicts the minimality of j. Thus we must have $a|b$. Let us write $b = ca$. Then, the relation $a^n - 1|b^n - 1$ implies $a^n - 1|c^n - 1$. And now we are finally done. Why? We have just seen that $a^n - 1|c^n - 1$ for all $n \geq 1$. But our previous argument applied to c instead of b shows that $a|c$. Thus, $c = ad$ and we deduce again that $a|d$. Since this process cannot be infinite, b must be a power of a.

It is worth saying that an even stronger result holds: it is enough to suppose that $a^n - 1|b^n - 1$ for infinitely many n. But this is a much more difficult problem and it follows from a 2003 result of Bugeaud, Corvaja and Zannier:

If integers $a, b > 1$ are multiplicatively independent in $\mathbb{Q}^*$ (that is $\log_a b \notin \mathbb{Q}$ or $a^n \neq b^m$ for $n, m \neq 0$), then for any $\varepsilon > 0$ there exists $n_0 = n_0(a, b, \varepsilon)$ such that $\gcd(a^n - 1, b^n - 1) < 2^{\varepsilon n}$ for all $n \geq n_0$. Unfortunately, the proof is too advanced to be presented here.

17.2 Practice problems

1. Let $(a_n)_{n\geq 1}$ be an increasing sequence of positive integers such that $a_n \mid a_1+a_2+\cdots+a_{n-1}$ for all $n \geq 2002$. Prove that there exists n_0 such that $a_n = a_1 + a_2 + \cdots + a_{n-1}$ for all $n \geq n_0$.

   Tournament of the Towns 2002

2. Let p be a polynomial with integer coefficients such that there exists a sequence of pairwise distinct positive integers $(a_n)_{n\geq 1}$ such that $p(a_1) = 0$, $p(a_2) = a_1$, $p(a_3) = a_2, \ldots$. Find the degree of this polynomial.

   Tournament of the Towns 2003

3. Find all pairs (a, b) of positive integers such that $an + b$ is triangular if and only if n is triangular.

   After a Putnam Competition problem

4. Let f and g be two real polynomials of degree 2 such that for any real number x, if $f(x)$ is integer, then so is $g(x)$. Prove that there are integers m, n such that $g(x) = mf(x) + n$ for all x.

   Bulgarian Olympiad

5. Let A, B be two finite sets of positive real numbers such that

$$\left\{\sum_{x\in A} x^n \mid n \in \mathbb{N}\right\} \subseteq \left\{\sum_{x\in B} x^n \mid n \in \mathbb{N}\right\}.$$

   Prove that there exists $k \in \mathbb{Z}$ such that $A = \{x^k \mid x \in B\}$.

   Gabriel Dospinescu

6. Let a and b be integers greater than 1. Prove that for any $k > 0$ there are infinitely many numbers n such that $\varphi(an + b) < kn$, where φ is Euler's totient function.

Gabriel Dospinescu

7. Prove that $\varphi(30n + 1) < \varphi(30n)$ for infinitely many positive integers n.

D.J.Newman

8. Prove that there are infinitely many positive integers n for which the equation

$$n = a^3 + b^5 + c^7 + d^9 + e^{11}$$

has no solutions in positive integers.

Belarus 2000

9. Prove that in any increasing sequence $(a_n)_{n \geq 1}$ of positive integers satisfying $a_n < 100n$ for all n, one can find infinitely many terms containing at least 1986 consecutive 1's.

Kvant

10. Prove that for each natural number m, there is a natural number N such that $s_b(N) > m$ for each $2 \leq b \leq 2010$. Here $s_b(n)$ is the sum of digits of n when written in base b.

Iranian TST 2010

11. Prove that there exists a positive integer n which is a 3-digit palindrome in base b for at least 2002 values of b.

Putnam 2002

12. Let a and b be positive integers such that for any n, the decimal representation of $a + bn$ contains a sequence of consecutive digits which form the decimal representation of n (for example, if $a = 600$, $b = 35$, $n = 16$ we have $600 + 16 \cdot 35 = 1160$). Prove that b is a power of 10.

Tournament of the Towns 2002

13. Suppose that a is a positive real number such that all numbers $1^a, 2^a, 3^a, \ldots$ are integers. Prove that a is also integer.

Putnam Competition

14. Let f be a complex polynomial such that for all positive integers n, the equation $f(x) = n$ has at least one rational solution. Prove that f has degree 1.

Mathlinks Contest

15. Let b be an integer greater than 4 and define the number

$$x_n = \underbrace{11\ldots1}_{n-1}\underbrace{22\ldots2}_{n}5$$

in base b. Prove that x_n is a perfect square for all sufficiently large n if and only if $b = 10$.

Laurenţiu Panaitopol, IMO Shortlist 2003

16. Find all monic polynomials with integer coefficients f such that $f(\mathbb{Z})$ is closed under multiplication.

Iranian TST 2007, Mohsen Jamali

17. Find all rational numbers $x = \frac{p}{q} > 1$ (p, q being integers such that $q > 0$ and $\gcd(p, q) = 1$) with the property: there exists a constant c such that for all sufficiently large n we have

$$|\{x^n\} - c| \leq \frac{1}{2(p+q)}.$$

Chinese TST 2007

18. The set of exponents of a positive integer is the unordered list of all the exponents of the primes appearing in its prime factorization. For example, the set of exponents of $300 = 2^2 \cdot 3 \cdot 5^2$ is $1, 2, 2$. Suppose that two arithmetical progressions $\left(a_n\right)_n$ and $\left(b_n\right)_n$ have the property that for any positive integer n, a_n and b_n have the same set of exponents. Prove that there is k such that $a_n = kb_n$ for all n.

Tuymaada Olympiad 2006

19. Find all triples (a, b, c) of integers such that $a \cdot 2^n + b$ is a divisor of $c^n + 1$ for any positive integer n.

Gabriel Dospinescu, Mathematical Reflections

20. (a) Find all increasing functions $f : \{1, 2, \ldots\} \to \mathbb{R}$ such that $f(ab) = f(a)f(b)$ for all positive integers a and b.

    (b) Find all increasing functions $f : \{1, 2, \ldots\} \to \mathbb{R}$ such that $f(ab) = f(a)f(b)$ for all relatively prime positive integers a and b.

Paul Erdős

21. Find all a, b, c such that $a \cdot 4^n + b \cdot 6^n + c \cdot 9^n$ is a perfect square for all sufficiently large n.

Vesselin Dimitrov

22. Let f be a polynomial with rational coefficients such that $f(2^n)$ is a perfect square for all positive integers n. Prove that there exists a polynomial g with rational coefficients such that $f = g^2$.

Gabriel Dospinescu

23. Let k be an integer greater than 1 and let f be a polynomial such that $\sqrt[k]{f(n)}$ is an integer for all sufficiently large n. Prove that there exists $g \in \mathbb{Q}[X]$ such that $f = g^k$.

24. Let f, g be two polynomials with real coefficients such that $f(\mathbb{Q}) = g(\mathbb{Q})$. Prove that there exist rational numbers a, b such that $f(X) = g(aX+b)$.

Miklos Schweitzer Competition

25. Given a polynomial $f(x)$ with rational coefficients, of degree at least 2, define the sets $f^0(\mathbb{Q}) = \mathbb{Q}$ and

$$f^n(\mathbb{Q}) = \{f(x) | x \in f^{n-1}(\mathbb{Q})\}.$$

Prove that $\bigcap_{n=0}^{\infty} f^n(\mathbb{Q})$ is a finite set.

Dan Schwarz, Romanian Masters in Mathematics 2010

26. Let A be a set of positive integers and let $b_1 < b_2 < \cdots$ be the positive integers which can be written as difference of two elements of A. If the sequence $b_{n+1} - b_n$ is unbounded, proved that A has density zero, that is

$$\lim_{n \to \infty} \frac{1}{n} \cdot |A \cap \{1, 2, \ldots, n\}| = 0.$$

Putnam 2004

27. Let A be a non-empty set of positive integers with the following property: there are positive integers $b_1, b_2, \ldots b_n$ and $c_1, c_2, \ldots, c_n$ such that the sets $b_iA + c_i = \{b_ia + c_i \colon a \in A\}$ are pairwise disjoint subsets of A.

Prove that
$$\frac{1}{b_1} + \frac{1}{b_2} + \ldots + \frac{1}{b_n} \leq 1.$$

IMO Shortlist 2002

28. Let $a_1, a_2, \ldots, a_n$ and $b_1, b_2, \ldots, b_n$ be positive integers such that any integer x satisfies at least one congruence $x \equiv a_i \pmod{b_i}$. Prove that there exists a nonempty subset I of $\{1, 2, \ldots, n\}$ such that $\sum_{i \in I} \frac{1}{b_i}$ is an integer.

M. Zhang

Chapter 18

18.1 Theory and examples

For a prime p, define the function $\left(\frac{a}{p}\right) : \mathbb{Z} \to \{-1, 1\}$ by $\left(\frac{a}{p}\right) = 1$ if the equation $x^2 = a$ has at least one solution in $\mathbb{Z}/p\mathbb{Z}$ and $\left(\frac{a}{p}\right) = -1$ otherwise. In the first case, we say that a is a quadratic residue modulo p; otherwise we say that it is a quadratic non-residue modulo p. This function is called Legendre's symbol and plays a fundamental role in number theory. We will unfold some easy properties of Legendre's symbol first, in order to prove a highly nontrivial result, Gauss's famous quadratic reciprocity law. First, let us present a useful theoretical (but not very practical) way of computing $\left(\frac{a}{p}\right)$ due to Euler.

Theorem 18.1. *The following identity is true provided $p \nmid a$:*

$$\left(\frac{a}{p}\right) \equiv a^{\frac{p-1}{2}} \pmod{p}.$$

In particular, we have $\left(\frac{-1}{p}\right) = (-1)^{\frac{p-1}{2}}$.

Proof. We will prove this result and many other simple facts concerning quadratic residues in what follows. First, let us assume that $\left(\frac{a}{p}\right) = 1$, and let x be a solution to the equation $x^2 = a$ in $\mathbb{Z}/p\mathbb{Z}$. Using Fermat's little theorem, we find that $a^{\frac{p-1}{2}} = x^{p-1} = 1 \pmod{p}$. Thus the equality $\left(\frac{a}{p}\right) = a^{\frac{p-1}{2}} \pmod{p}$ holds for all quadratic residues a modulo p. In addition, for any quadratic residue we have $a^{\frac{p-1}{2}} = 1 \pmod{p}$. Now, we will prove that there are exactly $\frac{p-1}{2}$ quadratic residues in $\mathbb{Z}/p\mathbb{Z} \setminus \{0\}$. This will enable us to conclude that quadratic residues are precisely the zeros of the polynomial $X^{\frac{p-1}{2}} - 1$ and also that non quadratic residues are exactly the zeros of the polynomial $X^{\frac{p-1}{2}} + 1$ (from Fermat's little theorem). Note that Fermat's little theorem implies that the polynomial $X^{p-1} - 1 = (X^{\frac{p-1}{2}} - 1)(X^{\frac{p-1}{2}} + 1)$ has exactly $p - 1$ zeros in the field $\mathbb{Z}/p\mathbb{Z}$. But in a field, the number of different zeros of a polynomial cannot exceed its degree. Thus each of the polynomials $X^{\frac{p-1}{2}} - 1$ and $X^{\frac{p-1}{2}} + 1$ has at most $\dfrac{p-1}{2}$ zeros in $\mathbb{Z}/p\mathbb{Z}$. These two observations show that in fact each of these polynomials has exactly $\dfrac{p-1}{2}$ zeros in $\mathbb{Z}/p\mathbb{Z}$. Let us observe next that there are at least $\dfrac{p-1}{2}$ quadratic residues modulo p. Indeed, all numbers $i^2 \pmod p$ with $1 \le i \le \dfrac{p-1}{2}$ are quadratic residues and they are all different (modulo p). This shows that there are exactly $\dfrac{p-1}{2}$ quadratic residues in $\mathbb{Z}/p\mathbb{Z}\setminus\{0\}$ and also proves Euler's criterion. □

Euler's criterion is a very useful result. Indeed, it allows a very quick proof of the fact that $\left(\dfrac{a}{p}\right) : \mathbb{Z} \to \{-1, 1\}$ is a group morphism. Indeed,

$$\left(\frac{ab}{p}\right) \equiv (ab)^{\frac{p-1}{2}} \equiv a^{\frac{p-1}{2}} b^{\frac{p-1}{2}} \equiv \left(\frac{a}{p}\right)\left(\frac{b}{p}\right) \pmod p.$$

The relation $\left(\dfrac{ab}{p}\right) = \left(\dfrac{a}{p}\right)\left(\dfrac{b}{p}\right)$ shows that while studying Legendre's symbol, it suffices to focus on the prime numbers only. Also, the same Euler's criterion implies that $\left(\dfrac{a}{p}\right) = \left(\dfrac{b}{p}\right)$ whenever $a \equiv b \pmod p$.

It is now time to discuss Gauss's celebrated quadratic reciprocity law. First of all, we will prove a lemma (also due to Gauss).

Lemma 18.2. *Let p be an odd prime and let $a \in \mathbb{Z}$ such that $gcd(a,p) = 1$. Define the least residue of $a \pmod n$ as the integer a' such that $a \equiv a' \pmod n$ and $-\frac{n}{2} < a' \le \frac{n}{2}$. Let a_j be the least residue of $aj \pmod p$ and l be the number of integers $1 \le j \le \frac{p-1}{2}$ for which $a_j < 0$. Then $\left(\frac{a}{p}\right) = (-1)^l$.*

Proof. The proof is not difficult at all. Observe that the numbers $|a_j|$ for $1 \leq j \leq r = \frac{p-1}{2}$ are a permutation of the numbers $1, 2, ..., r$. Indeed, we have $1 \leq |a_j| \leq r$ and $|a_j| \neq |a_k|$ (otherwise, we have either $p|a(j+k)$ or $p|a(j-k)$ which is impossible because $\gcd(a,p) = 1$ and $0 < j+k < p$). Therefore $a_1a_2\cdots a_r = (-1)^l|a_1||a_2|\cdots|a_r| = (-1)^l r!$. By the definition of a_j we also have $a_1a_2\cdots a_r \equiv a^r r! \pmod{p}$ and so $a^r \equiv (-1)^l \pmod{p}$. Using Euler's criterion, we deduce that $\left(\frac{a}{p}\right) = (-1)^l$. □

Using Gauss's lemma, the reader will enjoy the proof of the following classical results.

Theorem 18.3. *The identity* $\left(\frac{2}{p}\right) = (-1)^{\frac{p^2-1}{8}}$ *holds for any odd prime* p.

Proof. Let us take $a = 2$ in Gauss's lemma and observe that $l = \frac{p-1}{2} - \left\lfloor \frac{p}{4} \right\rfloor$. Indeed, we have $a_j = 2j$ if $1 \leq j \leq \left\lfloor \frac{p}{4} \right\rfloor$ and $a_j = 2j - p$ if $\left\lfloor \frac{p}{4} \right\rfloor < j \leq \frac{p-1}{2}$. Now, the conclusion follows, because $l = \frac{p-1}{2} - \left\lfloor \frac{p}{4} \right\rfloor$and $\frac{p^2-1}{8}$ have the same parity, as you can easily check.

□

But perhaps the most striking consequence of Gauss's lemma is the famous:

Theorem 18.4 (Quadratic reciprocity law). *For all distinct odd primes* p *and* q*, the following identity holds:*

$$\left(\frac{p}{q}\right)\left(\frac{q}{p}\right) = (-1)^{\frac{p-1}{2}\cdot\frac{q-1}{2}}.$$

Proof. The proof is a little bit more involved than that of the previous result. Consider R the rectangle defined by $0 < x < \frac{q}{2}$ and $0 < y < \frac{p}{2}$, and let $\left(\frac{p}{q}\right) = (-1)^l$ and $\left(\frac{q}{p}\right) = (-1)^m$, where l, m are defined as in Gauss's lemma. Observe that l is the number of lattice points (x, y) such that $0 < x < \frac{q}{2}$ and $-\frac{q}{2} < px - qy < 0$. These inequalities force $y < \frac{p+1}{2}$ and because y is an integer, it follows that $y < \frac{p}{2}$. Therefore l is the number of lattice points in R that satisfy $-\frac{q}{2} < px - qy < 0$ and similarly m is the number of lattice points in R that satisfy $-\frac{p}{2} < qy - px < 0$. Using Gauss's lemma, it is enough to prove that $\frac{(p-1)(q-1)}{4} - (l+m)$ is even. Because $\frac{(p-1)(q-1)}{4}$ is the number of lattice points in R, $\frac{(p-1)(q-1)}{4} - (l+m)$ is the number of lattice points in R that satisfy $px - qy \leq -\frac{q}{2}$ or $qy - px \leq -\frac{p}{2}$. These points determine two regions in R, which are clearly disjoint. Moreover, they have the same number of lattice points because $x = \frac{q+1}{2} - x', y = \frac{p+1}{2} - y'$ gives a one-to-one correspondence between the lattice points in the two regions. This shows that $\frac{(p-1)(q-1)}{4} - (l+m)$ is even and finishes the proof of this celebrated theorem. □

Using this powerful arsenal, we are now able to solve some interesting problems. Most of them are merely direct applications of the above results, but we think that they are still worthy, not necessarily because they appeared in various contests.

Example 1. Prove that the number $2^n + 1$ does not have prime divisors of the form $8k - 1$.

Vietnamese TST 2004

Solution. For the sake of contradiction, assume that p is a prime of the form $8k - 1$ that divides $2^n + 1$. Of course, if n is even, the contradiction is immediate, since in this case we have $-1 \equiv (2^{\frac{n}{2}})^2 \pmod{p}$ and so $-1 = (-1)^{\frac{p-1}{2}} = \left(\frac{-1}{p}\right) = 1$. Now, assume that n is odd. Then $-2 \equiv (2^{\frac{n+1}{2}})^2 \pmod{p}$ and so $\left(\frac{-2}{p}\right) = 1$. This can be also written in the form $\left(\frac{-1}{p}\right)\left(\frac{2}{p}\right) = 1$, or $(-1)^{\frac{p-1}{2} + \frac{p^2-1}{8}} = 1$. But if p is of the form $8k - 1$ the latter cannot hold and this is the contradiction that solves the problem.

Using the same idea and a bit more work, we obtain the following result.

Example 2. Prove that for any positive integer n, the number $2^{3^n}+1$ has at least n prime divisors of the form $8k+3$.

[Gabriel Dospinescu]

Solution. Using the result of the previous problem, we deduce that 2^n+1 does not have prime divisors of the form $8k+7$. We will prove that if n is odd, then it has no prime divisors of the form $8k+5$ either. Indeed, let p be a prime divisor of 2^n+1. Then $2^n \equiv -1 \pmod{p}$ and so $-2 \equiv (2^{\frac{n+1}{2}})^2 \pmod{p}$. Using the same argument as the one in the previous problem, we deduce that $\dfrac{p^2-1}{8}+\dfrac{p-1}{2}$ is even, which cannot happen if p is of the form $8k+5$. Now, let us solve the proposed problem. We assume $n>2$ (otherwise the verification is trivial). The essential observation is the identity

$$2^{3^n}+1=(2+1)(2^2-2+1)(2^{2\cdot 3}-2^3+1)\cdots(2^{2\cdot 3^{n-1}}-2^{3^{n-1}}+1)$$

Now, we prove that for all $1\le i<j\le n-1$, $\gcd(2^{2\cdot 3^i}-2^{3^i}+1,2^{2\cdot 3^j}-2^{3^j}+1)=3$. Indeed, assume that p is a prime number dividing $\gcd(2^{2\cdot 3^i}-2^{3^i}+1,2^{2\cdot 3^j}-2^{3^j}+1)$ We then have $p|2^{3^{i+1}}+1$. Thus,

$$2^{3^j}\equiv(2^{3^{i+1}})^{3^{j-i-1}}\equiv(-1)^{3^{j-i-1}}\equiv -1 \pmod{p},$$

implying

$$0\equiv 2^{2\cdot 3^j}-2^{3^j}+1\equiv 1-(-1)+1\equiv 3 \pmod{p}.$$

This cannot happen unless $p=3$. But since

$$v_3(\gcd(2^{2\cdot 3^i}-2^{3^i}+1,2^{2\cdot 3^j}-2^{3^j}+1))=1,$$

as you can immediately check, it follows that

$$\gcd(2^{2\cdot 3^i}-2^{3^i}+1,2^{2\cdot 3^j}-2^{3^j}+1)=3$$

and the claim is proved. It remains to show that each of the numbers $2^{2\cdot 3^i} - 2^{3^i} + 1$, with $1 \leq i \leq n-1$ has at least a prime divisor of the form $8k+3$, different from 3. From the previous remarks, it will follow that $2^{3^n}+1$ has at least $n-1$ distinct prime divisors of the form $8k+3$, and since it is also divisible by 3, the solution will be complete. Fix $i \in \{1, 2, \ldots, n-1\}$ and observe that any prime factor of $2^{2\cdot 3^i} - 2^{3^i} + 1$ is also a prime factor of $2^{3^n}+1$. Thus, from the first remark, this factor must be of one of the forms $8k+1$ or $8k+3$. Because $v_3(2^{2\cdot 3^i} - 2^{3^i} + 1) = 1$, all prime divisors of $2^{2\cdot 3^i} - 2^{3^i} + 1$ except for 3 are of the form $8k+1$, so $2^{2\cdot 3^i} - 2^{3^i} + 1 \equiv 8 \pmod 8$, which is clearly impossible. Thus at least a prime divisor of $2^{2\cdot 3^i} - 2^{3^i} + 1$ is different from 3 and is of the form $8k+3$. The claim is proved and the conclusion follows.

We have seen a beautiful proof of the following result in the chapter **Geometry and Numbers**. But there is another way to solve it, probably more natural and which turns out to be very useful in some other problems, too:

Example 3. Let n be a positive integer such that the equation $x^2+xy+y^2 = n$ has a solution in rational numbers. Prove that this equation also has a solution in integers.

Kömal

Solution. This looks quite familiar, especially after the discussion in chapter **Primes and Squares**. Indeed, let us start with a natural question: which primes can we expressed in the form x^2+xy+y^2 for some integers x, y? Suppose p is such a prime number. Then $4p = (2x+y)^2 + 3y^2$. This shows that $(2x+y)^2 \equiv -3y^2 \pmod p$. Now, if $p \neq 3$ then $y \not\equiv 0 \pmod p$ because otherwise $x \equiv 0 \pmod p$ and so $p^2 | p$, clearly false. The last relation implies therefore that $\left(\frac{-3}{p}\right) = 1$. Using the quadratic reciprocity law, we easily infer that this is equivalent to $\left(\frac{p}{3}\right) = 1$ and this happens precisely when $p \equiv 1 \pmod 3$. Therefore the primes that can be expressed as x^2+xy+y^2 are 3 and $p \equiv 1 \pmod 3$. We are not done yet, because we need to prove that all such primes can be written like that. For 3, there is no problem, but this is not

the case with arbitrary $p \equiv 1 \pmod 3$. Take such a prime number p. From the above arguments we know that $\left(\frac{-3}{p}\right) = 1$, which means that there exists a such that $a^2 = -3 \pmod p$. Now, recall Thue's lemma proved in chapter **Primes and Squares**: there exist integers $0 \leq x, y < \sqrt{p}$ not both zero such that $a^2x^2 \equiv y^2 \pmod p$. Therefore $p|3x^2+y^2$. Because $0 < 3x^2+y^2 < 4p$, we deduce that $3x^2+y^2$ is one of the numbers $p, 2p, 3p$. If it is p, then we obtain $p = (y-x)^2 + (y-x)\cdot 2x + (2\cdot x)^2$. If it is $3p$ then y must be a multiple of 3, say $y = 3z$ and then $p = x^2 + 3z^2$, thus we get the previous case. Finally, suppose that $2p = x^2+3y^2$. Then clearly x, y have the same parity. But then x^2+3y^2 is a multiple of 4, contradiction, because $2p$ is not divisible by 4. Thus this case is excluded and the proof of the first part is finished.
Now, we can attack the problem. Suppose that the equation $x^2+xy+y^2 = n$ has rational solutions, that is the equation $a^2+ab+b^2 = c^2n$ has integer solutions with $\gcd(a,b,c) = 1$. Take p a prime divisor of n and assume that $v_p(n)$ is odd. We claim that $p \equiv 3$ or $p \equiv 1 \pmod 3$. If not then $p|a$ and $p|b$ by the previous arguments, thus we can simplify by p^2 both members of the equation. Repeating this operation, we deduce in the end that $p|c$, which contradicts the fact that $\gcd(a,b,c) = 1$. Thus all prime divisors of the form $3k+2$ of n appear with even exponent. As we have already seen, all prime divisors of n not of the form $3k+2$ are of the form u^2+uv+v^2. Thus, all we need to prove now is that the product of two numbers of the form u^2+uv+v^2 is of the same form. But this is not difficult, because if $\epsilon = e^{\frac{2i\pi}{3}}$ then

$$(u^2+uv+v^2)(w^2+wt+t^2) = (u-\epsilon v)(u-\epsilon^2 v)(w-\epsilon t)(w-\epsilon^2 t)$$

that is $(A-\epsilon B)(A-\epsilon^2 B)$ for $A = uw - vt, B = ut + vw + vt$ and we are done. If you did find the above solution cumbersome, you are right! At first glance, the following problem seems trivial. It is actually very tricky, because brute force takes us nowhere. Yet, in the framework of the above results, this should not be so difficult.

Example 4. Find a number n between 100 and 1997 such that $n|2^n+2$.

APMO 1997

Solution. We will fail if we try to search for odd numbers (actually, this result was proved in the topic **Look at the Exponent!** and is due to Schinzel). So let us search for even numbers. The first attempt is to chose $n = 2p$, for some prime p. Unfortunately, this choice is ruled out by Fermat's little theorem. So let us set $n = 2pq$, for some different primes p and q. We need $pq|2^{2pq-1}+1$ and so we must have $\left(\frac{-2}{p}\right) = \left(\frac{-2}{q}\right) = 1$. Also, using Fermat's little theorem, $p|2^{2q-1}+1$ and $q|2^{2p-1}+1$. A simple case analysis shows that $q = 3, 5, 7$ are not good choices, so let us try $q = 11$. We find $p = 43$ and so it suffices to show that $pq|2^{2pq-1}+1$ for $q = 11$ and $p = 43$. This is not very hard: we have $p|2^{2q-1}+1$, implying $p|2^{p(2q-1)}+1 = 2^{2pq-p}+1$. Then $p|2^{2pq-1}+2^{p-1}$ and using Fermat's theorem ($p|2^{p-1}-1$) we get $p|2^{2pq-1}+1$ and an analogous reasoning shows that $q|2^{2pq-1}+1$, finishing the proof.

Are we wrong to present the following example? It apparently has no connection with quadratic reciprocity, but let us take a closer look.

Example 5. Let $f, g : \mathbb{N}^* \to \mathbb{N}^*$ be functions with the properties:

i) g is surjective;

ii) $2f(n)^2 = n^2 + g(n)^2$ for all positive integers n;

iii) $|f(n) - n| \le 2004\sqrt{n}$ for all n.

Prove that f has infinitely many fixed points.

[Gabriel Dospinescu] Moldova TST 2005

Solution. Let p_n be the sequence of prime numbers of the form $8k+3$ (the fact that there are infinitely many such numbers is a trivial consequence of Dirichlet's theorem, but we invite the reader to find an elementary proof). It is clear that for all n we have

$$\left(\frac{2}{p_n}\right) = (-1)^{\frac{p_n^2-1}{8}} = -1.$$

Using the condition i) we can find x_n such that $g(x_n) = p_n$ for all n. It follows that $2f(x_n)^2 = x_n^2 + p_n^2$, which yields $2f(x_n)^2 \equiv x_n^2 \pmod{p_n}$. Because

$\left(\frac{2}{p_n}\right) = -1$, the last congruence shows that $p_n | x_n$ and $p_n | f(x_n)$. Thus there exist sequences of positive integers a_n, b_n such that $x_n = a_n p_n$ and $f(x_n) = b_n p_n$ for all n. Clearly, ii) implies the relation $2b_n^2 = a_n^2 + 1$. Finally, using the property $|f(n) - n| \leq 2004\sqrt{n}$ we have

$$\frac{2004}{\sqrt{x_n}} \geq \left|\frac{f(x_n)}{x_n} - 1\right| = \left|\frac{b_n}{a_n} - 1\right|.$$

That is

$$\lim_{n\to\infty} \frac{\sqrt{a_n^2+1}}{a_n} = \sqrt{2}.$$

The last relation implies $\lim_{n\to\infty} a_n = 1$. Therefore, starting from a certain point, we have $a_n = 1 = b_n$, that is $f(p_n) = p_n$ and the conclusion follows.

We continue with a difficult classical result that often proves very useful. It characterizes the numbers that are quadratic residues modulo all sufficiently large prime numbers. Of course, perfect squares are such numbers, but how to prove that they are the only ones? Actually, this result has been extensively generalized, but all proofs are based on class field theory, a difficult series of theorems in algebraic number theory, that are far beyond the scope of this elementary book.

Example 6. Suppose that a is a non-square positive integer. Then $\left(\frac{a}{p}\right) = -1$ for infinitely many prime numbers p.

Solution. One may assume that a is square-free. Let us write $a = 2^e q_1 q_2 \dots q_n$, where q_i are different odd primes and $e \in \{0, 1\}$. Let us assume first that $n \geq 1$ (that is $a \neq 2$) and consider some odd distinct primes $r_1, r_2, \dots, r_k$, each of them different from $q_1, q_2, \dots, q_n$. We will show that there is a prime p, different from $r_1, r_2, \dots, r_k$, such that $\left(\frac{a}{p}\right) = -1$. Let s be a quadratic non-residue modulo q_n.

Using the Chinese remainder theorem, we can find a positive integer b such that

$$\begin{cases} b \equiv 1 \pmod{r_i}, & 1 \le i \le k \\ b \equiv 1 \pmod{8}, & \\ b \equiv 1 \pmod{q_i}, & 1 \le i < n \\ b \equiv s \pmod{q_n}. & \end{cases}$$

Now, write $b = p_1 \cdot p_2 \cdots p_m$, with p_i odd primes, not necessarily distinct. Using the quadratic reciprocity law, it follows that

$$\prod_{i=1}^{m} \left(\frac{2}{p_i}\right) = \prod_{i=1}^{m} (-1)^{\frac{p_i^2-1}{8}} = (-1)^{\frac{b^2-1}{8}} = 1$$

and

$$\prod_{j=1}^{m} \left(\frac{q_i}{p_j}\right) = \prod_{j=1}^{m} (-1)^{\frac{p_j-1}{2}\cdot\frac{q_i-1}{2}} \left(\frac{p_j}{q_i}\right) = (-1)^{\frac{q_i-1}{2}\cdot\frac{b-1}{2}} \left(\frac{b}{q_i}\right) = \left(\frac{b}{q_i}\right)$$

for all $i \in \{1, 2, \ldots, n\}$. Hence

$$\prod_{i=1}^{m} \left(\frac{a}{p_i}\right) = \left[\prod_{j=1}^{m} \left(\frac{2}{p_j}\right)\right]^e \prod_{i=1}^{n} \prod_{j=1}^{m} \left(\frac{q_i}{p_j}\right)$$

$$= \prod_{i=1}^{n} \left(\frac{b}{q_i}\right) = \left(\frac{b}{q_n}\right) = \left(\frac{s}{q_n}\right) = -1.$$

(We used the following observations in the above equalities: for any odd numbers $b_1, b_2, \ldots, b_m$, if $b = b_1 b_2 \cdots b_m$ then the numbers

$$\sum_{i=1}^{m} \frac{b_i^2 - 1}{8} - \frac{b^2 - 1}{8}$$

and

$$\sum_{i=1}^{m} \frac{b_i - 1}{2} - \frac{b - 1}{2}$$

are even. We leave to the reader this easy exercise, which can be handled by induction for instance.)

Thus, there exists $i \in \{1, 2, \ldots, m\}$ such that $\left(\frac{a}{p_i}\right) = -1$. Because $b \equiv 1 \pmod{r_i}$, $1 \le i \le k$, we also have $p_i \in \{1, 2, \ldots\} \setminus \{r_1, r_2, \ldots, r_k\}$ and the claim is proved.

The only case left is $a = 2$. But this is very simple, since it suffices to use Dirichlet's theorem to find infinitely many primes p such that $\frac{p^2-1}{8}$ is odd.

As in other units, we will now focus on some special cases. This time it is a problem almost trivial with the above framework but seemingly impossible to solve otherwise (we say this because there is a beautiful, but very difficult, solution using analytical tools, which we will not present here).

Example 7. Suppose that $a_1, a_2, \ldots, a_{2004}$ are nonnegative integers such that $a_1^n + a_2^n + \cdots + a_{2004}^n$ is a perfect square for all positive integers n. What is the least number of such integers that must equal 0?

[Gabriel Dospinescu] Mathlinks Contest

Solution. Suppose that $a_1, a_2, \ldots, a_k$ are positive integers such that $a_1^n + a_2^n + \cdots + a_k^n$ is a perfect square for all n. We will show that k is a perfect square. In order to prove this, we will use the above result and show that $\left(\frac{k}{p}\right) = 1$ for all sufficiently large primes p. This is not a difficult task. Indeed, consider a prime p, greater than any prime divisor of $a_1 a_2 \ldots a_k$. Using Fermat's little theorem, $a_1^{p-1} + a_2^{p-1} + \cdots + a_k^{p-1} \equiv k \pmod{p}$, and since $a_1^{p-1} + a_2^{p-1} + \cdots + a_k^{p-1}$ is a perfect square, it follows that $\left(\frac{k}{p}\right) = 1$. Thus k is a perfect square. And now the problem becomes trivial, since we must find the greatest perfect square less than 2004. A quick computation shows that this is $44^2 = 1936$ and so the desired minimal number is 68.

Here is another nice application of this idea. It is adapted after a problem given at the Saint Petersburg Olympiad. Actually, much more is true: Consider f a monic polynomial with integer coefficients, irreducible over $\mathbb{Q}$ and having degree greater than 1. Then there are infinitely many prime numbers p such that f has no root modulo p. For a proof of this result using (the difficult) Chebotarev's theorem and an elementary theorem of Jordan, as well as for many other aspects of this problem, the reader can consult Serre's beautiful paper *On a theorem of Jordan*, Bull.A.M.S 40 (2003).

Example 8. Suppose that $f \in \mathbb{Z}[X]$ is a second degree polynomial such that for any prime p there is at least one integer n for which $p|f(n)$. Prove that f has rational zeros.

Solution. Let $f(x) = ax^2 + bx + c$ be this polynomial. It suffices to prove that $b^2 - 4ac$ is a perfect square. This boils down to proving that it is a quadratic residue modulo any sufficiently large prime. Pick a prime number p and an integer n such that $p|f(n)$. Then

$$b^2 - 4ac \equiv (2an + b)^2 \pmod{p}$$

and so

$$\left(\frac{b^2 - 4ac}{p}\right) = 1.$$

This shows that our claim is true and finishes the solution.

Some of the properties of Legendre's symbol can also be found in the following problem.

Example 9. Let p be an odd prime and let

$$f(x) = \sum_{i=1}^{p-1} \left(\frac{i}{p}\right) X^{i-1}.$$

a) Prove that f is divisible by $X-1$ but not by $(X-1)^2$ if and only if $p \equiv 3 \pmod 4$;

b) Prove that if $p \equiv 5 \pmod 8$ then f is divisible by $(X-1)^2$ and not by $(X-1)^3$.

[Calin Popescu] Romanian TST 2004

Solution. The first question is not difficult at all. Observe that

$$f(1) = \sum_{i=1}^{p-1} \left(\frac{i}{p}\right) = 0$$

by the simple fact that there are exactly $\frac{p-1}{2}$ quadratic residues modulo p and $\frac{p-1}{2}$ quadratic non-residues in $\{1, 2, \dots p-1\}$. Also,

$$f'(1) = \sum_{i=1}^{p-1} (i-1) \left(\frac{i}{p}\right) = \sum_{i=1}^{p-1} i \left(\frac{i}{p}\right),$$

because $f(1) = 0$. The same idea of summing up in reversed order allows us to write:

$$\sum_{i=1}^{p-1} i \left(\frac{i}{p}\right) = \sum_{i=1}^{p-1} (p-i) \left(\frac{p-i}{p}\right)$$

$$= (-1)^{\frac{p-1}{2}} \sum_{i=1}^{p-1} (p-i) \left(\frac{i}{p}\right) = -(-1)^{\frac{p-1}{2}} f'(1)$$

(we used again the fact that $f(1) = 0$).
Hence for $p \equiv 1 \pmod 4$ we must also have $f'(1) = 0$. In this case f is divisible by $(X-1)^2$. On the other hand, if $p \equiv 3 \pmod 4$, then

$$f'(1) = \sum_{i=1}^{p-1} i \left(\frac{i}{p}\right) \equiv \sum_{i=1}^{p-1} i = \frac{p(p-1)}{2} \equiv 1 \pmod 2$$

and so f is divisible by $X-1$ but not by $(X-1)^2$.
The second question is much more technical, even though it uses the same main idea. Observe that

$$f''(1)=\sum_{i=1}^{p-1}(i^2-3i+2)\left(\frac{i}{p}\right)=\sum_{i=1}^{p-1}i^2\left(\frac{i}{p}\right)-3\sum_{i=1}^{p-1}i\left(\frac{i}{p}\right)$$

(once again we used the fact that $f(1)=0$). Observe that the condition $p\equiv 5 \pmod 8$ implies, by a), that f is divisible by $(X-1)^2$, so actually

$$f''(1)=\sum_{i=1}^{p-1}i^2\left(\frac{i}{p}\right).$$

Let us break this sum into two pieces and treat each of them independently. We have

$$\sum_{i=1}^{\frac{p-1}{2}}(2i)^2\left(\frac{2i}{p}\right)=4\left(\frac{2}{p}\right)\sum_{i=1}^{\frac{p-1}{2}}i^2\left(\frac{i}{p}\right).$$

Note that

$$\sum_{i=1}^{\frac{p-1}{2}}i^2\left(\frac{i}{p}\right)\equiv\sum_{i=1}^{\frac{p-1}{2}}i^2\equiv\sum_{i=1}^{\frac{p-1}{2}}i=\frac{p^2-1}{8}\equiv 1 \pmod 2,$$

so

$$\sum_{i=1}^{\frac{p-1}{2}}(2i)^2\left(\frac{2i}{p}\right)\equiv\pm 4 \pmod 8$$

(actually, using the fact that $\left(\frac{2}{p}\right)=(-1)^{\frac{p^2-1}{8}}$, we obtain that its value is -4). On the other hand,

$$\sum_{i=1}^{\frac{p-1}{2}}(2i-1)^2\left(\frac{2i-1}{p}\right)\equiv\sum_{i=1}^{\frac{p-1}{2}}\left(\frac{2i-1}{p}\right) \pmod 8.$$

If we prove that the last quantity is a multiple of 8, then the problem will be solved. But note that $f(1) = 0$ implies

$$0 = \sum_{i=1}^{\frac{p-1}{2}} \left(\frac{2i}{p}\right) + \sum_{i=1}^{\frac{p-1}{2}} \left(\frac{2i-1}{p}\right).$$

Also,

$$\sum_{i=1}^{\frac{p-1}{2}} \left(\frac{2i}{p}\right) = 1 + \sum_{i=1}^{\frac{p-3}{2}} \left(\frac{2i}{p}\right) = 1 + \sum_{i=1}^{\frac{p-3}{2}} \left(\frac{2\left(\frac{p-1}{2} - i\right)}{p}\right)$$

$$= 1 + \sum_{i=1}^{\frac{p-3}{2}} \left(\frac{2i+1}{p}\right) = \sum_{i=1}^{\frac{p-1}{2}} \left(\frac{2i-1}{p}\right).$$

Therefore $\sum_{i=1}^{\frac{p-1}{2}} \left(\frac{2i-1}{p}\right) = 0$ and the problem is finally solved.

There are more than 100 different proofs of the quadratic reciprocity law, each of them having a truly beautiful underlying idea. We decided not to present the proof using Gauss sums, which is probably the shortest one, as it needs some preparations concerning finite fields and their extensions. Instead, we present the following proof, which greatly simplifies the approach of V.A. Lebesgue.

Example 10. Let p and q be distinct odd primes. Prove that the equation

$$x_1^2 - x_2^2 + x_3^2 - x_4^2 + \cdots + x_p^2 = 1$$

has $q^{p-1} + q^{\frac{p-1}{2}}$ solutions in $(\mathbb{Z}/q\mathbb{Z})^p$. Deduce a new proof of the quadratic reciprocity law.

[Wouter Castryck]

Solution. For an odd number n let us define N_n, the number of solutions of the equation $x_1^2 - x_2^2 + x_3^2 - x_4^2 + ... + x_n^2 = 1$ in $(\mathbb{Z}/q\mathbb{Z})^n$. By replacing x_1 with $x_1 + x_2$ we obtain an equation with the same number of solutions:

$$x_1^2 + x_3^2 - \cdots + x_n^2 - 1 = -2x_1x_2.$$

There exist two kinds of solutions of this last equation: those in which $x_1 \neq 0$ and those in which $x_1 = 0$. The first case is very easy, because for any choice of $x_1 \neq 0$ and any choice of $x_3, ..., x_n$ there is precisely one x_2 such that $(x_1, x_2, ..., x_n)$ is a solution. Thus the first case gives $q^{n-2}(q-1)$ solutions of the equation. The second case is even easier, because the equation reduces to the corresponding one for $n-2$, so this second case gives qN_{n-2} new solutions (the factor q comes from the fact that any solution of

$$x_3^2 - x_4^2 + \cdots + x_n^2 = 1$$

gives q solutions of

$$x_1^2 + x_3^2 - \cdots + x_n^2 - 1 = -2x_1x_2,$$

x_2 being arbitrary). Therefore

$$N_n = q^{n-2}(q-1) + qN_{n-2}$$

and an immediate induction shows that $N_n = q^{n-1} + q^{\frac{n-1}{2}}$. The first part of the problem is now clear.

It is pretty clear that N_p can be written as

$$N_p = \sum_{a_1+a_2+\cdots+a_p=1} \left(1 + \left(\frac{a_1}{q}\right)\right) \cdot \left(1 + \left(\frac{-a_2}{q}\right)\right) \cdots \left(1 + \left(\frac{a_p}{q}\right)\right),$$

because the equation $x^2 = a$ has $1 + \left(\frac{a}{p}\right)$ solutions in $\mathbb{Z}/p\mathbb{Z}$ by definition of Legendre's symbol. On the other hand, imagine that we develop each product in the previous sum and collect terms. There will be a contribution of q^{p-1}

coming from 1 (because there are q^{p-1} solutions of the equation $a_1+a_2+\cdots+a_p=1$) and another contribution coming from the last product, namely

$$\left(\frac{(-1)^{\frac{p-1}{2}}}{q}\right)\cdot\sum_{a_1+\cdots+a_p=1}\left(\frac{a_1a_2\cdots a_p}{q}\right).$$

All other contributions are zero because $\sum_x\left(\frac{x}{p}\right)=0$. Thus

$$N_p=q^{p-1}+\left(\frac{(-1)^{\frac{p-1}{2}}}{q}\right)\cdot\sum_{a_1+\cdots+a_p=1}\left(\frac{a_1a_2...a_p}{q}\right).$$

Now, those p-tuples $(a_1,a_2,...,a_p)$ with $a_1+a_2+\cdots+a_p=1$ and not all a_i equal to p^{-1} can be collected in groups of size p and so, modulo p, the last quantity equals $1+\left(\frac{(-1)^{\frac{p-1}{2}}}{q}\right)\left(\frac{p^{-p}}{q}\right)$, which reduces to $1+(-1)^{\frac{p-1}{2}\cdot\frac{q-1}{2}}\cdot\left(\frac{p}{q}\right)$ (everything is taken mod p). On the other hand, the explicit value of N_p obtained in the first part shows that N_p is congruent to $1+\left(\frac{q}{p}\right)$ modulo p. Thus the two quantities must be equal modulo p, and since their values are -1 or 1 they are actually equal, which implies the quadratic reciprocity law.

Finally, a difficult problem.

Example 11. Find all positive integers n such that $2^n-1|3^n-1$.

[J. L. Selfridge] AMM

Solution. We will prove that $n=1$ is the only solution to the problem. Suppose that $n>1$ is a solution. Then 2^n-1 cannot be a multiple of 3, hence n is odd. Therefore, $2^n\equiv 8 \pmod{12}$. Because any odd prime different from 3 is of one of the forms $12k\pm1$ or $12k\pm5$ and since $2^n-1\equiv 7 \pmod{12}$, it follows that 2^n-1 has at least a prime divisor of the form $12k\pm5$, call it p. We must have $\left(\frac{3}{p}\right)=1$ (since $3^n\equiv 1 \pmod{p}$ and n is odd) and using the quadratic reciprocity law, we finally obtain $\left(\frac{p}{3}\right)=(-1)^{\frac{p-1}{2}}$. On the other hand, $\left(\frac{p}{3}\right) = \left(\frac{\pm 2}{3}\right) = -(\pm 1)$. Consequently, $-(\pm 1) = (-1)^{\frac{p-1}{2}} = \pm 1$, which is the desired contradiction. Therefore the only solution is $n = 1$.

18.2 Practice problems

1. Prove that for any odd prime p, the least positive quadratic non-residue modulo p is smaller than $1+\sqrt{p}$.

2. What is the number of solutions to the equation $a^2+b^2=1$ in $\mathbb{Z}/p\mathbb{Z} \times \mathbb{Z}/p\mathbb{Z}$? What about the equation $a^2-b^2=1$?

3. Let a and b be integers relatively prime to an odd prime p. Prove that

$$\sum_{i=1}^{p-1}\left(\frac{ai^2+bi}{p}\right)=-\left(\frac{a}{p}\right).$$

4. Let p be a prime of the form $4k+1$. Compute

$$\sum_{k=1}^{p-1}\left(\left\lfloor\frac{2k^2}{p}\right\rfloor-2\left\lfloor\frac{k^2}{p}\right\rfloor\right).$$

Korean TST 2000

5. Let $n \geq 0$ be an integer and let $p \equiv 7 \pmod 8$ be a prime number. Prove that

$$\sum_{k=1}^{p-1}\left\{\frac{k^{2^n}}{p}-\frac{1}{2}\right\}=\frac{p-1}{2}.$$

Calin Popescu, Romania TST 2005

6. Let A be the set of prime numbers dividing at least one of the numbers $2^{n^2+1}-3^n$. Prove that both A and $\mathbb{N}\backslash A$ are infinite.

Gabriel Dospinescu

7. Suppose that p is an odd prime and that A and B are two different non-empty subsets of $\{1,2,\ldots,p-1\}$ for which

(a) $A \cup B = \{1, 2, \ldots, p-1\}$;

(b) If a, b are both in A or both in B, then $ab \pmod p \in A$;

(c) If $a \in A$, $b \in B$, then $ab \pmod p \in B$.

Find all such subsets A and B.

Indian Olympiad

8. Let m, n be integers greater than 1 with n odd. Suppose that n is a quadratic residue mod p for any sufficiently large prime number $p \equiv -1 \pmod{2^m}$. Prove that n is a perfect square.

Ron Evans, AMM E 2627

9. Let a be a positive integer with the following property: for any positive integer n, $n^2 + a$ is a sum of two squares of integers. Prove that a is a perfect square.

Gabriel Dospinescu, Mathematical Reflections

10. Let a, b, c be positive integers such that $b^2 - 4ac$ is not a perfect square. Prove that for any $n > 1$ there are n consecutive positive integers, none of which can be written in the form $(ax^2 + bxy + cy^2)^z$ for some integers x, y, z with $z > 0$.

Gabriel Dospinescu, Gazeta Matematică

11. Let $p > 3$ be a prime and let a, b, c be integers with $a \neq 0$. Suppose that $ax^2 + bx + c$ is a perfect square for $2p - 1$ consecutive integers x. Prove that p divides $b^2 - 4ac$.

IMO Shortlist 1991

12. Let a and b be positive integers such that $a > 1$ and $a \equiv b \pmod{2}$. Prove that $2^a - 1$ is not a divisor of $3^b - 1$.

J.L.Selfridge, AMM E 3012

13. Let $p \equiv -1 \pmod{8}$ be a prime number. Prove that there exists an integer x such that $\frac{x^2-2}{p}$ is the square of an integer.

14. Prove that the numbers $3^n + 1$ have no divisor of the form $12k + 11$.

Fermat

15. Let m, n, A be positive integers such that $A = \frac{(m+3)^n+1}{3m}$. Prove that A is odd.

Bulgaria 1998

16. Let S be the set of all numbers of the form $2^{2^n} + 1$ or $6^{2^n} + 1$. Show that S contains infinitely many composite numbers.

Komal

17. Suppose that $\phi(5^m - 1) = 5^n - 1$ for a pair (m, n) of positive integers. Here ϕ is Euler's totient function. Prove that $\gcd(m, n) > 1$.

Taiwanese TST

18. Prove that for positive integers x, y, z the number $x^2 + y^2 + z^2$ is not divisible by $3(xy + yz + zx)$.

Mathlinks Contest

19. Find all positive integers a, b, c, d such that $a + b + d^2 = 4abc$.

Vietnamese TST

20. Let p be a prime of the form $4k+1$ such that $p^2 \mid 2^p - 2$. Prove that the largest prime divisor q of $2^p - 1$ satisfies the inequality $2^q > (6p)^p$.

Gabriel Dospinescu, Mathematical Reflections

21. Let $p = 4k + 3$ be a prime number. Find the number of different residues mod p of the numbers $(x^2 + y^2)^2$, where x, y run over the integers relatively prime to p.

Bulgarian TST 2007

22. Define the sequence $(a_n)_n$ by $a_0 = 4$ and $a_{n+1} = a_n^2 - 2$. Suppose that m is a positive integer and define $n = 2^m - 1$. Prove that n is a prime if and only if n divides a_{m-2}.

Lucas-Lehmer test

23. Prove that for any N there exists n_0 such that for any prime $p > n_0$ there are N consecutive quadratic residues mod p.

Brauer's theorem

24. Prove that for any $\varepsilon > 0$ there exists a prime p_0 with the property: for all primes $p > p_0$, the first quadratic non-residue in the interval $[1, p-1]$ is smaller than $p^{\frac{1}{2\sqrt{e}}+\varepsilon}$.

Vinogradov

Chapter 19

19.1 Theory and examples

Why are integrals pertinent for solving inequalities? When we say integral, we say in fact area which is a measurable concept, a comparable one. That is why there are plenty of inequalities which can be solved with integrals, some of them with a completely elementary statement. They seem elementary, but sometimes finding elementary solutions for them is a real challenge. Instead, there are beautiful and short solutions using integrals. The hard part is to find the integral that hides behind the elementary form of the inequality (and to be honest, the idea of using integrals to solve elementary inequalities is practically nonexistent in Olympiad books). Recall some basic properties.

- For all integrable functions $f, g : [a, b] \to \mathbb{R}$ and all real numbers α, β,
$$\int_a^b (\alpha f(x) + \beta g(x))dx = \alpha \int_a^b f(x)dx + \beta \int_a^b g(x) \text{ (linearity of integrals).}$$
- For all integrable functions $f, g : [a, b] \to \mathbb{R}$ such that $f \leq g$ we have
$$\int_a^b f(x)dx \leq \int_a^b g(x)dx \text{ (monotonicity for integrals).}$$
- For all integrable function $f : [a, b] \to \mathbb{R}$ we have
$$\int_a^b f^2(x)dx \geq 0.$$

Also, the well-known elementary inequalities of Cauchy-Schwarz, Chebyshev, Minkowski, Hölder, Jensen, and Young have corresponding integral inequalities, which are derived immediately from the algebraic inequalities (indeed, one just has to apply the corresponding inequalities for the numbers

$$f\left(a + \frac{k}{n}(b-a)\right), g\left(a + \frac{k}{n}(b-a)\right), \ldots \quad \text{where} \quad k \in \{1, 2, \ldots, n\}$$

and to use the fact that

$$\int_a^b f(x)dx = \lim_{n\to\infty} \frac{b-a}{n} \sum_{k=1}^{n} f\left(a + \frac{k}{n}(b-a)\right).$$

It seems at first glance that this is not a very intricate and difficult theory. Totally false! We will see how powerful this theory of integration is, and especially how hard it is to look beneath the elementary surface of a problem. To convince yourself of the strength of the integral, take a look at the following beautiful proof of the AM-GM inequality using integrals. This proof was found by H. Alzer and published in the American Mathematical Monthly.

Example 1. Prove that for any $a_1, a_2, \ldots, a_n \geq 0$ we have the inequality

$$\frac{a_1 + a_2 + \cdots + a_n}{n} \geq \sqrt[n]{a_1 a_2 \cdots a_n}.$$

Solution. Let us suppose that $a_1 \leq a_2 \leq \cdots \leq a_n$ and let

$$A = \frac{a_1 + a_2 + \cdots + a_n}{n}, \quad G = \sqrt[n]{a_1 a_2 \cdots a_n}.$$

Of course, we can find an index $k \in \{1, 2, \ldots, n-1\}$ such that $a_k \leq G \leq a_{k+1}$. Then it is immediate to see that

$$\frac{A}{G} - 1 = \frac{1}{n} \sum_{i=1}^{k} \int_{a_i}^{G} \left(\frac{1}{t} - \frac{1}{G} \right) dt + \frac{1}{n} \sum_{i=k+1}^{n} \int_{G}^{a_i} \left(\frac{1}{G} - \frac{1}{t} \right) dt$$

and the last quantity is clearly nonnegative, since each integral is nonnegative. Truly wonderful, is not it? This is also confirmed by the following problem, an absolute classic whose solution by induction can be a real nightmare.

Example 2. Let $a_1, a_2, \ldots, a_n$ be real numbers. Prove that

$$\sum_{i=1}^{n} \sum_{j=1}^{n} \frac{a_i a_j}{i + j} \geq 0.$$

Polish Mathematical Olympiad

Solution. Now we will see how easy this problem is if we manage to handle integrals. Note that

$$\frac{a_i a_j}{i+j} = \int_0^1 a_i a_j t^{i+j-1} dt.$$

We have translated the inequality into the language of integrals. The inequality

$$\sum_{i,j=1}^{n} \frac{a_i a_j}{i+j} \geq 0$$

is equivalent to

$$\sum_{i,j=1}^{n} \int_0^1 a_i a_j t^{i+j-1} dt \geq 0,$$

or, using the linearity of the integrals, to

$$\int_0^1 \left(\sum_{i,j=1}^{n} a_i a_j t^{i+j-1} \right) dt \geq 0.$$

This suggests finding an integrable function f such that

$$f^2(t) = \sum_{i,j=1}^{n} a_i a_j t^{i+j-1} dt.$$

This is not difficult, because the formula

$$\left(\sum_{i=1}^{n} a_i x_i \right)^2 = \sum_{i,j=1}^{n} a_i a_j x_i x_j$$

solves the task. We just have to take

$$f(x) = \sum_{i=1}^{n} a_i x^{i-\frac{1}{2}}$$

and we are done.

We continue the series of direct applications of classical integral inequalities with a problem which may also present serious difficulties if not attacked appropriately.

Example 3. Let $t \geq 0$ and define the sequence $(x_n)_{n\geq 1}$ by

$$x_n = \frac{1 + t + \cdots + t^n}{n+1}.$$

Prove that

$$x_1 \leq \sqrt{x_2} \leq \sqrt[3]{x_3} \leq \sqrt[4]{x_4} \leq \ldots$$

[Walther Janous] Crux Mathematicorum

Solution. It is clear that we have

$$x_n = \frac{1}{t-1}\int_1^t u^n du = \frac{1}{1-t}\int_t^1 u^n du$$

Using the first of these forms for $t > 1$ and the second for $t < 1$ the inequality to be proved (clear for $t = 1$) reduces to the more general inequality

$$\sqrt[k]{\frac{1}{b-a}\int_a^b f^k(x)dx} \leq \sqrt[k+1]{\frac{1}{b-a}\int_a^b f^{k+1}(x)dx}$$

for all $k \geq 1$ and any nonnegative integrable function $f : [a, b] \to \mathbb{R}$. And yes, this is a consequence of the Power Mean Inequality for integral functions.

The following problem has a long and quite complicated proof by induction. Yet using integrals it becomes trivial.

Example 4. Prove that for any positive real numbers x, y and any positive integers m, n

$$(n-1)(m-1)(x^{m+n} + y^{m+n}) + (m+n-1)(x^m y^n + x^n y^m)$$
$$\geq mn(x^{m+n-1}y + y^{m+n-1}x).$$

Solution. We transform the inequality as follows:

$$mn(x-y)(x^{m+n-1}-y^{m+n-1}) \geq (m+n-1)(x^m-y^m)(x^n-y^n) \Leftrightarrow$$

$$\frac{x^{m+n-1}-y^{m+n-1}}{(m+n-1)(x-y)} \geq \frac{x^m-y^m}{m(x-y)} \cdot \frac{x^n-y^n}{n(x-y)}$$

(we have assumed that $x > y$). The last relations can immediately be translated with integrals in the form

$$(x-y)\int_y^x t^{m+n-2}dt \geq \int_y^x t^{m-1}dt \int_y^x t^{n-1}dt.$$

And this follows from the integral form of Chebyshev inequality.

A nice blending of the arithmetic and geometric inequalities as well as integral calculus allows us to give a beautiful short proof of the following inequality.

Example 5. Let $x_1, x_2, \ldots, x_k$ be positive real numbers with $x_1x_2\cdots x_n \leq 1$ and m, n positive real numbers such that $n \leq km$. Prove that

$$m(x_1^n + x_2^n + \cdots + x_k^n - k) \geq n(x_1^m x_2^m \ldots x_k^m - 1).$$

IMO Shortlist 1985

Solution. Applying the AM-GM inequality, we find that

$$m(x_1^n + \cdots + x_k^n - k) \geq m(k\sqrt[k]{(x_1x_2\ldots x_k)^n} - k).$$

Let

$$P = \sqrt[k]{x_1x_2\ldots x_k} \leq 1.$$

We have to prove that

$$mkP^n - mk \geq nP^{mk} - n,$$

which is the same as

$$\frac{P^n-1}{n} \geq \frac{P^{mk}-1}{mk}.$$

This follows immediately from the fact that

$$\frac{P^x - 1}{x \ln P} = \int_0^1 P^{xt} dt$$

is decreasing as a function of positive x (for $P \leq 1$).

We have seen a rapid but difficult proof for the following problem, using the Cauchy-Schwarz inequality. Well, the problem originated by playing around with integral inequalities, and the following solution will show how one can create difficult problems starting from trivial ones.

Example 6. Prove that for any positive real numbers a, b, c such that $a + b + c = 1$ we have

$$(ab + bc + ca)\left(\frac{a}{b^2 + b} + \frac{b}{c^2 + c} + \frac{c}{a^2 + a}\right) \geq \frac{3}{4}.$$

[Gabriel Dospinescu]

Solution. As in the previous problem, the most important aspect is to translate the expression $\frac{a}{b^2+b} + \frac{b}{c^2+c} + \frac{c}{a^2+a}$ in the integral language. Fortunately, this isn't difficult, since it is just

$$\int_0^1 \left(\frac{a}{(x+b)^2} + \frac{b}{(x+c)^2} + \frac{c}{(x+a)^2}\right) dx.$$

Now, using the Cauchy-Schwarz inequality, we infer that (do not forget about $a + b + c = 1$):

$$\frac{a}{(x+b)^2} + \frac{b}{(x+c)^2} + \frac{c}{(x+a)^2} \geq \left(\frac{a}{x+b} + \frac{b}{x+c} + \frac{c}{x+a}\right)^2.$$

Using the same inequality again, we compare $\frac{a}{x+b} + \frac{b}{x+c} + \frac{c}{a+x}$ with $\frac{1}{x + ab + bc + ca}$. Consequently,

$$\frac{a}{(x+b)^2} + \frac{b}{(x+c)^2} + \frac{c}{(x+a)^2} \geq \frac{1}{(x + ab + bc + ca)^2},$$

and we can integrate this to find that

$$\frac{a}{b^2+b}+\frac{b}{c^2+c}+\frac{c}{a^2+a}\geq\frac{1}{(ab+bc+ca)(ab+bc+ca+1)}.$$

Now, all we have to do is to notice that

$$ab+bc+ca+1\leq\frac{4}{3}.$$

It seems a difficult challenge to find and prove a generalization of this inequality to n variables.

There is an important similarity between the following problem and example 2, yet here it is much more difficult to see the relation with integral calculus.

Example 7. Let $n \geq 2$ and let S be the set of all sequences $(a_1, a_2, \ldots, a_n) \subset [0,\infty)$ which satisfy

$$\sum_{i=1}^{n}\sum_{j=1}^{n}\frac{1-a_ia_j}{i+j}\geq 0.$$

Find the maximum value of the expression $\displaystyle\sum_{i=1}^{n}\sum_{j=1}^{n}\frac{a_i+a_j}{i+j}$ over all sequences from S.

[Gabriel Dospinescu]

Solution. Consider the function $f:\mathbb{R}\to\mathbb{R}$, $f(x)=a_1+a_2x+\cdots+a_nx^{n-1}$. Let us observe that

$$\sum_{i=1}^{n}\sum_{j=1}^{n}\frac{a_ia_j}{i+j}=\sum_{i=1}^{n}a_i\left(\sum_{j=1}^{n}\frac{a_j}{i+j}\right)=\sum_{i=1}^{n}a_i\int_0^1 x^if(x)dx$$

$$= \int_0^1 \left(xf(x) \sum_{i=1}^{n} a_i x^{i-1} \right) dx = \int_0^1 xf^2(x)dx.$$

So, if we denote $M = \sum_{1 \le i,j \le n} \frac{1}{i+j}$, we infer (using the hypothesis) that

$$\int_0^1 xf^2(x)dx \le M.$$

On the other hand, we have the identity

$$\sum_{i=1}^{n}\sum_{j=1}^{n} \frac{a_i + a_j}{i+j} = 2\left(\frac{a_1}{2} + \cdots + \frac{a_n}{n+1} + \cdots + \frac{a_1}{n+1} + \cdots + \frac{a_n}{2n} \right)$$

$$= 2\int_0^1 (x + x^2 + \cdots + x^n) f(x)dx.$$

Now, the problem becomes easy, since we must find the maximal value of

$$2\int_0^1 (x + x^2 + \cdots + x^n) f(x)dx$$

where

$$\int_0^1 xf^2(x)dx \le M.$$

The Cauchy-Schwarz inequality for integrals is the way to go:

$$\left(\int_0^1 (x + x^2 + \cdots + x^n) f(x)dx \right)^2$$

$$= \left(\int_0^1 \sqrt{xf^2(x)} \sqrt{x(1 + x + \cdots + x^{n-1})^2} dx \right)^2$$

$$= \int_0^1 xf^2(x)dx \int_0^1 (1 + x + \cdots + x^{n-1})^2 dx \le M^2.$$

This shows that $\sum_{i=1}^{n}\sum_{j=1}^{n}\frac{a_i+a_j}{i+j} \leq 2M$ and now the conclusion easily follows: the maximal value is $2\sum_{1\leq i,j\leq n}\frac{1}{i+j}$, attained for $a_1 = a_2 = \cdots = a_n = 1$.

We already said that grouping terms was a mathematical crime. It is time to say it again. We present a new method of solving inequalities involving fractions. The next examples show that bunching could be a great pain.

Example 8. Let a, b, c be positive real numbers. Prove that

$$\frac{1}{3a}+\frac{1}{3b}+\frac{1}{3c}+\frac{3}{a+b+c} \geq \frac{1}{2a+b}+\frac{1}{2b+a}+\frac{1}{2b+c}$$
$$+\frac{1}{2c+b}+\frac{1}{2c+a}+\frac{1}{2a+c}.$$

[Gabriel Dospinescu]

Solution. Of course, the reader has noticed that this is stronger than Popoviciu's inequality, so it seems that classical methods will have no chance. And what if we say that this is Schur's inequality revisited? Indeed, let us write Schur's inequality in the form:

$$x^3+y^3+z^3+3xyz \geq x^2y+y^2x+y^2z+z^2y+z^2x+x^2z$$

where $x = t^{a-\frac{1}{3}}$, $y = t^{b-\frac{1}{3}}$, $z = t^{c-\frac{1}{3}}$ and integrate the inequality as t ranges between 0 and 1. And surprise... since what we get is exactly the desired inequality.

In the same category, here is another application of this idea.

Example 9. Prove that for any positive real numbers a, b, c the following inequality holds:

$$\frac{1}{3a}+\frac{1}{3b}+\frac{1}{3c}+2\left(\frac{1}{2a+b}+\frac{1}{2b+c}+\frac{1}{2c+a}\right)$$

$$\geq 3\left(\frac{1}{a+2b}+\frac{1}{b+2c}+\frac{1}{c+2a}\right).$$

[Gabriel Dospinescu]

Solution. If the previous problem could be solved using bunching (or not? anyway, we haven't tried), this one is surely impossible to solve in this manner. With the experience from the previous problem, we see that the problem asks us in fact to prove that

$$x^3+y^3+z^3+2(x^2y+y^2z+z^2x) \geq 3(xy^2+yz^2+zx^2)$$

for any positive real numbers x, y, z.
Let us assume that $x = \min(x, y, z)$ and write $y = x + m$, $z = x + n$ for some nonnegative real numbers m, n. Simple computations show that the inequality is equivalent to

$$2x(m^2-mn+n^2)+(n-m)^3+m^3 \geq (n-m)m^2.$$

Therefore, it suffices to prove that

$$(n-m)^3+m^3 \geq (n-m)m^2,$$

which is the same as $t^3+1 \geq t$ for all $t \geq -1$ (via the substitution $t = \dfrac{n-m}{m}$), which is immediate.

At the start of this topic we said that there is a deep relation between integrals and areas, but in the sequel we seemed to neglect the last concept. We ask the reader to accept our apologies and bring to their attention two mathematical gems, in which they will surely have the occasion to play around with areas. If only this was easy to see... In fact, these problems are discrete forms of Young and Steffensen inequalities for integrals.

Example 10. Let $a_1 \geq a_2 \geq \cdots \geq a_{n+1} = 0$ and let $b_1, b_2, \ldots, b_n \in [0, 1]$. Prove that if

$$k = \left\lfloor \sum_{i=1}^{n} b_i \right\rfloor + 1,$$

then

$$\sum_{i=1}^{n} a_i b_i \leq \sum_{i=1}^{k} a_i.$$

St. Petersburg Olympiad, 1996

Solution. The very experienced reader will have already seen a resemblance to Steffensen's inequality: for any continuous functions $f, g : [a, b] \to \mathbb{R}$ such that f is decreasing and $0 \leq g \leq 1$ we have

$$\int_a^{a+k} f(x)dx \geq \int_a^b f(x)g(x)dx,$$

where

$$k = \int_a^b g(x)dx.$$

So, probably an argument using areas (this is how we avoid integrals and argue with their discrete forms, areas!!!) could lead to a neat solution. Let us consider a coordinate system XOY and let us draw the rectangles $R_1, R_2, \ldots, R_n$ such that the vertices of R_i are the points $(i-1, 0)$, $(i, 0)$, $(i-1, a_i)$, (i, a_i) (we need n rectangles of heights $a_1, a_2, \ldots, a_n$ and horizontal sides 1, so as to view $\sum_{i=1}^{k} a_i$ as a sum of areas) and the rectangles $S_1, S_2, \ldots, S_n$, where the vertices of S_i are the points $\left(\sum_{j=1}^{i-1} b_j, 0\right)$, $\left(\sum_{j=1}^{i} b_j, 0\right)$, $\left(\sum_{j=1}^{i-1} b_j, a_i\right)$, $\left(\sum_{j=1}^{i} b_j, a_i\right)$ (where $\sum_{j=1}^{0} b_j = 0$). We have made this choice because we need two sets of pairwise disjoint rectangles with the same heights and areas $a_1, a_2, \ldots, a_n$ and $a_1b_1, a_2b_2, \ldots, a_nb_n$ respectively, so that we can compare the areas of the unions of the rectangles in the two sets. Thus, we have to show that the set of rectangles $S_1, S_2, ..., S_n$ can be covered by the rectangles $R_1, R_2, ..., R_{k+1}$. This is quite obvious, by drawing a picture, but let us make it rigorous. Since the width of the union of $S_1, S_2, \ldots, S_n$ is $\sum_{j=1}^{n} b_j < k+1$ (and the width of $R_1, R_2, \ldots, R_{k+1}$ is $k+1$), it is enough to prove this for any horizontal line. But if we consider a horizontal line $y = p$ and an index r such that $a_r \geq p > a_{r+1}$, then the corresponding width for the set $R_1, R_2, \ldots, R_{k+1}$ is p, which is at least $b_1 + b_2 + \cdots + b_p$, the width for $S_1, S_2, \ldots, S_n$. And the problem is solved. And now the second problem, given this time in a Balkan Mathematical Olympiad.

Example 11. Let $(x_n)_{n\geq 0}$ be an increasing sequence of nonnegative integers such that for all $k \in \mathbb{N}$ the number of indices $i \in \mathbb{N}$ for which $x_i \leq k$ is $y_k < \infty$. Prove that for all $m, n \in \mathbb{N}$,

$$\sum_{i=0}^{m} x_i + \sum_{j=0}^{n} y_j \geq (m+1)(n+1).$$

Balkan Mathematical Olympiad 1999

Solution. Again, the experienced reader will see immediately a similarity with Young's inequality: for any strictly increasing one-to-one map $f : [0, A] \to [0, B]$ and any $a \in (0, A)$, $b \in (0, B)$ we have the inequality

$$\int_0^a f(x)dx + \int_0^b f^{-1}(x)dx \geq ab.$$

Indeed, it suffices to take the given sequence $(x_n)_{n\geq 0}$ as the one-to-one increasing function in Young's inequality and the sequence $(y_n)_{n\geq 0}$ as the inverse of f. Just view $\sum_{i=0}^{m} x_i$ and $\sum_{j=0}^{m} y_j$ as the corresponding integrals, and the similarity will be obvious. Thus, probably a geometrical solution is hiding behind some rectangles again. Indeed, consider the vertical rectangles with width 1 and heights $x_0, x_1, \ldots, x_m$ and the rectangles with width 1 and heights

$y_0, y_1, \ldots, y_n$. Then in a similar way one can prove that the set of these rectangles covers the rectangle of sides $m+1$ and $n+1$. Thus the sum of their areas is at least the area of this rectangle.

It will be difficult to solve the following problems using integrals, since the idea is very well hidden. Yet there is such a solution, and it is more than beautiful.

Example 12. Prove that for any $a_1, a_2, \ldots, a_n$ and $b_1, b_2, \ldots, b_n \geq 0$ the following inequality holds

$$\sum_{1 \leq i < j \leq n} (|a_i - a_j| + |b_i - b_j|) \leq \sum_{1 \leq i,j \leq n} |a_i - b_j|.$$

Poland 1999

Solution. Let us define the functions $f_i, g_i : [0, \infty) \to \mathbb{R}$,

$$f_i(x) = \begin{cases} 1, & t \in [0, a_i], \\ 0, & t > a_i \end{cases} \quad \text{and } g_i(x) = \begin{cases} 1, & x \in [0, b_i], \\ 0, & x > b_i. \end{cases}$$

Also, let us define

$$f(x) = \sum_{i=1}^{n} f_i(x), \quad g(x) = \sum_{i=1}^{n} g_i(x).$$

Now, let us compute $\int_0^\infty f(x)g(x)dx$. We see that

$$\int_0^\infty f(x)g(x)dx = \int_0^\infty \left(\sum_{1 \leq i,j \leq n} f_i(x)g_j(x) \right) dx$$

$$= \sum_{1 \leq i,j \leq n} \int_0^\infty f_i(x)g_j(x)dx = \sum_{1 \leq i,j \leq n} \min(a_i, b_j).$$

A similar computation shows that

$$\int_0^\infty f^2(x)dx = \sum_{1\le i,j\le n} \min(a_i, a_j)$$

and

$$\int_0^\infty g^2(x)dx = \sum_{1\le i,j\le n} \min(b_i, b_j).$$

Since

$$\int_0^\infty f^2(x)dx + \int_0^\infty g^2(x)dx = \int_0^\infty (f^2(x)+g^2(x))dx \ge 2\int_0^\infty f(x)g(x)dx,$$

we find that

$$\sum_{1\le i,j\le n} \min(a_i, a_j) + \sum_{1\le i,j\le n} \min(b_i, b_j) \ge 2 \sum_{1\le i,j\le n} \min(a_i, b_j).$$

Now, remember that $2\min(x,y) = x+y-|x-y|$ and the last inequality becomes

$$\sum_{1\le i,j\le n} |a_i - a_j| + \sum_{1\le i,j\le n} |b_i - b_j| \le 2 \sum_{1\le i,j\le n} |a_i - b_j|$$

and since

$$\sum_{1\le i,j\le n} |a_i - a_j| = 2 \sum_{1\le i<j\le n} |a_i - a_j|,$$

the problem is solved.

Using the same idea, here is a difficult problem, whose elementary solution is awful and which has a three-line solution. Of course, this is easy to find for the author of the problem, but in a contest things change!

Example 13. Let $a_1, a_2, \ldots, a_n > 0$ and let $x_1, x_2, \ldots, x_n$ be real numbers such that

$$\sum_{i=1}^n a_i x_i = 0.$$

a) Prove that the inequality $\sum_{1\le i<j\le n} x_i x_j |a_i - a_j| \le 0$ holds;

b) Prove that we have equality in the above inequality if and only if there exist a partition $A_1, A_2, \ldots, A_k$ of the set $\{1, 2, \ldots, n\}$ such that for all $i \in \{1, 2, \ldots, k\}$ we have $\sum_{j\in A_i} x_j = 0$ and $a_{j_1} = a_{j_2}$ if $j_1, j_2 \in A_i$.

[Gabriel Dospinescu] Mathlinks Contest

Solution. Let λ_A be the characteristic function of an arbitrary set A. Let us consider the function

$$f : [0, \infty) \to \mathbb{R}, \quad f = \sum_{i=1}^{n} x_i \lambda_{[0,a_i]}.$$

Now, let us compute

$$\begin{aligned}\int_0^\infty f^2(x)dx &= \sum_{1\le i,j\le n} x_i x_j \int_0^\infty \lambda_{[0,a_i]}(x)\lambda_{[0,a_j]}(x)dx \\ &= \sum_{1\le i,j\le n} x_i x_j \min(a_i, a_j).\end{aligned}$$

Then

$$\sum_{1\le i,j\le n} x_i x_j \min(a_i, a_j) \ge 0.$$

Since

$$\min(a_i, a_j) = \frac{a_i + a_j - |a_i - a_j|}{2}$$

and

$$\sum_{1\le i,j\le n} x_i x_j (a_i + a_j) = 2\left(\sum_{i=1}^{n} x_i\right)\left(\sum_{i=1}^{n} a_i x_i\right) = 0,$$

we conclude that

$$\sum_{1\le i<j\le n} x_i x_j |a_i - a_j| \le 0.$$

Let us suppose that we have equality. We find that

$$\int_0^{\infty} f^2(x)dx = 0$$

and so $f(x) = 0$ almost anywhere. Now, let $b_1, b_2, \ldots, b_k$ the distinct numbers that appear among $a_1, a_2, \ldots, a_n > 0$ and let $A_i = \{j \in \{1, 2, \ldots, n\} | \ a_j = b_i\}$. Then $A_1, A_2, \ldots, A_k$ is a partition of the set $\{1, 2, \ldots, n\}$ and we also have

$$\sum_{i=1}^{k} \left(\sum_{j \in A_i} x_j \right) \lambda_{[0,b_i]} = 0$$

almost anywhere, from which we easily conclude that

$$\sum_{i \in A_i} x_j = 0 \text{ for all } i \in \{1, 2, \ldots, k\}.$$

The conclusion follows.

Because we have proved the nice inequality

$$\sum_{1 \le i,j \le n} x_i x_j \min(a_i, a_j) \ge 0$$

for all $x_1, x_2, \ldots, x_n, a_1, a_2, \ldots, a_n > 0$ let us take a further step and give the magnificent proof found by Ravi Boppana for one of the most difficult inequalities ever given in a contest. The solution is based on the above result.

Example 14. Prove the following inequality

$$\sum_{1 \le i,j \le n} \min(a_i a_j, b_i b_j) \le \sum_{1 \le i,j \le n} \min(a_i b_j, a_j b_i)$$

for all nonnegative real numbers $a_1, \ldots, a_n$ and $b_1, \ldots, b_n$.

[G. Zbaganu] USAMO 1999

Solution. Let us define the numbers $r_i = \dfrac{\max(a_i, b_i)}{\min(a_i, b_i)} - 1$ and $x_i = \operatorname{sgn}(a_i - b_i)$ (if, by any chance, one of $a_i, b_j = 0$, we can simply put $r_i = 0$). The crucial observation is the following identity:

$$\min(a_i b_j, a_j b_i) - \min(a_i a_j, b_i b_j) = x_i x_j \min(r_i, r_j).$$

Proving this relation can be achieved by distinguishing four cases, but let us observe that actually we may assume that $a_i \geq b_i$ and $a_j \geq b_j$, which leaves us with only two cases. The first one is when at least one of the two inequalities $a_i \geq b_i$ and $a_j \geq b_j$ becomes an equality. This case is trivial, so let us assume the contrary. Then

$$x_i x_j \min(r_i, r_j) = b_i b_j \min\left(\frac{a_i}{b_i} - 1, \frac{a_j}{b_j} - 1\right) = b_i b_j \left(\min\left(\frac{a_i}{b_i}, \frac{a_j}{b_j}\right) - 1\right)$$
$$= \min(a_i b_j, a_j b_i) - b_i b_j = \min(a_i b_j, a_j b_i) - \min(a_i a_j, b_i b_j).$$

Now, we can write

$$\sum_{1 \leq i,j \leq n} \min(a_i b_j, a_j b_i) - \sum_{1 \leq i,j \leq n} \min(a_i a_j, b_i b_j) = \sum_{i,j} x_i x_j \min(r_i, r_j) \geq 0,$$

the last inequality being the main ingredient of the preceding problem.

Finally, a problem, which is a consequence of this last hard inequality. Consider this a hint and try to solve it, since otherwise the problem is really hard.

Example 15. Let $x_1, x_2, \ldots, x_n$ be some positive real numbers such that

$$\sum_{1 \leq i,j \leq n} |1 - x_i x_j| = \sum_{1 \leq i,j \leq n} |x_i - x_j|.$$

Prove that $\displaystyle\sum_{i=1}^{n} x_i = n.$

[Gabriel Dospinescu]

Solution. Consider $b_i = 1$ in the inequality from example 14. We obtain:

$$\sum_{1\le i,j\le n} \min(x_i, x_j) \ge \sum_{1\le i,j\le n} \min(1, x_i x_j).$$

Now, use the formula $\min(u, v) = \dfrac{u + v - |u - v|}{2}$ and rewrite the above inequality in the form

$$2n\sum_{i=1}^{n} x_i - \sum_{1\le i,j\le n} |x_i - x_j| \ge n^2 + \left(\sum_{i=1}^{n} x_i\right)^2 - \sum_{1\le i,j\le n} |1 - x_i x_j|.$$

Taking into account that

$$\sum_{1\le i,j\le n} |1 - x_i x_j| = \sum_{1\le i,j\le n} |x_i - x_j|,$$

we obtain

$$2n\sum_{i=1}^{n} x_i \ge n^2 + \left(\sum_{i=1}^{n} x_i\right)^2,$$

which can be rewritten as $\left(\sum_{i=1}^{n} x_i - n\right)^2 \le 0$. Therefore $\sum_{i=1}^{n} x_i = n$.

19.2 Practice problems

1. Prove that for any $x > 0$ and any positive integer n,

$$\frac{\binom{2n}{0}}{x} - \frac{\binom{2n}{1}}{x+1} + \frac{\binom{2n}{2}}{x+2} - \cdots + \frac{\binom{2n}{2n}}{x+2n} > 0.$$

Kömal

2. Let $x_0 = 0$ and $x_i > 0, i = 1, 2, \ldots, n$ such that $\sum_i^n x_i = 1$. Prove that

$$\sum_{i=1}^{n} \frac{x_i}{\sqrt{1 + x_0 + x_1 + \cdots + x_{i-1}}\sqrt{x_i + \cdots + x_n}} < \frac{\pi}{2}$$

China 1996

3. Prove that for all real numbers $a_1, a_2, \ldots, a_n$

$$\sum_{i,j=1}^{n} \frac{ij}{i+j-1} a_i a_j \geq \left(\sum_{i=1}^{n} a_i \right)^2.$$

4. Let $a_1, a_2, \ldots, a_n \in \mathbb{R}$ and $c, x_1, x_2, \ldots, x_n > 0$. Prove that

$$\sum_{i=1}^{n} \sum_{j=1}^{n} \frac{a_i a_j}{(x_i + x_j)^c} \geq 0.$$

Komal

5. Prove that for any positive real numbers a, b, c such that $a + b + c = 1$,

$$\left(1 + \frac{1}{a}\right)^b \left(1 + \frac{1}{b}\right)^c \left(1 + \frac{1}{c}\right)^a \geq 1 + \frac{1}{ab + bc + ca}.$$

Marius and Sorin Radulescu

6. Let $k \in \mathbb{N}$ and $a_1, a_2, \ldots, a_{n+1}$ nonnegative real numbers such that $a_{n+1} = a_1$. Prove that

$$\sum_{\substack{1 \le i \le n \\ 1 \le j \le k}} a_i^{k-j} a_{i+1}^{j-1} \ge \frac{k}{n^{k-2}} \left(\sum_{i=1}^{n} a_i \right)^{k-1}.$$

Hassan A. Shah Ali, Crux Mathematicorum

7. Prove that for all $a_1, a_2, \ldots, a_n, b_1, b_2, \ldots, b_n \ge 0$

$$\left(\sum_{1 \le i,j \le n} \min(a_i, a_j) \right) \left(\sum_{1 \le i,j \le n} \min(b_i, b_j) \right) \ge \left(\sum_{1 \le i,j \le n} \min(a_i, b_j) \right)^2.$$

Don Zagier

8. Consider vectors $a_1, a_2, \ldots, a_n$ and $b_1, b_2, \ldots, b_m$ in the plane. For every line through the origin, let the projection of the vectors onto the line be $A_1, A_2, \ldots, A_n$ and $B_1, B_2, \ldots, B_m$. Suppose that for any such line we have

$$|A_1| + |A_2| + \cdots + |A_n| \ge |B_1| + |B_2| + \cdots + |B_m|.$$

Prove that

$$|a_1| + |a_2| + \cdots + |a_n| \ge |b_1| + |b_2| + \cdots + |b_m|,$$

where $|v|$ is the length of the vector v.

9. Prove that for any $x_1 \ge x_2 \ge \cdots \ge x_n > 0$ we have

$$\sum_{i=1}^{n} \sqrt{\frac{x_i^2 + x_{i+1}^2 + \cdots + x_n^2}{i}} \le \pi \sum_{i=1}^{n} x_i.$$

Adapted after an IMC 2000 problem

10. Prove that for any positive real numbers $x_1, x_2, \ldots, x_n$ such that

$$\sum_{i=1}^{n} \frac{1}{1+x_i} = \frac{n}{2}$$

we have the inequality

$$\sum_{1 \le i,j \le n} \frac{1}{x_i + x_j} \ge \frac{n^2}{2}.$$

Gabriel Dospinescu

11. Prove that the function $f : [0,1) \to \mathbb{R}$ defined by

$$f(x) = \log_2(1-x) + x + x^2 + x^4 + x^8 + \cdots$$

is bounded.

Komal

12. Let $x_1, x_2, \ldots, x_n$ and $y_1, y_2, \ldots, y_n$ be positive real numbers such that for all $t > 0$ there are at most $\frac{1}{t}$ pairs (i, j) satisfying $x_i + y_j \ge t$. Prove that

$$(x_1 + x_2 + \cdots + x_n) \cdot (y_1 + y_2 + \cdots + y_n) \le \max_{1 \le i,j \le n} (x_i + y_j).$$

Gabriel Dospinescu

13. The sides and diagonals of a (not necessarily convex) polygon have length at most 1. Prove that the area of the polygon is less than $\frac{\pi}{4}$.

14. Let $x_1, x_2, \ldots, x_m, y_1, y_2, \ldots, y_n$ be positive real numbers and let

$$X = \sum_{i=1}^{m} x_i, Y = \sum_{j=1}^{n} y_j.$$

Prove that

$$2XY\sum_{i=1}^{m}\sum_{j=1}^{n}|x_i-y_j| \geq X^2\sum_{i=1}^{n}\sum_{j=1}^{n}|y_i-y_j| + Y^2\sum_{i=1}^{m}\sum_{j=1}^{m}|x_i-x_j|.$$

Chinese TST 2009

15. Prove that for any complex numbers $z_1, z_2, \ldots, z_n$ one can find a nonempty subset $I \subset \{1, 2, \ldots, n\}$ such that

$$\left|\sum_{i\in I} z_i\right| \geq \frac{1}{\pi}\sum_{i=1}^{n}|z_i|.$$

Is the constant $\frac{1}{\pi}$ optimal?

16. Find the best constant k such that for any $n \geq 2$ and any nonnegative real numbers $x_1, \ldots, x_n$ we have

$$(x_1 + 2x_2 + \cdots + nx_n)(x_1^2 + x_2^2 + \cdots + x_n^2) \geq k(x_1 + x_2 + \cdots + x_n)^3.$$

17. Let $a_1, a_2, \ldots, a_n$ be positive real numbers and let $S = a_1 + a_2 + \cdots + a_n$ be their sum. Prove that

$$\frac{1}{n}\cdot\sum_{i=1}^{n}\frac{1}{a_i} + \frac{n(n-2)}{S} \geq \sum_{i\neq j}\frac{1}{S + a_i - a_j}.$$

Gabriel Dospinescu

18. Prove that for any real numbers $a_1, a_2, \ldots, a_n$,

$$\sum_{i=1}^{n}\sum_{j=1}^{n}\frac{a_i \cdot a_j}{1 + |i-j|} \geq 0.$$

19. Prove that for any real numbers $a_1, a_2, \ldots, a_n$,

$$\sum_{1 \le i,j \le n} \frac{a_i a_j}{i+j} \le \pi \sum_{i=1}^{n} a_i^2.$$

Prove that π is the best constant.

Hilbert's inequality

Chapter

20

20.1 Theory and examples

It is very difficult to imagine a completely trivial mathematical statement which has absolutely nontrivial applications. And if there is such a candidate, then surely the pigeonhole principle will be the winner: what could be easier than the observation that if we put more than n objects in n boxes, there will be a box containing at least two objects? Yet, this observation, combined with some trivial variations, turn out to be a completely revolutionary idea in mathematics. Quantitative results such as Siegel's lemma, or the fact that the class group of a number field is finite, are fundamental results in number theory, and are all consequences of this principle. There is also an enormous quantity of difficult Ramsey-type (and other) results in combinatorics, all based on this little observation. The purpose of this chapter is to present some of these applications of the pigeonhole principle, most of them elementary.

Let us begin with some combinatorial statements in which the use of the pigeonhole principle is more or less clear. But the reader must pay attention, because what is easy to state is not necessarily easy to write! This is why even the easiest problems of this chapter will have some subtle parts, and the reader should not expect straightforward applications of the pigeonhole principle.

Example 1. Let $A_1, A_2, ..., A_{50}$ subsets of a finite set A such that any subset has more than half of the number of elements of A. Prove that there exists a subset of A with at most 5 elements that has nonempty intersection with each of the 50 subsets.

United Kingdom 1976

Solution. Let $A = \{a_1, a_2, ..., a_n\}$ and define $f(i)$ to be the number the subsets among $A_1, A_2, ..., A_{50}$ that contain a_i. Then clearly

$$f(1) + f(2) + \cdots + f(n) = |A_1| + |A_2| + \cdots + |A_{50}| > 25n.$$

Thus there exists an i such that $f(i) \geq 26$, which implies the existence of an a_x in at least 26 subsets, let them be $A_{25}, A_{26}, ..., A_{50}$. Working with the remaining 24 subsets only and using the same argument we deduce the existence of an element a_y which belongs to at least 13 subsets among $A_1, A_2, ..., A_{24}$, let them be $A_{12}, A_{13}, ..., A_{24}$. Similarly, there exists a_z which belongs to at least 6 subsets among $A_1, ..., A_{11}$, let them be $A_6, A_7, ..., A_{11}$ and if we continue this process we define similarly a_u and a_v. It is clear that the set of a_x, a_y, a_z, a_u, a_v satisfies all conditions of the problem.

The strange statement of the following problem should not mislead the reader: after all, we have said that all problems of this chapter are based on the pigeonhole principle, but we haven't said where this idea hides. After reading the solution, the reader will surely say: but it was obvious! Yes, it is obvious, but only if we proceed correctly...

Example 2. Let $A = \{1, ..., 100\}$ and let $A_1, A_2, ..., A_m$ be subsets of A, each with 4 elements, any two of them having at most 2 elements in common. Prove that if $m \geq 40425$ then there exist 49 subsets among the chosen ones such that their union is A, but the union of any 48 subsets (among the 49) is not A.

[Gabriel Dospinescu]

Solution. Let us consider the collection of all two-element subsets of each $A_1, A_2, ..., A_m$. We obtain a collection of $6m$ two-element subsets of A. But the number of distinct subsets of cardinal 2 in A is 4950. Thus, by the pigeonhole principle, there exist distinct elements $x, y \in A$ which belong to at least 49 subsets. Let these subsets be $A_1, A_2, ..., A_{49}$. Then the conditions of the problem imply that the union of these subsets has $2 + 49 \times 2 = 100$ elements, so the union is A. However, the union of any 48 subsets among these 49 has at most $2 + 2 \times 48 = 98$ elements, so it is different of A.

The following example is, in a certain sense, typical for problems involving estimations of trigonometric sums. Its presence as the last problem in an international contest for undergraduate students shows that it is more difficult than it looks, even though the solution is again a pure application of the pigeonhole principle.

Example 3. Let A be a subset of $\mathbb{Z}_n$ with at most $\frac{\ln n}{1.7}$ elements. Prove that there exists a nonzero integer r such that $\left|\sum\limits_{s\in A} e^{\frac{2i\pi}{n}sr}\right| \geq \frac{|A|}{2}$.

IMC 1999

Solution. Let $A = \{a_1, a_2, ..., a_k\}$ and define $g(t) = \left(e^{\frac{2i\pi}{n}a_1 t}, ..., e^{\frac{2i\pi}{n}a_k t}\right)$ for $0 \leq t \leq n-1$. If we divide the unit circle into 6 equal arcs then these k-tuples are divided into 6^k classes. Because $n > 6^k$, there are two k-tuples in the same class, that is there exist $t_1 < t_2$ such that $g(t_1)$ and $g(t_2)$ are in the same class. Observe now that if we consider $r = t_2 - t_1$ then $\operatorname{Re}\left(e^{\frac{2i\pi}{n}ra_j}\right) = \cos\left(\frac{2\pi a_j(t_2-t_1)}{n}\right) \geq \cos\left(\frac{\pi}{3}\right)$. Therefore $|f(r)| \geq \operatorname{Re}(f(r)) \geq \frac{|A|}{2}$ and the problem is solved.

Sometimes, even the completely obvious observation that an infinite sequence taking only a finite number of values must have (at least) two equal terms (actually, an infinite constant subsequence) can be really useful. This is shown by the following extension of a difficult problem given in a Romanian TST in 1996:

Example 4. Let $x_1, x_2, ..., x_k$ be real numbers such that $A = \{\cos(n\pi x_1) + \cos(n\pi x_2) + \cdots + \cos(n\pi x_k) | n \in \mathbb{N}^*\}$ is finite. Prove that x_i are all rational numbers.

[Vasile Pop]

Solution. The beautiful idea is that if the sequence $a_n = \cos(n\pi x_1) + \cos(n\pi x_2) + \cdots + \cos(n\pi x_k)$ takes a finite number of distinct values, then so does the sequence in $\mathbb{R}^k$ defined by $u_n = (a_n, a_{2n}, ..., a_{kn})$. Thus there exist $m < n$ such that $a_n = a_m, a_{2n} = a_{2m}, ..., a_{kn} = a_{km}$. Let us analyze these relations more closely. We know that $\cos(nx)$ is a polynomial of degree n with integer coefficients in $\cos(x)$. If $A_i = \cos(n\pi x_i)$ and $B_i = \cos(m\pi x_i)$ then the previous relations combined with this observation, show that $A_1^j + A_2^j + \cdots + A_k^j = B_1^j + B_2^j + \cdots + B_k^j$ for all $j = 1, 2, ..., k$. Using Newton's formula, we deduce that the polynomials having zeros $A_1, A_2, ..., A_k$ and $B_1, B_2, ..., B_k$ are equal. Thus there exists a permutation σ of $1, 2, ..., n$ such that $A_i = B_{\sigma(i)}$. Thus $\cos(n\pi x_i) = \cos(m\pi x_{\sigma(i)})$, which means that $nx_i - mx_{\sigma(i)}$ is a rational number for all i. This easily implies that all x_i are rational numbers.

The same idea can be used with success when dealing with remainders of recursive sequences modulo certain positive integers. This kind of problem has become quite classical, being present in lots of mathematical competitions .

Example 5. Consider the sequence $(a_n)_{n\geq 1}$ defined by $a_1 = a_2 = a_3 = 1$ and $a_{n+3} = a_{n+1}a_{n+2} + a_n$. Prove that any positive integer has a multiple which is a term of this sequence.

[Titu Andreescu, Dorel Mihet] Revista Matematică Timişoara

<u>**Solution.**</u> Consider a positive integer N and let the first term of the extended sequence to be $a_0 = 0$. Now, look at the sequence of triples (a_n, a_{n+1}, a_{n+2}) reduced mod N. This sequence takes at most N^3 distinct values because there are N possible remainders mod N. Thus we can find two positive integers $i < j$ such that $a_i \equiv a_j \pmod{N}, a_{i+1} \equiv a_{j+1} \pmod{N}$ and $a_{i+2} \equiv a_{j+2} \pmod{N}$. Using the recursive relation, we deduce that the sequence becomes periodic mod N with period $j - i$. Indeed, it follows immediately from the recursive relation that $a_k \equiv a_{k+j-i} \pmod{N}$ for all $k \geq i$, and using the fact that $a_n = a_{n+3} - a_{n+1}a_{n+2}$ we can proceed backwards with an inductive argument to prove that $a_k \equiv a_{k+j-i} \pmod{N}$ for all $k \leq i$. In particular, it follows that a_{j-i} is a multiple of N, so N divides at least one term of the sequence.

A classical application of the pigeonhole principle is to prove that for any coloring of the lattice points in a plane with a finite number of colors, there are rectangles having all vertices of the same color. We advise the reader who does not know this problem to solve it first and then to proceed to the following similar problem.

Example 6. Let m, n be positive integers and let A be a set of lattice points in the plane such that any open disc of radius m contains at least one point of A. Prove that no matter how we color the points in A with n colors there exist four points of the same color in A which are vertices of a rectangle.

Romanian TST 1996

Solution. Consider first a huge square of side-length a (to be determined later) and having sides parallel to the coordinate axes. Divide it into smaller squares of side-length $2m$ and inscribe a circle in each such smaller square. We find at least $\left\lfloor \frac{a^2}{4m^2} \right\rfloor$ circles of radius m inside this huge square, and thus at least as many points of A. But these points lie on $a - 1$ vertical lines. By the pigeonhole principle, there exists a vertical line containing at least $n+1$ points of A if a is suitably chosen (for instance, any multiple of $4nm^2$). Again by the pigeonhole principle, two of these points have the same color. This shows that in any such huge square there exists a vertical line and two points on it that have the same color. Because there are finitely many positions of these pairs of points on a segment of finite length and because we can put infinitely many huge squares consecutively on the Ox axis, there will be two squares in which the points of the same color and on the same vertical line have identical positions and same color. These points will determine a monochromatic rectangle.

It is time to consider some more involved problems in which the use of the pigeonhole principle is far from obvious. Several articles in the American Mathematical Monthly were dedicated to the following problem, which shows that it is not surprising that only a few students solved it when it was proposed for the Putnam Competition (in a weaker form than the example below):

Example 7. Let S_a be the set of numbers of the form $\lfloor na \rfloor$ for some positive integer n. Prove that if a, b, c are positive real numbers, then the three sets S_a, S_b, S_c cannot be pairwise disjoint.

Solution. Let us pick an integer N and consider the triples $(\{\frac{i}{a}\},\{\frac{i}{b}\},\{\frac{i}{c}\})$ for $i = 0, 1, ..., N^3$. These points lie in the unit cube $[0,1]^3$ so by the pigeonhole principle there are two points lying in a cube of side-length $\frac{1}{N}$, that is there exist $i > j$ such that for some integers m, n, p we have $\left|\frac{i-j}{a} - m\right| \le \frac{1}{N}, \left|\frac{i-j}{b} - n\right| \le \frac{1}{N}, \left|\frac{i-j}{c} - p\right| \le \frac{1}{N}$. This can be written as $|i - j - ma| < 1, |i - j - nb| < 1$ and $|i - j - pc| < 1$. Therefore all numbers $\lfloor ma \rfloor, \lfloor nb \rfloor, \lfloor pc \rfloor$ are equal to $i - j$ or $i - j - 1$, which shows that some two of them are equal, and so two of the sets S_a, S_b, S_c intersect.

We continue with a very beautiful problem from an Iranian Olympiad, where there are some traps in applying the pigeonhole principle.

Example 8. Let m be a positive integer and $n = 2^m+1$. Consider $f_1, f_2, ..., f_n : [0,1] \to [0,1]$ to be increasing functions such that $f_i(0) = 0$ and $|f_i(x) - f_i(y)| \le |x - y|$ for all $1 \le i \le n$ and all $x, y \in [0,1]$. Prove that there exist $1 \le i < j \le n$ such that $|f_i(x) - f_j(x)| \le \frac{1}{m+1}$ for all $x \in [0,1]$.

Iran 2001

Solution. This time, everything is clear: the solution of this problem should use the pigeonhole principle. But how? Looking at the graph of such functions, we observe that the points of a regular subdivision of $[0,1]$ play a special role in their behavior. Therefore, let us concentrate more on these points, so let us associate to each function f_i an $(m+1)$-tuple $(a_1(i), a_2(i), ..., a_{m+1}(i))$, where $a_j(i)$ is the smallest integer k such that $f_i(\frac{j}{m+1}) \in [\frac{k}{m+1}, \frac{k+1}{m+1}]$. In this way, we can control the behavior of the function f_i very well at all points of the regular distribution $\left(0, \frac{1}{m+1}, \frac{2}{m+1}, ..., 1\right)$. Because f_i is increasing, it is clear that $a_{j+1}(i) \ge a_j(i)$. Also, the inequality $f_i\left(\frac{j+1}{m+1}\right) - f_i\left(\frac{j}{m+1}\right) \le \frac{1}{m+1}$ assures us that $a_{j+1}(i) \le a_j(i) + 1$. Furthermore, note that

$$0 \le f_i\left(\frac{1}{m+1}\right) = f_i\left(\frac{1}{m+1}\right) - f_i(0) \le \frac{1}{m+1},$$

so $a_1(i) = 0$ for all i. Therefore there are at most 2^m such sequences that can be associated with $f_1, f_2, ..., f_n$. By the pigeonhole principle, two functions f_i, f_j with $i < j$ must be associated with the same $(m+1)$-tuple. This shows that we can find some integers $b_0, b_1, ..., b_{m+1}$ and two indices $i < j$ such that $f_i(\frac{k}{m+1})$ and $f_j(\frac{k}{m+1})$ are both in $[\frac{b_k}{m+1}, \frac{b_k+1}{m+1}]$ for all $k = 0, 1, ..., m+1$.

Now we are almost done, because we have found our candidates i, j. What remains to be verified is straightforward. Indeed, consider $x \in [0,1]$ and k an integer to be such that $x \in [\frac{k}{m+1}, \frac{k+1}{m+1}]$. We know that $0 \le b_{k+1} - b_k \le 1$ by the previous observations. Now, we have two cases. First let $b_{k+1} = b_k$, so

$$f_i(x) \le f_i\left(\frac{k+1}{m+1}\right) \le \frac{b_k+1}{m+1} \le \frac{1}{m+1} + f_j\left(\frac{k}{m+1}\right) \le f_j(x) + \frac{1}{m+1},$$

and by a similar argument we also obtain $f_j(x) \le f_i(x) + \frac{1}{m+1}$. So, assume that $b_{k+1} = b_k + 1$. Then

$$f_i(x) \le f_i\left(\frac{k}{m+1}\right) + x - \frac{k}{m+1} \le x + \frac{b_k - k + 1}{m+1},$$

while

$$f_j(x) \ge f_j\left(\frac{k+1}{m+1}\right) + x - \frac{k+1}{m+1} \ge x + \frac{b_k - k}{m+1},$$

from where $f_i(x) - f_j(x) \le \frac{1}{m+1}$. Analogously we obtain $f_j(x) - f_i(x) \le \frac{1}{m+1}$, which shows that in both cases $|f_i(x) - f_j(x)| \le \frac{1}{m+1}$.

In the same category of difficult (or very difficult) problems can be included the next example, too. Here it is absolutely not obvious how to use the pigeonhole principle. The solution presented here was given by Gheorghe Eckstein:

Example 9. 49 students take a test consisting of 3 problems, marked from 0 to 7. Show that there are two students A and B such that A scores at least as many as B for each problem.

IMO 1988 Shortlist

Solution. Let us consider the set of triples (a, b, c) where each component can be $0, 1, ...7$. We define an order on these triples by saying that (a, b, c) is greater than or equal to (x, y, z) if $a \geq x, b \geq y$, and $c \geq z$. A similar order is defined for pairs (a, b). We need to prove that among any 49 triples there are two that are comparable. Supposing the contrary, it is clear that such a set A of triples cannot contain two triples with the same first two coordinates. Now, consider the following chains:

$$(1) \quad (0,0) < (0,1) < (0,2) < (0,3) < (0,4) < (0,5) < (0,6) < (0,7) < (1,7) < (2,7) < (3,7) < (4,7) < (5,7) < (6,7) < (7,7)$$

$$(2) \quad (1,0) < (1,1) < (1,2) < (1,3) < (1,4) < (1,5) < (1,6) < (2,6) < (3,6) < (4,6) < (5,6) < (6,6) < (7,6)$$

$$(3) \quad (2,0) < (2,1) < (2,2) < (2,3) < (2,4) < (2,5) < (3,5) < (4,5) < (5,5) < (6,5) < (7,5)$$

$$(4) \quad (3,0) < (3,1) < (3,2) < (3,3) < (3,4) < (4,4) < (5,4) < (6,4) < (7,4).$$

Note that no such chain can contain more than 8 pairs of the first two coordinates of some triples in A (otherwise there are two with the same last coordinate among them and so they are comparable). On the other hand, there are 48 pairs (a, b) with $0 \leq a, b \leq 7$ covered by these four chains. Therefore there are $64 - 48 = 16$ remaining pairs of two elements which are not covered by the chains. Each such pair corresponds to at most one element of A. Therefore A has at most $4 \times 8 + 16 = 48$ elements, a contradiction. Note that the above construction shows that the property fails with only 48 students.

A highly nontrivial example of how the pigeonhole principle can be used in combinatorial problems is the following example. The solution was given by Andrei Jorza.

Example 10. The 2^n rows of a $2^n \times n$ table are filled with all the different n-tuples of 1 and -1. After that, some numbers are replaced by zeros. Prove that there exists a nonempty set of rows such that their sum is the zero vector.

Tournament of the Towns 1996

Solution. Take any numbering $L_1, L_2, ..., L_{2^n}$ of the rows before the replacement of some numbers by 0, in such a way that L_1 is the vector with all coordinates equal to 1 and L_{2^n} is the vector $(-1, -1, ..., -1)$. Define $f(L)$ to be the new line, obtained by (possibly) replacing some numbers by 0. For any row L that now contains some zeros, let $g(L)$ be the corresponding row in the initial table, obtained by the following rule: any 1 in L becomes the value -1 in $g(L)$ and any 0 or -1 in L becomes a 1 in $g(L)$. Now, define the following sequence: $x_0 = (0, 0, ..., 0)$, $x_1 = f(L_1)$ and $x_{r+1} = x_r + f(g(x_r))$. We claim that all terms of this sequence have all coordinates equal to 0 or 1. This is clear if $n = 1$. Assuming that it holds for x_r, observe that the only places in which the value -1 can appear in $f(g(x_r))$ are those on which x_r has a 1, thus all coordinates of x_{r+1} are nonnegative. Also, the places on which a 1 appears on $f(g(x_r))$ must be among the places on which x_r had a 0. This proves that x_{r+1} also has all coordinates equal to 0 or 1. Now, it follows from the pigeonhole principle that for some $i > j$ we have $x_i = x_j$, which can be also written as

$$f(g(x_j)) + f(g(x_{j+1})) + \cdots + f(g(x_{i-1})) = 0$$

and this means precisely that a sum of rows in the new table is zero.
There is no trace of the pigeonhole principle in the following problem. At least at first glance. However, a very clever argument based on the pigeonhole principle allows an elegant proof:

Example 11. Let $(a_n)_{n \geq 1}$ be an increasing sequence of positive integers such that $a_{n+1} - a_n \leq 2001$ for all n. Prove that there are infinitely many pairs (i, j) with $i < j$ such that $a_i | a_j$.

Solution. Let us construct an infinite matrix A with 2001 columns in the following way: the first line consists of the numbers $a_1+1, a_1+2, ..., a_1+2001$. Now, suppose we constructed the first k lines and the kth line is $x_1+1, x_1+2, ..., x_1+2001$. Define the $(k+1)$st line to be $N+x_1+1, N+x_1+2, ..., N+x_1+2001$ where $N=(x_1+1)(x_1+2)\cdots(x_1+2001)$. The way in which this matrix is constructed ensures (an inductive argument doing the job) that for any two elements situated on the same column, one divides the other. Now, pick any 2002 consecutive lines. On each line there is at least one term of the sequence, because $(a_n)_{n\geq 1}$ is increasing and $a_{n+1}-a_n \leq 2001$. Thus there are at least 2002 terms of the sequence on the matrix formed by the selected lines. By the pigeonhole principle, there exist two terms of the sequence on some of the 2001 columns. Those terms will form a good pair. Thus for each choice of 2002 consecutive lines we find a good pair. Because the numbers on each column are increasing, it is enough to apply this procedure to the first 2002 lines, then to the next 2002 lines and so on. This will produce infinitely many good pairs.

The following example was taken from an article called "24 Times the Pigeonhole Principle". We must confess we did not count exactly how many times this phrase appears in the following solution, but we do warn the reader that this will normally take a considerable amount of time.

Example 12. Let $n \geq 10$. Prove that for any coloring with red and blue of the edges of the complete graph with n vertices there exist two vertex-disjoint triangles having all six edges colored with the same color.

[Ioan Tomescu]

Solution. Have courage, this is going to be long! First, we will establish a very useful result, that will be repeatedly used in the solution:

Lemma 20.1. *Every coloring with two colors of the complete graph with six vertices induces a monochromatic triangle. The only coloring with two colors of the complete graph with five vertices that does not induce monochromatic triangles has the form: there exists a pentagon with edges red and diagonals blue.*

Proof. Consider first the case of a complete graph with five vertices. It is clear that with every vertex there are at least two incident edges having the same color. If for a vertex at least three of them have the same color, it can be easily argued that a monochromatic triangle appears. So, suppose that every vertex is incident with two red and two blue edges. Let x be an arbitrary vertex and suppose that xy and xz are red. Then yz is blue. Now, let t be a vertex distinct from x, y, z and suppose that the edge connecting y and t is red and the edge connecting x and t is blue. Let w be the fifth vertex of the graph. Then the edges wz and wt are red, while wx and wy are blue. Similarly, zt is blue and so we can consider the pentagon $xytwz$ which has red edges and blue diagonals.

The case of the complete graph with six vertices is much easier: pick a vertex x. There exist three edges having the same color (say red) leaving from x (again the pigeonhole principle). Let y, z, t be their extremities. If yzt is blue, we are done. Otherwise, assume that yz is red. Then xyz is a monochromatic triangle. The lemma is proved.

□

Now, choose six vertices of the graph. They clearly induce a complete subgraph with six vertices. By the lemma, there exists a monochromatic triangle xyz. If we consider six of the remaining seven vertices, we find another monochromatic triangle uvw, whose set of vertices is disjoint from the set of vertices of xyz. If the two triangles have the same color, we are done. Otherwise, suppose that xyz is red and uvw is blue. Because there are nine edges between the two triangles, by the pigeonhole principle at least five edges have the same color, say blue. By the same principle, there exists a vertex of xyz, call it x, which is incident with at least two blue edges having the other extremity in triangle uvw. Suppose without loss of generality that these vertices are u, v. Thus two triangles xyz and xuv appear with x as a common vertex, the edges of xyz being red and the edges of xuv blue. Look at the remaining five vertices, which form a complete graph with five vertices. If this graph contains a monochromatic triangle, we are done. Otherwise, by the lemma the remaining five vertices form a pentagon $abcde$ with red sides and blue diagonals. By the pigeonhole principle, there exist three edges among those connecting x to vertices of $abcde$ that have the same color. Now we have two cases.

In the first case, vertices y and z are joined by at least three edges having the same color with vertices of $abcde$. If, for instance, the color corresponding to y is blue, then we can consider two blue edges joining y with $abcde$. Then no blue triangle with a vertex in y appears if and only if the two blue edges join y with two consecutive vertices of the pentagon, for example with a and b. But there is still a third blue edge joining y with one of c, d, e, and this shows the existence of a blue triangle with vertices y and the two extremities of a diagonal of the pentagon. So, two blue triangles with disjoint sets of vertices appear. Let us now consider the case when y and z are each joined by at least three red edges with the vertices of the pentagon. So, there is a red triangle with vertices x and two neighboring vertices of the pentagon, say a and b. Consider now y, z, c, d, e. If the induced complete graph with five vertices contains a monochromatic triangle, we are done, because we still have the red triangle xab and the blue triangle xuv. Otherwise, again using the lemma, yz, cd and de are red, so either ze, yc are red or zc, ye are red. In both cases all other edges of the complete graph induced are blue. Let us consider just the first subcase (ze, yc red), the second one being similarly treated. Then y is joined by at least three red edges with vertices of $abcde$, and since yd and ye are diagonals in $ycdez$ (thus they are blue), it follows that ya and yb are red. Similarly we find that za, zb are red and so we have two good triangles zae and xyb.

Finally, let us consider the second case. Actually, all we have to do is to argue as in the first case, by considering vertices u, v joined each by at least three edges of the same color with vertices of $abcde$. So we are done.

The following problems are more computational, but contain much more mathematics than the previous examples. The first one is a famous example due to Behrend, concerning subsets with large cardinality containing no three elements in arithmetic progression. This is related to an even more famous (but notoriously difficult) theorem of Roth: the maximum cardinality of a subset of $\{1, 2, ..., n\}$ having no three elements in arithmetic progression is at most $C\frac{n}{\ln(\ln n)}$ for an absolute constant C. This was refined by Bourgain to

$Cn\sqrt{\frac{\ln(\ln n)}{\ln n}}$. The proofs of these results are very deep, but finding a lower bound for the maximum cardinality of such a set is is not so difficult, if you use the pigeonhole principle. Only easy when compared to the proofs of the mentioned theorems, of course...

Example 13. There exists an absolute constant $c > 0$ such that for all sufficiently large integers N there exists a subset A of $\{1, 2, ..., N\}$ with at least $Ne^{-c\sqrt{\ln N}}$ elements and such that no three elements of A form an arithmetic progression.

Behrend's theorem

Solution. The beautiful idea that provides an elegant proof of this result is the observation that a line cuts a sphere of $\mathbb{R}^n$ in at most two points. For $n = 3$, this is immediate geometrically, and for larger n this follows from the Cauchy-Schwarz inequality: if $||x|| = ||y|| = ||\alpha x + (1-\alpha)y|| = r$ for some $\alpha \in (0, 1)$, it easily follows by squaring the last relation that $\langle x, y\rangle = ||x|| \cdot ||y||$, where $\langle\rangle$ is the natural inner product and $||\cdot||$ the Euclidean norm. By the Cauchy-Schwarz inequality, the last relation implies that x, y are colinear and from here the conclusion easily follows. Now, define $F(n, M, r)$ to be the set of vectors x all of whose coordinates $x_1, x_2, ..., x_n$ are in $\{1, 2, ..., M\}$ and such that $x_1^2 + x_2^2 + \cdots + x_n^2 = r^2$. Fix n, M and observe that as r^2 varies from n to nM^2, the sets $F(n, M, r)$ cover the set of vectors with all coordinates in $\{1, 2, ..., M\}$. Using the pigeonhole principle it follows that there exists some r such that $\sqrt{n} \le r \le M\sqrt{n}$ for which $F(n, M, r)$ has at least $\frac{M^n}{n(M^2-1)} > \frac{M^{n-2}}{n}$ elements. Let us now define the function f from $F(n, M, r)$ to $\{1, 2, ..., N\}$ by $f(x_1, x_2, ..., x_n) = \sum_{i=1}^{n} (2M)^{i-1} x_i$. We claim that if $f(x), f(y), f(z)$ form an arithmetic progression, then $x = y = z$. Indeed, it follows that

$$\sum_{i=1}^{n} (x_i + y_i - 2z_i)(2M)^{i-1} = 0.$$

Put $\alpha_i = x_i + y_i - 2z_i$. Then $|\alpha_i| \le 2M - 1$ and the last relation easily implies that $\alpha_i = 0$ for all i (indeed, $|\sum_{i=1}^{n-1} \alpha_i (2M)^{i-1}| < (2M)^{n-1}$, so $\alpha_n = 0$; now, use an inductive argument to finish the proof). Therefore $x+y=2z$ and because x, y, z lie on a sphere, the observation made in the beginning of the solution shows that $x=y=z$. Also, f is injective: if $f(x)=f(y)$, then $f(x)+f(y)=2f(y)$ and from the above argument, $x+y=2y$, thus $x=y$. Finally,

$$|f(x_1, x_2, \dots, x_n)| \leq M\frac{(2M)^n-1}{2M-1} \leq (2M)^n.$$

Therefore, if we consider M the largest integer such that $(2M)^n \leq N$, then $f(F(n, M, r))$ is a subset of $\{1, 2, \dots, N\}$ which has no arithmetic progressions of length three. Now we need to choose some n as to obtain an optimal cardinality for $f(F(n, M, r))$. But this cardinality is the same as that of $F(n, M, r)$ (because f is injective), which is at least $\frac{M^{n-2}}{n}$ by the choice of r. But

$$\frac{M^{n-2}}{n} > \frac{N^{\frac{n-2}{n}}}{4^{n-2}n}.$$

So choose n the integer part of $\sqrt{\ln N}$ to see that $f(F(n, M, r))$ has at least $Ne^{-c\sqrt{\ln N}}$ elements and has no three elements in arithmetic progression.

We now pass to another revolutionary result, the famous Siegel's lemma. The applications of this theorem are so numerous and important that they would fill a book by themselves. We leave the interested reader to search in the huge literature of transcendental number theory for variations of the following result and for applications, among them the difficult Thue-Siegel-Roth theorem (do not kid yourselves, these require much more than Siegel's lemma alone!).

Example 14. Let $1 \leq m < n$ be integers and let $A = (a_{ij})_{1\leq i\leq n, 1\leq j\leq m}$ be a matrix with integer entries. Suppose that for all $1 \leq j \leq m$, the number $A_j = \sum_{i=1}^{n} |a_{ij}|$ is nonzero. Prove that there exist integers $x_1, x_2, \dots, x_n$, not all zero, such that $|x_i| \leq \sqrt[n-m]{A_1A_2\dots A_m}$ for all $1 \leq i \leq n$ and $\sum_{i=1}^{n} a_{ij}x_i = 0$ for all $1 \leq j \leq m$.

Siegel's lemma

Solution. The idea is the following: for a nonnegative real number M, we will prove that the quantity $\sum_{i=1}^{n} a_{ij}x_i$ cannot take too many distinct values when $(x_1, x_2, ..., x_n)$ goes through the set of vectors with integer coordinates, all of them between 0 and M. It will follow that the image of the function

$$f(x_1, x_2, ..., x_n) = \left(\sum_{i=1}^{n} a_{i1}x_i, ..., \sum_{i=1}^{n} a_{im}x_i\right)$$

is not too big and we will be able to use the pigeonhole principle as long as $([M]+1)^n$ is greater than the image of f.

Consider integers $a_1, a_2, ..., a_n$ and suppose that $a_1, a_2, ..., a_p$ are nonnegative and $a_{p+1}, a_{p+2}, ..., a_n$ are negative. Then it is clear that for any integers x_i such that $0 \le x_i \le M$ we have

$$\lfloor M \rfloor (a_{p+1} + \cdots + a_n) \le a_1x_1 + a_2x_2 + \cdots + a_nx_n \le (a_1 + \cdots + a_p)\lfloor M \rfloor.$$

Thus there are at most $1 + (|a_1| + |a_2| + \cdots + |a_n|)\lfloor M \rfloor$ values taken by $a_1x_1 + a_2x_2 + \cdots + a_nx_n$, which means that the image of f has at most

$$(1 + \lfloor M \rfloor A_1)(1 + \lfloor M \rfloor A_2) \cdots (1 + \lfloor M \rfloor A_m) \le A_1A_2 \cdots A_m(1 + \lfloor M \rfloor)^m$$

elements. Because there are $(1 + \lfloor M \rfloor)^n$ vectors in $\mathbb{Z}^n$ all of whose coordinates are between 0 and M, it follows that f is not injective if we take $M = \sqrt[n-m]{A_1A_2 \cdots A_m}$. Thus there exist two distinct vectors x, y such that $f(x) = f(y)$. It is clear that the vector $v = x - y$ satisfies all the desired conditions.

And here is a surprising, yet very challenging, application of Siegel's lemma, inspired by a USAMO problem:

Example 15. Let $C > 0$ and $A < e^{\frac{2}{11}}$ be two real numbers and let $f : \{1, 2, ...\} \to \{1, 2, ...\}$ be a function satisfying $f(n) < CA^n$ for all n. Suppose that $f(n + p - 1) - f(n)$ is a multiple of

p for any prime number p and any n. Prove that there are integers $r_1, r_2, ..., r_s$ not all zero such that for all n we have

$$r_1 f(n) + r_2 f(n+1) + \cdots + r_s f(n+s-1) = 0.$$

[Gabriel Dospinescu, Vesselin Dimitrov]

Solution. Let us consider positive integers m, n such that $m = \left\lfloor \frac{n}{2} \right\rfloor$ (this is usually the best choice in Siegel's lemma) and let us define $a_{ij} = f(i+j)$ for $1 \le i \le n$ and $1 \le j \le m$. We claim that we can choose some m such that if $x_1, x_2, ..., x_n$ is a solution of the system given by Siegel's lemma, then $x_1 f(j+1) + x_2 f(j+2) + \cdots + x_n f(j+n) = 0$ holds for all positive integers j. For this, we will need some preparation, which will be done in the next paragraph.

Take x_j to be any solution given by Siegel's lemma, and observe that the desired relation holds for $j \le m$. Assume that it fails for some $k > m$ and let $k+1$ be the smallest index for which it fails (thus it holds for all $j \le k$ and $k \ge m$). Consider p any prime smaller than $k+2$. Then $1 \le k+2-p \le k$ and so

$$x_1 f(1+k+2-p) + \cdots + x_n f(n+k+2-p) = 0.$$

But this last sum is congruent (mod p) to

$$\lambda = x_1 f(1+(k+1)) + \cdots + x_n f(n+(k+1)) \qquad (20.1)$$

which is nonzero by the choice of k. This shows that the last quantity λ is actually a multiple of the product of all primes up to $k+1$. The desired contradiction will follow from the fact that Siegel's lemma and the hypothesis on f ensure that λ is small enough and thus cannot be divisible by the product of all p with $p \le k+1$. Let us estimate first the growth of x_j. Using the notations of Siegel's lemma, we have

$$A_j \le C(A^{j+1} + \cdots + A^{j+n}) < C_1 A^{n+j},$$

where $C_1 > 1$ depends only on A, C. Thus

$$|x_j| \leq (A_1 ... A_m)^{\frac{1}{n-m}} < C_1^{1/2} \cdot A^{\frac{mn}{n-m} + \frac{m(m+1)}{2(n-m)}} < C_2 A^{5n/4},$$

for some $C_2 > 0$ depending only on A, C. Therefore

$$|x_1 f(1 + (k+1)) + \cdots + x_n f(n + (k+1))| \leq$$
$$(\max(|x_j|) \cdot C(A^{k+1+1} + \cdots + A^{n+k+1}) < C_3 A^{9n/4+k},$$

where C_3 is again a constant depending only on A, C.

Now, we can prove the claim and thus end the solution. Suppose that the statement does not hold, so for infinitely many k (remember that for each m the corresponding k was at least m) we will have

$$\prod_{p \leq k} p \leq C_3 A^{9n/4+k}.$$

Because $k \geq m > n/2 - 1$, we have

$$A^{9n/4+k} C_3 < A^{11k/2} C_4.$$

Thus for infinitely many k one must have

$$\frac{11k}{2} \cdot \ln A + \ln C_4 > \sum_{p \leq k} \ln p$$

and this forces, from the prime number theorem, $A \geq e^{\frac{2}{11}}$, a contradiction with the choice of A.

We end this chapter with a very challenging problem concerning the growth of coefficients of divisors of a polynomial whose coefficients are $0, 1$ or -1. This type of problems, concerning the multiplicity of roots of polynomials with coefficients $-1, 0, 1$ has been subject to extensive research, but seems to be a quite difficult problem. One estimation in the following problem can easily be obtained using the pigeonhole principle; the other requires a beautiful theorem of Landau.

Example 16. For $n \geq 2$, let A_n be the set of polynomial divisors of all polynomials of degree n with coefficients in $\{-1, 0, 1\}$. Let $C(n)$ be the largest coefficient of a polynomial with integer coefficients that belongs to A_n. Prove that for any $\varepsilon > 0$ there exists a k such that for all $n > k$,

$$2^{n^{\frac{1}{2}-\varepsilon}} < C(n) < 2^n.$$

Solution. Let us start with the left hand side inequality: $C(n) > 2^{n^{\frac{1}{2}-\varepsilon}}$. For a polynomial f with coefficients 0 or 1 and degree at most n define the function $\phi(f) = (f(1), f'(1), ..., f^{N-1}(1))$. Taking into account that all coefficients of f are 0 or 1, we can immediately deduce that $f^{(j)}(1) \leq (1+n)^{j+1}$ for all j, thus the image of f has at most $(1+n)^{1+2+\cdots+N} < (1+n)^{N^2}$ elements. On the other hand, f is defined on a set of 2^{n+1} elements. So, if $2^{n+1} > (1+n)^{N^2}$ then by the pigeonhole principle two polynomials f, g will have the same image and thus their difference will have all coefficients $-1, 0$ or 1 and degree at most n. Also, $f - g$ will be divisible by $(X-1)^N$. Thus $C(n) \geq \binom{2N}{N}$, because the largest coefficient of $(X-1)^N$ is $\binom{2N}{N}$. Because $\binom{2N}{N}$ is the largest binomial coefficient among $\binom{2N}{k}$, we have $\binom{2N}{N} > \frac{4^N}{2N+1} > 2^N$ for $N > N_0$. By taking $N = \left\lfloor \sqrt{\frac{n}{\log_2(n+1)}} \right\rfloor$, we have $(1+n)^{N^2} < 2^{n+1}$, thus $C(n) > 2^N$ and it is easy to see that $N > n^{\frac{1}{2}-\varepsilon}$ for n large enough.

The other part, $C(n) < 2^n$, is much more subtle. For a polynomial $f(X) = a_nX^n + \cdots + a_1X + a_0$ with real coefficients (everything that follows applies verbatim for complex coefficients), define its Mahler measure by

$$M(f) = |a_n| \prod_{i=1}^{n} \max(1, |x_i|) \tag{20.2}$$

where x_i are the roots of f. The following inequality is due to Landau:

Lemma 20.2. $M(f) \le \sqrt{a_0^2 + a_1^2 + \cdots + a_n^2}$.

Proof. There are many proofs of this lemma, but we particularly like the following one, which we haven't encountered in the literature. Consider $N > n$ and let $z_1, z_2, ..., z_N$ be the N-th roots of unity. A simple computation, based on the fact that $\sum_{j=1}^N z_j^k = N$ if $N|k$ and 0 otherwise, shows that

$$\sum_{j=1}^N |f(z_j)|^2 = \sum_{j=1}^N \left(\sum_{i=0}^n a_i z_j^i\right)\left(\sum_{i=0}^n a_i z_j^{-i}\right) =$$

$$\sum_{u,v=0}^n a_u a_v \cdot \sum_{j=1}^N z_j^{u-v} = N \cdot \sum_{i=0}^n a_i^2.$$

Now, applying the AM-GM inequality, we obtain that

$$\sum_{i=0}^n a_i^2 \ge \sqrt[N]{|f(z_1)f(z_2)\cdots f(z_N)|^2}.$$

On the other hand, the identity $(X - z_1)(X - z_2)\cdots(X - z_N) = X^N - 1$ and the fact that $f(X) = a_n(X - x_1)(X - x_2)\cdots(X - x_n)$ imply the identity

$$|f(z_1)f(z_2)\cdots f(z_N)| = |a_n|^N |1 - x_1^N||1 - x_2^N|\cdots|1 - x_n^N|,$$

which, combined to the previous inequality, implies

$$\sqrt{\sum_{i=0}^n a_i^2} \ge |a_n| \cdot \prod_{i=1}^n \sqrt[N]{|1 - x_i^N|}. \tag{20.3}$$

Now, it is pretty clear that $\lim_{N\to\infty} \sqrt[N]{|1-z^N|} = \max(1,|z|)$ whenever $|z| \neq 1$. Thus the inequality is proved whenever all roots of f lie outside the unit circle. What happens in the opposite case? It really does not matter! Actually, Viéte's formulae show that the inequality $M(f) \leq \sqrt{a_0^2 + a_1^2 + \cdots + a_n^2}$ reduces to an inequality involving only absolute values of polynomials in x_i. If this inequality holds whenever the variables x_i are not on the unit circle, it also holds in the other cases, by continuity. Therefore the lemma is proved. □

The previous lemma shows that polynomials with all coefficients of absolute value at most 1 have Mahler measure at most $\sqrt{n+1}$. Take now any divisor f of a polynomial g with all coefficients $-1, 0, 1$ and write $g = hf$. Suppose that f has integer coefficients. It is easy to see that $M(g) = M(h)M(f) \geq M(f)$. Thus $M(f) \leq \sqrt{n+1}$. Now, observe that by Viéte's formula, the triangular inequality and the obvious fact that $|x_{i_1}x_{i_2}...x_{i_s}| \leq M(f)$ for all distinct $i_1, ..., i_s$ and all s, we have that any coefficient of f is bounded in absolute value by the fact that

$$\binom{n}{\lfloor \frac{n}{2} \rfloor} M(f) \leq \sqrt{n+1} \cdot \binom{n}{\lfloor \frac{n}{2} \rfloor} < 2^n$$

for sufficiently large n. Thus the conclusion follows.

20.2 Practice problems

1. Prove that for infinitely many positive integers A the equation $\lfloor x\sqrt{x}\rfloor + \lfloor y\sqrt{y}\rfloor = A$ has at least 1980 solutions in positive integers.

   Russia 1980

2. Find the largest positive integer n with the property: there exist nonnegative integers $x_1, x_2, \ldots, x_n$, not all of them equal to 0 and such that n^3 does not divide any of the numbers $a_1x_1 + a_2x_2 + \cdots + a_nx_n$ with $a_1, a_2, \ldots, a_n \in \{-1, 0, 1\}$, not all of them equal to 0.

   Dorel Mihet, Romanian TST 1996

3. Consider a set of 2000 positive integers not exceeding 10^{100}. Prove that this set has two nonempty disjoint subsets with the same size, the same sum of elements, and the same sum of squares of the elements.

   Poland 2001

4. Let r, n be positive integers. For a set A, let $\binom{A}{r}$ be the set of subsets of A having r elements. Let A be an infinite set and let $f : \binom{A}{r} \to \{1, 2, \ldots, n\}$ be a map. Prove that there exists an infinite subset B of A such that $f(X) = f(Y)$ for all $X, Y \in \binom{B}{r}$.

   IMO 1987 Shortlist

5. Let $n \geq 3$ be an integer. On a circle of length 1 consider finitely many pairwise disjoint arcs, whose sum of lengths is greater than $1 - \frac{1}{n}$. Prove that there exists a regular n-gon having vertices on these arcs.

   Marius Cavachi, Romanian TST 1993

6. Prove that given any n^2 integers, we can always put them in an $n \times n$ matrix whose determinant is divisible by $n^{\lfloor \frac{n-1}{2} \rfloor}$.

Titu Andreescu, Revista Matematică Timisoara

7. Prove that any integer k greater than 1 has a multiple smaller than k^4 which has at most four distinct digits.

Ioan Tomescu, Romanian TST 1989

8. Prove that no matter how we choose more than $\frac{2^{n+1}}{n}$ points in $\mathbb{R}^n$, all of whose coordinates are ± 1, there exists an equilateral triangle with vertices in three of these points.

Putnam 2000

9. Let n be a positive integer. What is the size of the largest subset of $\{-n, -n+1, \ldots, n-1, n\}$ which does not contain three elements a, b, c (not necessarily distinct) satisfying $a + b + c = 0$?

USAMO 2009

10. Find the least n with the following property: there exists a partition with n classes of $\{1, 2, \ldots, 40\}$ such that whenever a, b, c (not necessarily distinct) are in the same class, we have $a \neq b + c$.

Belarus 2000

11. Prove that any sequence of $mn + 1$ real numbers contains an increasing subsequence with $m + 1$ terms or a decreasing subsequence with $n + 1$ terms.

Erdős-Szekeres's theorem

12. Consider n points in the plane. Prove that we can choose $[\sqrt{n}]$ of them such that no three are vertices of an equilateral triangle.

Romanian Contest

13. Let A be the set of the first $2^m \cdot n$ positive integers and let S be a subset of A with $(2^m - 1)n + 1$ elements. Prove that there exist $a_0, a_1, \ldots, a_m$ distinct elements of S such that $a_0 \mid a_1 \mid \cdots \mid a_m$.

Romanian TST 2006

14. Consider an 11×11 chess board whose unit squares are colored using three colors. Prove that there exists an $m \times n$ rectangle with $2 \leq m, n \leq 11$ whose vertices are in squares having the same color.

Ioan Tomescu, Romanian TST 1988

15. In a competition, 50 students are to solve the same 8 problems. A total of 171 correct solutions were received. Prove that there are at least three problems that were solved by at least three students.

Valentin Vornicu, Radu Gologan, Mathlinks Contest

16. Let k be an integer, and let $a_1, a_2, \ldots, a_n$ be integers which give at least $k+1$ distinct remainders when divided by $n+k$. Prove that some of these n numbers add up to a multiple of $n+k$.

Kömal

17. Let $P_0, P_1, \ldots, P_{n-1}$ be some points on the unit circle. Also let $A_1A_2 \ldots A_n$ be a regular polygon inscribed on this circle. Fix an integer k, with $1 \leq k \leq \frac{n}{2}$. Prove that one can find i, j such that $A_iA_j \geq A_1A_k \geq P_iP_j$.

AMM

18. Prove that for all N there exists a k such that more than N prime numbers can be written in the form T^2+k for some integer T. Prove the same result with T^2+k replaced by any nonconstant monic polynomial $f \in \mathbb{Z}[X]$.

Sierpinski

19. Let A be a set of n remainders modulo n^2. Prove the existence of a set B of n remainders modulo n^2 such that $A+B=\{a+b \pmod{n^2} \mid a \in A, b \in B\}$ has at least $\frac{n^2}{2}$ elements.

IMO Shortlist 1999

20. Let a be a real number with $0<a<\frac{1}{2}$ and let $(a_n)_{n\geq 1}$ be an increasing sequence of positive integers such that for all sufficiently large n there are at least $n \cdot a$ terms of the sequence smaller than n. Prove that for all $k>\frac{1}{a}$ there are infinitely many terms of the sequence that can be written as the sum of at most k other terms of the sequence.

Paul Erdős, AMM

21. There are $(n+1)^2$ points in the interior of a square of side-length 1. Prove that one can choose three of them such that the area of the triangle determined by them is at most $\frac{1}{2}$.

Dan Schwarz, Romanian Masters in Mathematics 2008

22. Let $f(n)$ be the largest prime divisor of n and let $(a_n)_{n\geq 1}$ be an increasing sequence of positive integers. Prove that the set of all $f(a_i+a_j)$ (for all $i \neq j$) is unbounded.

G. Grunwald, D. Lazar

23. Let $n \geq 3$ and consider $3n^2$ positive integers smaller than or equal to n^3. Prove that among them one can find nine distinct numbers $a_1, a_2, \ldots, a_9$ and one can find nonzero integers x, y, z such that $a_1x + a_2y + a_3z = 0$, $a_4x + a_5y + a_6z = 0$ and $a_7x + a_8y + a_9z = 0$.

Marius Cavachi, Romanian TST 1996

24. Let $a < b < c$ be positive integers. Prove that there exist integers x, y, z, not all zero, such that $ax + by + cz = 0$ and

$$\max(|x|, |y|, |z|) \leq 1 + \frac{2}{\sqrt{3}}c.$$

Miklos Schweitzer Competition

25. For a pair a, b of integers with $0 < a < b < 1000$, the subset S of $\{1, 2, \ldots, 2003\}$ is called a skipping set for (a, b) if $|s_1 - s_2| \notin \{a, b\}$ for any $(s_1, s_2) \in S^2$. Let $f(a, b)$ be the maximum size of a skipping set for (a, b). Determine the maximum and minimum values of f.

Zuming Feng, USA TST 2003

26. A frog stays at the origin $(0, 0)$ of the plane. At every second, if the frog is at the point (x, y), it can jump to one of the points $(x + 1, y)$ or $(x, y + 1)$. Suppose that the frog jumps infinitely many times. Prove that for any $n \geq 3$ there are n collinear points on the frog's path.

T.C.Brown, AMM

Chapter

21

21.1 Theory and examples

It is notoriously difficult to decide whether a given polynomial is irreducible over a certain field. There exist a variety of criteria that allow us to prove that a certain polynomial is irreducible, but unfortunately they are very limited, and their hypotheses are usually not satisfied. Furthermore, there are not many elementary techniques: a few classical irreducibility criteria and the study of roots of polynomials are practically the only ideas that we will discuss in this chapter. But, as you can easily see, even those are not trivial, and some of the problems can be extremely difficult, even though they have elementary solutions. We will discuss a very useful irreducibility criterion, Capelli's theorem, which is really not as well known as it should be, and we will see some striking consequences of this result. Also, we will insist on the method of studying the roots of polynomials, because it gives elegant solutions for problems of this type: Perron's criterion and Rouche's theorem are discussed, as well as some applications. Finally, we will see that working with reductions of polynomials modulo primes can often give precious information about their irreducibility properties. In this chapter, we will assume that the reader is familiar with notions of algebraic number theory, but those will not exceed the results discussed in the chapter **A Brief Introduction to Algebraic Number Theory**.

We will begin the discussion with the most elementary method, which is the study of roots of polynomials. Let us observe from the beginning two quite useful results: if a monic polynomial with integer coefficients f has a nonzero free term (constant term) and exactly one root of absolute value greater than 1, then f is irreducible in $\mathbb{Q}[X]$. Indeed, if $f = gh$ for some nonconstant polynomials g, h with integer coefficients, we may assume that g has all roots of absolute value smaller than 1. Then $|g(0)| < 1$, because it is just the product of the absolute values of the roots of g. Because $|g(0)|$ is an integer, it follows that $g(0) = 0$ and thus $f(0) = 0$, contradiction.

The second result is very similar: if f is monic and all roots of f are outside the closed unit disc and $|f(0)|$ is a prime number, then f is irreducible in $\mathbb{Q}[X]$. Indeed, with the same notations, we may assume that $|g(0)| = 1$. Because

$|g(0)|$ is the product of the absolute values of the roots of g, it follows that at least one root of g is within the unit disc. But then f has at least one root in the closed unit disc, which is a contradiction. Here are some examples, the first two extremely simple, but useful, and the others more and more difficult.

Example 1. Let $f(X) = a_0+a_1X+\cdots+a_nX^n$ be a polynomial with integer coefficients such that a_0 is prime and $|a_0| > |a_1| + |a_2| + \cdots + |a_n|$. Prove that f is irreducible in $\mathbb{Z}[X]$.

Solution. By previous arguments, it is enough to prove that all zeros of f are outside the closed unit disk of the complex plane. But this is not difficult, because if z is a zero of f and if $|z| \leq 1$ then

$$|a_0| = |a_1z + a_2z^2 + \cdots + a_nz^n| \leq |a_1| + |a_2| + \cdots + |a_n|,$$

which contradicts the hypothesis of the problem.

The previous example may look a bit artificial, but it is quite powerful for theoretic purposes. For example, it immediately implies the Goldbach theorem for polynomials with integer coefficients: any such polynomial can be written as the sum of two irreducible polynomials. Actually, it proves much more: for any polynomial f with integer coefficients there are infinitely many positive integers a such that $f + a$ is irreducible in $\mathbb{Z}[X]$.

We have already discussed algebraic numbers and some of their properties. We will see that they play a fundamental role in proving the irreducibility of a polynomial. However, we will work with an extension of the notion of algebraic number: for any field $K \subset \mathbb{C}$, we say that the number $z \in \mathbb{C}$ is algebraic over K if there exists a polynomial $f \in K[X]$ such that $f(z) = 0$. Exactly the same arguments as those presented for algebraic numbers over $\mathbb{Q}$ allow us to deduce the same properties of the minimal polynomial of an algebraic number over K. Also, α is an algebraic number, the set $K[\alpha]$ of numbers of the form $g(\alpha)$ with $g \in K[X]$ is a field included in $\mathbb{C}$. The following fundamental result is frequently used.

Example 2. Let K be a subfield of $\mathbb{C}$, p a prime number, and $a \in K$. The polynomial $X^p - a$ is reducible in $K[X]$ if and only if there exists $b \in K$ such that $a = b^p$.

Solution. One implication being obvious, let us concentrate on the more difficult part. Suppose that $X^p - a$ is reducible in $K[X]$, and consider α such that $\alpha^p = a$. Let f be the minimal polynomial of α over K and let $m = \deg(f)$. Clearly $m < p$. Let $f(X) = (X - \alpha_1)(X - \alpha_2)\cdots(X - \alpha_m)$ and introduce the numbers $r_1 = \alpha, r_i = \frac{\alpha_i}{\alpha}$ for $i \geq 2$. Because f divides $X^p - a$, we have $r_i^p = 1$. Hence $(-1)^m f(0) = c\alpha^m$ for some c, a root of unity of order p. Since $m < p$, there exist integers u, v such that $um + vp = 1$. It follows that $(-1)^{um} f^u(0) = c^u \alpha^{1-vp}$. Combining this observation with the fact that $\alpha^p = a$, we deduce that $c^u\alpha = (-1)^{mu} f(0)^u a^v = b \in K$, thus $a = \alpha^p = b^p$. This finishes the proof of the hard part of the problem.

We continue with a very beautiful result, the celebrated Cohn's theorem. It shows how to produce lots of irreducible polynomials: just pick prime numbers, write them in any base you want and make a polynomial with the digits in that base!

Example 3. Let $b \geq 2$ and let p be a prime number. Write $p = a_0 + a_1 b + \cdots + a_n b^n$ with $0 \leq a_i \leq b - 1$. Then the polynomial $f(X) = a_n X^n + a_{n-1} X^{n-1} + \cdots + a_1 X + a_0$ is irreducible in $\mathbb{Q}[X]$.

[Cohn's theorem]

Solution. It is clear that $\gcd(a_0, a_1, \ldots, a_n) = 1$, so by Gauss's lemma it is enough to prove that f is irreducible over $\mathbb{Z}$. First, we will discuss the case $b > 3$, the case $b = 2$ being, as we will see, much more difficult. Suppose that $f(X) = g(X)h(X)$ is a nontrivial factorization of f. Because p is a prime, one of the numbers $g(b)$ and $h(b)$ is equal to 1 or -1. Let this number be $g(b)$ and let $x_1, x_2, \ldots, x_n$ be the zeros of f. There exists a subset A of the set

$\{1, 2, \ldots, n\}$ such that $g(X) = a \prod_{i \in A} (X - x_i)$. We now prove a helpful result.

Lemma 21.1. *Each complex zero of f has either nonpositive real part or an absolute value smaller than $\frac{1+\sqrt{4b-3}}{2}$.*

Proof. The proof is rather tricky, but not complicated. It is enough to observe that if $|z| > 1$ and $\mathrm{Re}(z) > 0$ then $\mathrm{Re}\left(\frac{1}{z}\right) > 0$ and so by the triangle inequality

$$\left|\frac{f(z)}{z^n}\right| \geq \left|a_n + \frac{a_{n-1}}{z}\right| - (b-1)\left(\frac{1}{|z|^2} + \cdots + \frac{1}{|z|^n}\right)$$

$$> \mathrm{Re}\left(a_n + \frac{a_{n-1}}{z}\right) - \frac{b-1}{|z|^2 - |z|} \geq \frac{|z|^2 - |z| - (b-1)}{|z|^2 - |z|}.$$

Therefore if $f(z) = 0$ and $\mathrm{Re}(z) > 0$ then either $|z| \leq 1$ or $|z| < \frac{1+\sqrt{4b-3}}{2}$ and this establishes the lemma. □

It remains now to cleverly apply this result. We claim that for any zero x_i of f we have $|b - x_i| > 1$. Indeed, if $\mathrm{Re}(x_i) \leq 0$, everything is clear. Otherwise, $|b - x_i| \geq b - |x_i| > b - \frac{1+\sqrt{4b-3}}{2} \geq 1$, as you can easily verify if $b \geq 3$. Now, everything is clear, because this result implies that $|g(b)| > 1$, a contradiction.

Now let us deal with the very difficult case $b = 2$. We will present a very beautiful solution communicated by Alin Bostan. The idea is to prove that $|2 - x_i| > |1 - x_i|$ for any zero x_i of f. Keeping the previous notations, we will deduce that $1 = |g(2)| > |g(1)|$ and so $g(1) = 0$. This implies $f(1) = 0$, which is clearly impossible. Now take x to be a zero of f and observe that if $|2 - x| < |1 - x|$ then $\mathrm{Re}(x) > \left(\frac{3}{2}\right)$, and so if $y = \frac{1}{x}$ we have $|y| < 1$ and y satisfies a relation of the form

$$y^n + \left(\frac{1}{2} \pm \frac{1}{2}\right) y^{n-1} + \cdots + \left(\frac{1}{2} \pm \frac{1}{2}\right) y + 1 = 0.$$

Multiplying by y^{n+1} and adding the two relations, we find another relation of the same type (but with n increased) and by repeating this argument we deduce that there are infinitely many N for which y satisfies the relation

$$y^N + \left(\frac{1}{2} \pm \frac{1}{2}\right) y^{N-1} + \cdots + \left(\frac{1}{2} \pm \frac{1}{2}\right) + 1 = 0.$$

This can be also written as

$$1 + \frac{1}{2} \cdot \left(y + y^2 + \cdots + y^N\right) = \frac{1}{2} \cdot \left(\pm y \pm y^2 \pm \cdots \pm y^N\right).$$

The triangle inequality implies

$$\left|\frac{2 - y - y^{N+1}}{2(1-y)}\right| \le \frac{|y| - |y|^{N+1}}{2(1-|y|)}$$

and this for infinitely many N. Taking into account the fact that $|y| < 1$, we deduce from the above inequality that

$$\left|\frac{2-y}{1-y}\right| \le \frac{|y|}{1-|y|}.$$

Finally, the last inequality implies

$$\left|\frac{2x-1}{x-1}\right| \le \frac{1}{|x|-1} \le 2$$

and thus $|2x-1| \le |2x-2|$, which is impossible for $\mathrm{Re}(x) \ge \frac{3}{2}$. This finishes the proof of the claim and also the solution of this difficult problem.

We end this part of the chapter with a very beautiful criterion due to Perron and with a difficult theorem of Selmer. Perron's criterion is quite similar to the first example, but much more difficult to prove: it states that if a coefficient is "too large", then the polynomial is irreducible. Here is the precise statement:

Example 4. Let $f(X) = X^n + a_{n-1}X^{n-1} + \cdots + a_1 X + a_0$ be a polynomial with integer coefficients. If $|a_{n-1}| > 1 + |a_0| + |a_1| + \cdots + |a_{n-2}|$ and $a_0 \neq 0$ then f is irreducible in $\mathbb{Q}[X]$.

[Perron]

Solution. We will prove that f has exactly one zero outside the closed unit disk of the complex plane. This will show that f is irreducible in $\mathbb{Z}[X]$, and by Gauss's lemma it will also be irreducible in $\mathbb{Q}[X]$. It is quite clear that no zero of f is on the unit circle, because if z is such a zero, then

$$|a_{n-1}| = |a_{n-1}z^{n-1}| = |z^n + a_{n-2}z^{n-2} + \cdots + a_1 z + a_0| \le 1 + |a_0| + \cdots + |a_{n-2}|,$$

a contradiction. On the other hand, $|f(0)| \ge 1$, so by Viéte's formula at least one zero of f lies outside the unit disk. Call this zero x_1 and let $x_2, \ldots, x_n$ be the other zeros of f. Let

$$g(X) = X^{n-1} + b_{n-2}X^{n-2} + \cdots + b_1 X + b_0 = \frac{f(X)}{X - x_1}.$$

By identifying coefficients in the formula $f(X) = (X - x_1)g(X)$, we deduce that

$$a_{n-1} = b_{n-2} - x_1,\ a_{n-2} = b_{n-3} - b_{n-2}x_1,\ \ldots, a_1 = b_0 - b_1 x_1,\ a_0 = -b_0 x_1.$$

Therefore the hypothesis $|a_{n-1}| > 1 + |a_0| + |a_1| + \cdots + |a_{n-2}|$ can be rewritten as

$$|b_{n-2} - x_1| > 1 + |b_{n-3} - b_{n-2}x_1| + \cdots + |b_0 x_1|.$$

Taking into account that $|b_{n-2}| + |x_1| \ge |b_{n-2} - x_1|$ and

$$|b_{n-3} - b_{n-2}x_1| \ge |x_1||b_{n-2}| - |b_{n-3}|,\ \ldots,\ |b_0 - b_1 x_1| \ge |b_1||x_1| - |b_0|,$$

we deduce that $|x_1| - 1 > (|x_1| - 1)(|b_0| + |b_1| + \cdots + |b_{n-2}|)$ and since $|x_1| > 1$, it follows that $|b_0| + |b_1| + \cdots + |b_{n-2}| < 1$. Using an argument based on the triangle inequality, similar to the one in the first example, we immediately infer that g has only zeros inside the unit disk, which shows that f has exactly one zero outside the unit disk. This finishes the proof of this criterion.

The above elegant solution, due to Laurenţiu Panaitopol, shows that deep theorems can be avoided even when this seems impossible. The classical proof of this criterion uses Rouché's theorem. Because this is also a very powerful tool, we prefer to prove it in a very particular, but very common, case for polynomials and circles.

Theorem 21.2 (Rouché's theorem). *Let P, Q be two polynomials with complex coefficients and let R be a positive real number. If P, Q satisfy the inequality $|P(z) - Q(z)| < |Q(z)|$ for all z on the circle of radius R, centered at the origin, then the two polynomials have the same number of zeros inside the circle, multiplicities being counted.*

Proof. The proof of this theorem is not elementary, but with a little bit of integral calculus it can be proved in a very elegant way. Let L be the set of all curves $\gamma : [0, 2\pi] \to \mathbb{C}$ which are differentiable, with continuous derivative, such that $\gamma(0) = \gamma(2\pi)$ and γ does not vanish. The index of $\gamma \in L$ is defined as

$$I(\gamma) = \frac{1}{2i\pi} \cdot \int_0^{2\pi} \frac{\gamma'(t)}{\gamma(t)} dt \qquad (21.1)$$

We claim that $I(\gamma)$ is an integer. Indeed, consider $K(t) = e^{\int_0^t \frac{\gamma'(x)}{\gamma(x)} dx}$ and note that K is differentiable and that $K'(t) = K(t) \cdot \frac{\gamma'(t)}{\gamma(t)}$. This shows that $\frac{K(t)}{\gamma(t)}$ is a constant function. Therefore, because $\gamma(0) = \gamma(2\pi)$, we must have $K(0) = K(2\pi)$, which says exactly that $I(\gamma)$ is an integer. The following result is essential in the proof:

Lemma 21.3. *The index of a curve $\gamma \in L$* contained in a disc that does not contain the origin is 0.

Proof. Let $B(x, r)$ be the open disc of center x and radius $r > 0$ and suppose that γ is contained in $B(\omega, s)$, a disc that does not contain the origin (thus $s < |\omega|$), that is $|\gamma(t) - \omega| < s$ for all t. The idea is to make a continuous deformation of γ, keeping the index unchanged, and such that at a certain moment the index of the new curve can be trivially computed. In order to do this, take $u \in [0, 1]$ and consider the application $f_u(t) = u\gamma(t) + (1 - u)\omega$, defined on $[0, 2\pi]$. The triangle inequality shows that $f_u \in L$ and also that this curve is contained in $B(\omega, s)$. On the other hand, we claim that the mapping $\phi(u) = I(f_u)$ is continuous. Because it takes only integer values (by the previous remark), it will be constant. Therefore, $I(\gamma) = I(f_1) = I(f_0) = 0$. So, let us prove that $I(f_u)$ is continuous with respect to u. Indeed, note that

$$\left|\frac{f_u'(t)}{f_u(t)} - \frac{f_v'(t)}{f_v(t)}\right| = \left|\frac{\omega \cdot (u-v) \cdot \gamma'(t)}{(u\gamma(t) + (1-u)\omega)(v\gamma(t) + (1-v)\omega)}\right|$$
$$\leq \frac{|\omega| \cdot |u-v| \cdot |\gamma'(t)|}{(|\omega| - s)^2}$$

because $|u\gamma(t) + (1-u)\omega| \geq |\omega| - |u||\gamma(t) - \omega| \geq |\omega| - s$. This inequality shows by integration that $I(f_u)$ satisfies

$$|I(f_u) - I(f_v)| \leq \frac{|\omega|}{2\pi(|\omega| - s)^2} \cdot \int_0^{2\pi} |\gamma'(t)| dt \cdot |u - v|,$$

which proves that $I(f_u)$ is continuous, and finishes the proof of the lemma. □

This lemma implies that two curves in L sufficiently close have equal index. Indeed, suppose that γ_1 and γ_2 are in L and satisfy $|\gamma_1(t) - \gamma_2(t)| < |\gamma_2(t)|$ for all t. Then the curve $\gamma(t) = \frac{\gamma_1(t)}{\gamma_2(t)}$ satisfies $|\gamma(t) - 1| < 1$ for all t. Because $|\gamma(t) - 1|$ is also continuous on the compact interval $[0, 2\pi]$, it follows that its maximum is smaller than 1, that is, there exists a disc that does not contain the origin and which contains γ. By the lemma, γ has index 0. But a quick computation shows that $I(\gamma) = I(\gamma_1) - I(\gamma_2)$. Thus γ_1 and γ_2 have the same index. Finally, let us prove this particular case of Rouché's theorem. Consider the curves $\gamma_1(t) = P(Re^{it})$ and $\gamma_2(t) = Q(Re^{it})$. Observe that the inequality $|P(z) - Q(z)| < |Q(z)|$ implies that γ_i does not vanish on $[0, 2\pi]$. Thus γ_1, γ_2 are in L and also $|\gamma_1(t) - \gamma_2(t)| < |\gamma_2(t)|$. Thus the two curves have the same index. But for a polynomial P one can easily compute the index of the associated curve! Indeed, suppose that $P(z) = a(z - z_1)(z - z_2) \cdots (z - z_n)$, where z_i are not necessarily distinct. Then it is well known that

$$\frac{P'(z)}{P(z)} = \sum_{i=1}^{n} \frac{1}{z - z_i} \tag{21.2}$$

This shows that if $\gamma(t) = P(Re^{it})$, then

$$I(\gamma) = \frac{R}{2\pi} \cdot \sum_{j=1}^{n} \int_0^{2\pi} \frac{e^{it}dt}{Re^{it} - z_j}.$$

Now, we have seen that $|z_j| \neq R$. Suppose that $|z_j| < R$. Then

$$\int_0^{2\pi} \frac{e^{it}dt}{Re^{it} - z_j} = \frac{1}{R} \int_0^{2\pi} \frac{dt}{1 - \frac{z_j}{R}e^{-it}} = \frac{2\pi}{R}.$$

Indeed,

$$\frac{1}{1 - \frac{z_j}{R}e^{-it}} = 1 + \sum_{m \geq 1} \frac{z_j^m}{R^m} e^{-imt},$$

and the mean value of e^{-imt} over $[0, 2\pi]$ is 0 for all $m \geq 1$. It is enough to change the order of integral and summation (which is legal, because of the uniform convergence with respect to t) in order to see that

$$\frac{1}{R} \int_0^{2\pi} \frac{dt}{1 - \frac{z_j}{R}e^{-it}} = \frac{2\pi}{R}.$$

Now, in exactly the same way, you can prove that

$$\int_0^{2\pi} \frac{e^{it}dt}{Re^{it} - z_j} = 0$$

if $|z_j| > R$. Thus $I(\gamma)$ is exactly the number of zeros of P inside the circle of radius R centered at the origin. This finishes the proof of Rouché's theorem.

□

Observe that Perron's criterion instantly solves the following old IMO problem: the polynomial $X^n + 5X^{n-1} + 3$ is irreducible in $\mathbb{Q}[X]$, just because "five is greater than four!" Here are two nicer examples, where this criterion turns out to be extremely efficient. The solution to the first problem is due to Mikhail Leipnitski.

Example 5. Let $f_1, f_2, \ldots, f_n$ be polynomials with integer coefficients. Prove that there exists a reducible polynomial $g \in \mathbb{Z}[X]$ such that all polynomials $f_1 + g, f_2 + g, \ldots, f_n + g$ are irreducible in $\mathbb{Q}[X]$.

Iranian Olympiad

Solution. Using Perron's criterion, it is clear that if M is sufficiently large and m is greater than $\max(\deg(f_1), \deg(f_2), \ldots, \deg(f_n))$, the polynomials $X^{m+1} + MX^m + f_i(X)$ are all irreducible in $\mathbb{Q}[X]$. Therefore we can choose $g(X) = X^{m+1} + MX^m$.

Example 6. Let $(f_n)_{n\geq 0}$ be the Fibonacci sequence, defined by $f_0 = 0, f_1 = 1$ and $f_{n+1} = f_n + f_{n-1}$. Prove that for any $n \geq 2$ the polynomial $X^n + f_n f_{n+1} X^{n-1} + \cdots + f_2 f_3 X + f_1 f_2$ is irreducible in $\mathbb{Q}[X]$.

[Valentin Vornicu] Mathlinks Contest

Solution. By Perron's criterion, it suffices to verify the inequality

$$f_{n+1} f_n > f_n f_{n-1} + \cdots + f_2 f_1 + 1$$

for all $n \geq 3$. For $n = 3$ it is obvious. Supposing the inequality true for n, we have

$$f_{n+1} f_n + f_n f_{n-1} + \cdots + f_2 f_1 + 1 < f_{n+1} f_n + f_{n+1} f_n < f_{n+2} f_{n+1},$$

because this is equivalent to $2f_n < f_{n+2} = f_{n+1} + f_n$ and this one is obvious. The inductive step is proved and so is the proof for $n \geq 3$.

Finally, a very difficult example of an irreducibility problem that can be solved by studying the roots of polynomials. It generalizes a classical result stating that $X^p - X - 1$ is irreducible over the field of rational numbers if p is a prime number.

Example 7. Prove that $X^n - X - 1$ is irreducible in $\mathbb{Q}[X]$ for all n.

Selmer's theorem

Solution. Let us consider a factorization $X^n - X - 1 = f(X)g(X)$ for some integer nonconstant polynomials f, g. It is not difficult to check that $X^n - X - 1$ has distinct complex roots. Thus f will have some roots $z_1, z_2, \ldots, z_s$ of $X^n - X - 1$, which are pairwise distinct. The essential observation is the following estimation:

Lemma 21.4. *For each root z of $X^n - X - 1$ one has*

$$2Re\left(z - \frac{1}{z}\right) > \frac{1}{|z|^2} - 1.$$

Proof. By writing $z = re^{it}$, the inequality comes down to $(1+2r\cos t)(r^2-1) > 0$. However, $r^{2n} = |z|^{2n} = |z+1|^2 = 1 + 2r\cos t + r^2$, so what we need is $(r^{2n} - r^2)(r^2 - 1) > 0$, which is clear. $\square$

Using the lemma, it follows that

$$2\text{Re}\left(z_1 - \frac{1}{z_1}\right) + 2\text{Re}\left(z_2 - \frac{1}{z_2}\right) + \cdots + 2\text{Re}\left(z_s - \frac{1}{z_s}\right) >$$

$$> \frac{1}{|z_1|^2} + \frac{1}{|z_2|^2} + \cdots + \frac{1}{|z_s|^2} - s \geq 0,$$

by the AM-GM inequality, because the product of $|z_i|$ is just $|f(0)| = 1$. Thus

$$\operatorname{Re}\left(z_1 - \frac{1}{z_1}\right) + \operatorname{Re}\left(z_2 - \frac{1}{z_2}\right) + \cdots + \operatorname{Re}\left(z_s - \frac{1}{z_s}\right) > 0.$$

On the other hand, because f is monic and has integer coefficients,

$$\operatorname{Re}\left(z_1 - \frac{1}{z_1}\right) + \operatorname{Re}\left(z_2 - \frac{1}{z_2}\right) + \cdots + \operatorname{Re}\left(z_s - \frac{1}{z_s}\right)$$

is an integer, so it is actually at least 1. Working similarly with g, we deduce that $\operatorname{Re}\left(z_1 - \frac{1}{z_1} + z_2 - \frac{1}{z_2} + \cdots + z_n - \frac{1}{z_n}\right) \geq 2$, where $z_1, z_2, \ldots, z_n$ are all roots of $X^n - X - 1$. However, this is impossible, because by Viete's formula $z_1 - \frac{1}{z_1} + z_2 - \frac{1}{z_2} + \cdots + z_n - \frac{1}{z_n} = 1$. This shows that any such factorization is impossible, and so $X^n - X - 1$ is irreducible in $\mathbb{Z}[X]$. All we need now is to apply Gauss's lemma to obtain a complete proof.

We pass now to a proof of the celebrated Capelli's theorem. As we will immediately see, this is a very powerful criterion for the irreducibility of compositions of polynomials, even though the proof is really easy. However, this does not seem to be well known, especially in the world of mathematical competitions. We thank Marian Andronache for showing us this striking result and some of its consequences.

Example 8. Let K be a subfield of $\mathbb{C}$ and $f, g \in K[X]$. Let α be a complex root of f and assume that f is irreducible in $K[X]$ and $g(X) - \alpha$ is irreducible in $K[\alpha][X]$. Then $f(g(X))$ is irreducible in $K[X]$.

Capelli's theorem

Solution. Define $h(X) = g(X) - \alpha$ and consider β a zero of the polynomial h. Because $f(g(\beta)) = f(\alpha) = 0$, β is algebraic over K. Let $\deg(f) = n, \deg(h) = m$ and let s be the minimal polynomial of β over K. If we manage to prove that $\deg(s) = mn$, then we are done, since s is irreducible over K and s divides

$f(g(X))$, which has degree mn. So, let us suppose the contrary. By using a repeated division algorithm, we can write $s = r_{n-1}g^{n-1}+r_{n-2}g^{n-2}+\cdots+r_1g+r_0$, where $\deg(r_i) < m$. Hence $r_{n-1}(\beta)\alpha^{n-1}+\cdots+r_1(\beta)\alpha+r_0(\beta) = 0$. By grouping terms according to increasing powers of β, we deduce from the last relation an equation $k_{m-1}(\alpha)\beta^{m-1} + \cdots + k_1(\alpha)\beta + k_0(\alpha) = 0$. Here the polynomials k_i have coefficients in K and have degree at most $n-1$. Because h is irreducible in $K(\alpha)[X]$, the minimal polynomial of β over $K(\alpha)$ is h and thus it has degree m. Therefore the last relation implies $k_{m-1}(\alpha) = \cdots = k_1(\alpha) = k_0(\alpha) = 0$. Now, because f is irreducible in $K[X]$, the minimal polynomial of α has degree n, and since $\deg(k_i) < n$, we must have $k_{m-1} = \cdots = k_1 = k_0 = 0$. This shows that $r_{n-1} = \cdots = r_1 = r_0 = 0$ and thus $s = 0$, which is clearly a contradiction. This shows that s has degree mn, and thus it is equal (up to a multiplicative constant) to $f(g(X))$ and this polynomial is irreducible.

This previous proof could have been written in a much shorter and conceptual form, using some basic facts of extensions of fields. Namely, let β be a zero of $g - \alpha$. Then $[K(\alpha,\beta) : K(\alpha)] = \deg(g)$ because $g - \alpha$ is irreducible, and thus the minimal polynomial of β over $K(\alpha)$. On the other hand, f being irreducible over K, it is the minimal polynomial of α over K. Thus $[K(\alpha) : K] = \deg(f)$. Thus, by multiplicativity of degrees in extensions, $[K(\alpha,\beta) : K] = \deg(f) \cdot \deg(g)$. On the other hand, $\alpha = g(\beta)$, thus $K(\alpha,\beta) = K(\beta)$, so the degree of β over K is at least $\deg(f) \cdot \deg(g) = \deg(f(g(X))$. Because $f(g(X))$ has β as zero, it follows that it is the minimal polynomial of β over K and so it is irreducible over K.

Using the previous result, we obtain a generalization (and a more general statement) of two difficult problems given in recent Romanian TST's:

Example 9. Let f be a monic polynomial with integer coefficients and let p be a prime number. If f is irreducible in $\mathbb{Z}[X]$ and $\sqrt[p]{(-1)^{\deg(f)}f(0)}$ is irrational, then $f(X^p)$ is also irreducible in $\mathbb{Z}[X]$.

Solution. Consider α a complex zero of f and let $n = \deg(f)$ and $g(X) = X^p$ and $h = g - \alpha$. Using previous results, it suffices to prove that h is irreducible

in $\mathbb{Q}[\alpha][X]$. Because $\mathbb{Q}[\alpha]$ is a subfield of $\mathbb{C}$, it suffices to prove that α is not the p-th power of an element of $\mathbb{Q}[\alpha]$. Suppose there is $u \in \mathbb{Q}[X]$ of degree at most $n-1$ such that $\alpha = u^p(\alpha)$. Let $\alpha_1, \alpha_2, \dots, \alpha_n$ be the zeros of f. Because f is irreducible and α is one of its zeros, f is the minimal polynomial of α, so f must divide $u^p(X) - X$. Therefore $\alpha_1 \cdot \alpha_2 \cdots \alpha_n = (u(\alpha_1) \cdot u(\alpha_2) \cdots u(\alpha_n))^p$. Finally, using the fundamental theorem of symmetric polynomials, $u(\alpha_1) \cdot u(\alpha_2) \cdots u(\alpha_n)$ is rational. But $\alpha_1 \cdot \alpha_2 \cdots \alpha_n = (-1)^n f(0)$, implies $\sqrt[p]{(-1)^n f(0)} \in \mathbb{Q}$, a contradiction.
A direct application of Capelli's theorem solves the following problem, which is not as easy otherwise:

Example 10. Prove that for each positive integer n the polynomial $f(X) = (X^2 + 1^2)(X^2 + 2^2) \cdots (X^2 + n^2) + 1$ is irreducible in $\mathbb{Z}[X]$.

Japan 1999

Solution. Consider the polynomial $g(X) = (X + 1^2)(X + 2^2) \cdots (X + n^2) + 1$. Let us prove first that this polynomial is irreducible in $\mathbb{Z}[X]$. Suppose that $g(X) = F(X)G(X)$ with $F, G \in \mathbb{Z}[X]$ nonconstant. Then $F(-i^2)G(-i^2) = 1$ for any $1 \le i \le n$. Therefore $F(-i^2)$ and $G(-i^2)$ are equal to 1 or -1 and since their product is 1, we must have $F(-i^2) = G(-i^2)$ for all $1 \le i \le n$. This means that $F - G$ is divisible by $(X + 1^2)(X + 2^2) \cdots (X + n^2)$ and because it has degree at most $n - 1$, it must be the zero polynomial. Therefore $g = F^2$ and so $(n!)^2 + 1 = g(0)$ must be a perfect square. This is clearly impossible, so g is irreducible. All we have to do now is to apply the result in example 4.

Sophie Germain's identity $m^4 + 4n^4 = (m^2 - 2mn + 2n^2)(m^2 + 2mn + 2n^2)$ shows that the polynomial $X^4 + 4a^4$ is reducible in $\mathbb{Z}[X]$ for all integers a. However, finding an irreducibility criterion for polynomials of the form $X^n + a$ is not an easy task. The following result, even though very particular, shows that this problem is not an easy one. Actually, there exists a general criterion, also known as Capelli's criterion: for rational a and $m \ge 2$, the polynomial $X^m - a$ is irreducible in $\mathbb{Q}[X]$ if and only if $\sqrt[p]{a}$ is irrational for any prime p dividing m and also, if $4|m$, a is not of the form $-4b^4$ with b rational.

Example 11. Let $n \geq 2$ be an integer and let K be a subfield of $\mathbb{C}$. If the polynomial $f(X) = X^{2^n} - a \in K[X]$ is reducible in $K[X]$, then either there exists $b \in K$ such that $a = b^2$ or there exists $c \in K$ such that $a = -4c^4$.

Solution. Suppose the contrary, that $X^2 - a$ is irreducible in $K[X]$. Let α be a zero of this polynomial. First, we will prove that $X^4 - a$ is irreducible in $K[X]$. Using the result in example 8, it is enough to prove that $X^2 - \alpha$ is irreducible in $K[\alpha][X]$. If this is not true, then there are $u, v \in K$ such that $\alpha = (u + \alpha v)^2$, which can be also written as $v^2\alpha^2 + (2uv - 1)\alpha + u^2 = 0$. Because $\alpha^2 \in K$ and α is not in K, it follows that $2uv = 1$ and $u^2 + av^2 = 0$. Thus $a = -4u^4$ and we can take $c = u$, a contradiction. Therefore $X^2 - \alpha$ is irreducible in $K[\alpha][X]$ and $X^4 - a$ is irreducible in $K[X]$. Now, we will prove by induction on n the following assertion: for any subfield K of $\mathbb{C}$ and any $a \in K$ not of the form b^2 or $-4c^4$ with $b, c \in K$, the polynomial $X^{2^n} - a$ is irreducible in $K[X]$. Assume it is true for $n - 1$ and take α a zero of $X^2 - a$. Let K^t be the set of x^t when $x \in K$. Then with the same argument as above one can prove that α does not belong to $-K^2[\alpha]$ (thus it is not in $-4K^4[\alpha]$) and it does not belong to $K^2[\alpha]$. Therefore $X^{2^{n-1}} - \alpha$ is irreducible over $K[\alpha]$. In the same way we prove that $X^{2^{n-1}} + \alpha$ is irreducible over $K(\alpha)$. Now, observe that $X^{2^n} - a = (X^{2^{n-1}} - \alpha)(X^{2^{n-1}} + \alpha)$, so it has at most two irreducible factors over K. If it is not irreducible over K, then one of its irreducible factors over K will be $X^{2^{n-1}} + \alpha$ or $X^{2^{n-1}} - \alpha$, thus one of these polynomials would have coefficients in K. This would imply that $\alpha \in K$, which means that a is a square in K. This is a contradiction which finishes the proof.

The following example is a notoriously difficult problem given a few years ago in a Romanian Team Selection Test.

Example 12. Prove that the polynomial $(X^2 + X)^{2^n} + 1$ is irreducible in $\mathbb{Q}[X]$ for all integers $n \geq 0$.

[Marius Cavachi] Romanian TST 1997

Solution. Using Capelli's theorem, it is enough to prove that if α is a root of $f(X) = X^{2^n} + 1$ (which is clearly irreducible in $\mathbb{Q}[X]$ by Eisenstein's theorem applied to $f(X+1)$), then $X^2 + X - \alpha$ is irreducible in $\mathbb{Q}[\alpha][X]$ (this is also immediate from the previous problem). But this is not difficult, because a polynomial of degree 2 (or 3) is reducible over a field if and only if it has roots in that field. Here, it is enough to prove that we cannot find a polynomial $g \in \mathbb{Q}[X]$ such that $g(\alpha)^2 + g(\alpha) = \alpha$. Suppose by contradiction that g is such a polynomial. Then, if $\alpha_1, \alpha_2, \ldots, \alpha_{2^n}$ are the roots of f it follows from the irreducibility of f that $\left(g(\alpha_i) + \frac{1}{2}\right)^2 = \alpha_i + \frac{1}{4}$ for all i. By multiplying these relations, we deduce that $f\left(-\frac{1}{4}\right)$ is the square of a rational number (the argument is always the same, based on the theorem of symmetric polynomials). But this means that $4^{2^n} + 1$ is a perfect square, which is clearly impossible.

A very efficient method for proving that a certain polynomial is irreducible is working modulo p for suitable prime numbers p. There are several criteria involving this idea, and Eisenstein's criterion is probably the easiest to state and verify. It asserts that if $f(X) = a_nX^n + a_{n-1}X^{n-1} + \cdots + a_1X + a_0$ is a polynomial with integer coefficients for which there exists a prime p such that p divides all coefficients except a_n and p^2 does not divide a_0 then f is irreducible in $\mathbb{Q}[X]$. The proof is not complicated. Observe first of all that by dividing f by the greatest common divisor of its coefficients, the resulting polynomial is primitive and has the same property. Therefore we may assume that f is primitive and so it is enough to prove the irreducibility in $\mathbb{Z}[X]$. Suppose that $f = gh$ for some nonconstant integer polynomials g, h and look at this equality in the field $\mathbb{Z}/p\mathbb{Z}$. Let f^* be the polynomial f reduced modulo p. We have $g^*h^* = a_nX^n$ (by convention, a_n will also denote $a_n \pmod{p}$). This implies that $g^*(X) = bX^r$ and $h^*(X) = cX^{n-r}$ for some $0 \le r \le n$, with $bc = a_n$. Suppose first that $r = 0$. Then $h(X) = cX^n + pu(X)$ for a certain polynomial with integer coefficients u. Because p does not divide a_n, it does not divide c and so $\deg(h) \ge n$, contradiction. This shows that $r > 0$ and similarly $r < n$. Thus there exist polynomials u, v with integer coefficients such that $g(X) = bX^r + pu(X)$ and $h(X) = cX^{n-r} + pv(X)$. This shows that $a_0 = f(0) = p^2u(0)v(0)$ is a multiple of p^2, contradiction.

Before passing to the next example, note two important consequences of Ei-

senstein's criterion. First, if p is a prime number, then $f(X) = 1 + X + X^2 + \cdots + X^{p-1}$ is irreducible in $\mathbb{Q}[X]$. This follows from Gauss's lemma and the observation that $f(X+1) = \frac{1}{X}((1+X)^p - 1)$ satisfies the conditions of Eisenstein's criterion. Second, for all n there is a polynomial of degree n which is irreducible in $\mathbb{Q}[X]$. Indeed, for $X^n - 2$, Eisenstein's criterion can be applied with $p = 2$ and the result follows from Gauss's lemma.
The following example is more general than Eisenstein's criterion. And older!

Example 13. Let $k = f^n + pg$ with $n \geq 1$, p a prime, and f and g polynomials with integer coefficients such that $\deg(f^n) > \deg(g)$, k is primitive, and there exists a prime p such that f^* is irreducible in $\mathbb{Z}/p\mathbb{Z}[X]$ and f^* does not divide g^*.
Then k is irreducible in $\mathbb{Q}[X]$.

[Schönemann's criterion]

Solution. Suppose that $k = k_1 k_2$ is a nontrivial factorization in polynomials with integer coefficients. By passing to $\mathbb{Z}/p\mathbb{Z}[X]$ we deduce that $k_1^* k_2^* = (f^*)^n$. From the hypothesis and this equality, it follows that there exist nonnegative integers u, v with $u + v = n$ and polynomials with integer coefficients g_1, g_2 such that $k_1 = f^u + pg_1$ and $k_2 = f^v + pg_2$, with $\deg(g_1) < u \deg(f)$ and $\deg(g_2) < v \deg(f)$. From here we infer that $g = f^u g_2 + f^v g_1 + pg_1 g_2$. Because k_1 is not identical 1, we have $u > 0$ and $v > 0$. Let us assume, without loss of generality, that $u \leq v$. From the previous relation there exists a polynomial h with integer coefficients such that $g = f^u h + pg_1 g_2$. It is enough to pass again in $\mathbb{Z}/p\mathbb{Z}[X]$ this last relation to deduce that f^* divides g^*, which contradicts the hypothesis. Therefore F is irreducible.

Here is an application of the above criterion, hardly approachable otherwise:

Example 14. Let p be a prime of the form $4k+3$ and let a, b be integers such that $\min(v_p(a), v_p(b-1)) = 1$. Prove that the polynomial $X^{2p} + aX + b$ is irreducible in $\mathbb{Z}[X]$.

[Laurenţiu Panaitopol, Doru Ştefănescu]

Solution. Indeed, the fact that $p = 3 \pmod 4$ ensures that $X^2 + 1$ is irreducible in $\mathbb{Z}/p\mathbb{Z}[X]$ (indeed, being of degree 2, it is enough to prove that it has no roots in $\mathbb{Z}/p\mathbb{Z}$, which was proved for instance in the chapter **Primes and Squares**). Let us try to write $X^{2p} + aX + b$ as $(X^2+1)^p + pg(X)$, just as in the previous example. It is enough to take

$$g(X) = \frac{a}{p}X + \frac{b-1}{p} + \frac{1}{p}\cdot\left[\binom{p}{1}X^{2(p-1)} + \binom{p}{2}X^{2(p-2)} + \cdots + \binom{p}{p-1}X^2\right].$$

Now it is immediate that all conditions of Schonemann's criterion are satisfied, so the problem is solved.

Now let us see a beautiful proof of the irreducibility of the cyclotomic polynomials. This is not an easy problem, as the reader can immediately observe. But for the reader who is not so familiar with these polynomials, let us make a (very small) introduction. Let n be a positive integer. If $n = 1$ we define $\phi_1(X) = X - 1$ and if $n > 1$ we put

$$\phi_n(X) = \prod_{\gcd(k,n)=1, 1\le k\le n} \left(X - e^{\frac{2ik\pi}{n}}\right) \tag{21.3}$$

From this definition, it is not even clear why this polynomial has integer coefficients. Actually, one can easily prove the identity $\prod_{d|n} \phi_d(X) = X^n - 1$, which allows a direct proof by induction of the fact that $\phi_n(X) \in \mathbb{Z}[X]$. Indeed, just observe that $X^n - 1$ has no repeated zero, that clearly the left-hand side divides $X^n - 1$ because every zero of it is a zero of $X^n - 1$ (it is clear from the definition that ϕ_n has no repeated zeros and also that ϕ_n and ϕ_m are relatively prime for distinct m, n) and finally that the degree of $\prod_{d|n} \phi_d(X)$ is n because of the identity $\sum_{d|n} \varphi(d) = n$ (proved in the chapter **The Smaller, the Better**).

Now, let us prove the following important result:

Example 15. The polynomial ϕ_n is irreducible in $\mathbb{Q}[X]$ for each positive integers n.

Solution. Let α be a primitive root of unity of order n and let p be a prime number relatively prime to n. Let f and g be the minimal polynomials of α and α^p over the field of rational numbers. Because α is an algebraic integer, f, g have integer coefficients. Also, because $g(\alpha^p) = 0$, it follows that f divides $g(X^p)$. The idea is that in $\mathbb{Z}/p\mathbb{Z}$ we have $g(X^p) = g(X)^p$ and so if f^* and g^* are the polynomials f, g reduced modulo p, then f^* divides $(g^*)^p$ in $\mathbb{Z}/p\mathbb{Z}[X]$. Thus, if r is a root of f^* in some algebraic closure of $\mathbb{Z}/p\mathbb{Z}$, then $g^*(r) = 0$. Now, suppose that $f \neq g$. Both f and g divide ϕ_n in $\mathbb{Z}[X]$, because α^p is also a primitive root of unity. Because $f \neq g$ are irreducible, they are relatively prime and fg divides ϕ_n in $\mathbb{Z}[X]$, thus f^*g^* divides $X^n - 1$ (seen as a polynomial in $\mathbb{Z}/p\mathbb{Z}[X]$). But this is impossible, because it would follow that r is a root of multiplicity at least 2 of the polynomial $X^n - 1$ modulo p, that is we also have $nr^{n-1} = 0$ in that algebraic extension. Because n and p are relatively prime, this implies that $r = 0$, which is impossible, because $r^n = 1$.

The above contradiction shows that $f = g$, that is α and α^p have the same minimal polynomial for all prime numbers p relatively prime to n. This immediately implies that α and α^k have the same minimal polynomial for all k relatively prime to n. Thus, the minimal polynomial of α must have as roots all primitive roots of unity of order n and thus degree at least $\varphi(n)$, which means that it is ϕ_n, that is ϕ_n is irreducible.

There exists another beautiful proof of this result, but which uses the difficult (and non elementary) Dirichlet's theorem on primes in arithmetic progressions. Let ω be a primitive root of unity of order n and let $s = \varphi(n) = \deg(\phi_n)$. Also, let f be an irreducible factor of ϕ_n with integer coefficients, which has ω as a zero. Then the zeros of f (which, as we have seen in chapter **A Brief Introduction to Algebraic Number Theory**, are called the conjugates of ω) are of the form ω^t. Also, if ϕ_n is not irreducible then the number of zeros of f is smaller than s. Now, take p to be a prime number. Because f is monic and has all zeros of absolute value 1, it follows that $|f(\omega^t)| \leq 2^s$. But because $f(\omega) = 0$, it follows that $\frac{f(\omega^p)}{p}$ is an algebraic integer (this result is not obvious, but it has been discussed in the same chapter). Its conjugates are also algebraic integers of the form $\frac{f(\omega^{tp})}{p}$. Thus if we choose $p > 2^s$, then all conjugates of the algebraic integer $\frac{f(\omega^p)}{p}$ are inside the unit disc of the complex plane, thus $f(\omega^p) = 0$ (indeed, if $x = \frac{f(\omega^p)}{p}$ and g is the minimal polynomial of x, then by Gauss's lemma g has integer coefficients, and thus the product of the absolute values of all conjugates of x is just $|g(0)|$; if all conjugates are inside the unit disc, then $g(0) = 0$ and because g is irreducible, $g(X) = X$, thus $x = 0$). Therefore, for any prime number $p > 2^s$, ω^p is a zero of f. All we need to observe now is that Dirichlet's theorem assures us of the existence of infinitely many primes $p \equiv r \pmod{n}$ for any r such that $\gcd(r, n) = 1$. Therefore all ω^r with $\gcd(r, n) = 1$ are zeros of f, which shows that $\deg(f) \geq \deg(\phi_n)$ and proves the irreducibility of ϕ_n.

21.2 Practice problems

1. Let a and n be integers and p be a prime such that $p > |a| + 1$. Prove that $X^n + aX + p$ is irreducible in $\mathbb{Z}[X]$.

Laurenţiu Panaitopol, Romanian TST 1999

2. Let $p > 3$ be a prime number and m, n be positive integers. Prove that $X^m + X^n + p$ is irreducible in $\mathbb{Z}[X]$.

Laurenţiu Panaitopol

3. Prove that for all positive integers d there is a monic polynomial f of degree d such that $X^n + f(X)$ is irreducible in $\mathbb{Z}[X]$ for all n.

4. Let k, d be integers greater than 1. Prove that for any partition of the set of positive integers into k classes there is a class and infinitely many polynomials of degree d with all coefficients in that class and which are irreducible in $\mathbb{Z}[X]$.

Marian Andronache, Ion Savu, Unesco Contest 1995

5. Find the number of irreducible polynomials of the form $X^p + pX^k + pX^l + 1$, where $p > 2$ is a fixed prime number and k, l are subject to the conditions $1 \leq l < k \leq p - 1$.

Valentin Vornicu, Romanian TST 2006

6. Let p and q be distinct prime numbers and $n \geq 3$. Find all integers a for which $X^n + aX^{n-1} + pq$ is reducible in $\mathbb{Z}[X]$.

Chinese TST 1994

7. Let n and r be positive integers. Prove the existence of a polynomial f with integer coefficients and degree n such that for any polynomial g with integer coefficients and degree at most n, if the coefficients of $f-g$ have absolute values at most r, then g is irreducible in $\mathbb{Q}[X]$.

Miklos Schweitzer Competition

8. Prove that for any $n \geq 2$ there exists a polynomial f with integer nonzero coefficients, irreducible over the rational numbers and such that $|f(x)|$ is composite for any integer x.

Chinese TST 2008

9. Prove that for any positive integer n, the polynomial

$$(X^2+2)^n + 5(X^{2n-1} + 10X^n + 5)$$

is irreducible in $\mathbb{Z}[X]$.

Laurenţiu Panaitopol, Doru Ştefănescu

10. Let p be a prime of the form $4k+3$ and let n be a positive integer. Prove that $(X^2+1)^n + p$ is irreducible in $\mathbb{Z}[X]$.

N Popescu, Gazeta Matematică

11. Does there exist an infinite sequence of pairwise relatively prime positive integers $(a_i)_{i\geq 0}$ such that all polynomials $a_0 + a_1X + \cdots + a_nX^n$ (with $n \geq 1$) are irreducible over the rational numbers?

Omid Hatami, Iran TST

12. Let p be a prime number and let k be an integer not divisible by p. Prove that $X^p - X + k$ is irreducible in $\mathbb{Z}[X]$.

13. Let p, q be odd primes such that q does not divide $p-1$ and let $a_1, a_2, \dots, a_n$ be distinct integers such that q divides $a_i - a_j$ for all i, j. Prove that the polynomial $(X-a_1)(X-a_2)\cdots(X-a_n) - p$ is irreducible in $\mathbb{Q}[X]$.

Ivan Borsenco, Mathematical Reflections

14. Let $f(X) = a_m X^m + a_{m-1}X^{m-1} + \cdots + a_1 X + a_0$ be a polynomial of degree m in $\mathbb{Z}[X]$ and define $H = \max_{0 \le i \le m-1} |\frac{a_i}{a_m}|$. If $f(n)$ is prime for some integer $n \ge H+2$ then f is irreducible in $\mathbb{Z}[X]$.

Ram Murty, AMM

15. Let p be a prime number, $n_1 > n_2 \cdots > n_p$ be positive integers and $d = \gcd(n_1, n_2, \dots, n_p)$. Prove that the polynomial

$$P(X) = \frac{X^{n_1} + X^{n_2} + \cdots + X^{n_p} - p}{X^d - 1}$$

is irreducible in $\mathbb{Q}[X]$.

Romanian TST 2010

16. Let f be a primitive polynomial with integer coefficients of degree n for which there exist distinct integers $x_1, x_2, \dots, x_n$ such that

$$0 < |f(x_i)| < \frac{\lfloor \frac{n+1}{2} \rfloor !}{2^{\lfloor \frac{n+1}{2} \rfloor}}.$$

Prove that f is irreducible in $\mathbb{Z}[X]$.

17. Find all quadratic polynomials $f \in \mathbb{Z}[X]$ for which one can find $n \ge 2$ such that $f^{2^n} + 1$ is reducible in $\mathbb{Q}[X]$.

Gabriel Dospinescu, Marian Tetiva

18. Let $f \in \mathbb{Z}[X]$ be a polynomial of degree greater than 1 with the property that $f(X^2+aX)$ is reducible in $\mathbb{Q}[X]$ for infinitely many integers a. Does it follow that f is reducible in $\mathbb{Q}[X]$?

Gabriel Dospinescu, Mathematical Reflections

19. Let f be an irreducible polynomial in $\mathbb{Q}[X]$ of degree p, where $p > 2$ is prime. Let $x_1, x_2, \ldots, x_p$ be the zeros of f. Prove that for any nonconstant polynomial g with rational coefficients, of degree smaller than p, the numbers $g(x_1), g(x_2), \ldots, g(x_p)$ are pairwise distinct.

Toma Albu, Romanian TST 1983

20. Let $f \in \mathbb{Z}[X]$ be a monic polynomial irreducible in $\mathbb{Z}[X]$, and suppose that there exists a positive integer m such that $f(X^m)$ is reducible in $\mathbb{Z}[X]$. Show that for any prime p dividing $f(0)$ we have $v_p(f(0)) \geq 2$.

Marian Andronache

21. Let $d > 1$ be an integer and let $f(n)$ be the probability that a polynomial of degree d with all coefficients bounded by n in absolute value is reducible in $\mathbb{Z}[X]$. Prove that $f(n) \cdot \frac{n}{\ln n}$ is a bounded sequence.

22. Let a be a nonzero integer. Prove that the polynomial

$$X^n + aX^{n-1} + \cdots + aX^2 + aX - 1$$

is irreducible in $\mathbb{Z}[X]$.

Marian Andronache, Ion Savu, Romanian Olympiad 1990

23. Let f be a monic polynomial with integer coefficients and having distinct integer roots. Prove that $f^2 + 1$ and $f^4 + 1$ are irreducible in $\mathbb{Q}[X]$.

Schur

24. Is there a polynomial f with rational coefficients such that $f(1) \neq -1$ and $X^n f(X) + 1$ is reducible over the rational numbers for all $n \geq 1$?

Schinzel

25. Let $p_1, p_2, \ldots, p_n$ be distinct prime numbers. Prove that the polynomial

$$f(X) = \prod_{e_1, e_2, \ldots, e_n = \pm 1} (X + e_1\sqrt{p_1} + e_2\sqrt{p_2} + \cdots + e_n\sqrt{p_n})$$

is irreducible in $\mathbb{Z}[X]$.

Chapter

22

22.1 Theory and examples

After a very elementary chapter about extremal properties of graphs, it is time to see how the study of their cycles can give valuable information in combinatorial problems. We will assume in this chapter some familiarity with basic concepts of graph theory that can be found in practically any book of combinatorics. We prefer to do so, because recalling all definitions would require a large digression and would largely diminish the quantity of examples presented. And since the topic is very subtle and the problems are in general difficult, we think it is better to present many examples. We would like to thank Adrian Zahariuc for the large quantity of interesting results and solutions that he communicated to us.

We start with a simple, but important result. It was extended by Erdős in a much more difficult to prove statement: if the number of edges of a graph on n vertices is at least $\frac{(n-1)k}{2}$ then there exists a cycle of length at least $k+1$ (if $k>1$). Let us remain modest and prove the following much easier result :

Example 1. In a graph G with n vertices, every vertex has degree at least k. Prove that G has a cycle of length at least $k+1$.

Solution. The shortest solution uses the extremal principle. Consider the longest chain $x_0, x_1, \dots, x_r$ in G and observe that this maximality property ensures that all vertices adjacent to x_0 are in this longest chain. Or, the degree of x_0 being at least k, we deduce that there is a vertex x_i adjacent to x_0 such that $k \le i \le r$. Therefore $x_0, x_1, \dots, x_i, x_0$ is a cycle of length at least $k+1$.

Any graph with n vertices and at least n edges must have a cycle. The following problem is an easy application of this fact:

Example 2. Suppose $2n$ points of an $n \times n$ grid are marked. Prove that there exists a $k>1$ and $2k$ distinct marked points $a_1, a_2, \dots, a_{2k}$ such that for all i, a_{2i-1} and a_{2i} are in the same row, while a_{2i} and a_{2i+1} are in the same column.

IMC 1999

Solution. Here it is not difficult to discover the graph to work on. It is enough to look at the n lines and n columns as the two classes of a bipartite graph. We connect two vertices if the intersection of the corresponding row and column is marked. Clearly, this graph has $2n$ vertices and $2n$ edges, so there must exist a cycle. But the existence of a cycle is equivalent (by the definition of the graph) to the conclusion of the problem.

The following example is an extremal problem in graph theory, of the same kind as Turan's theorem. This type of problem can go from easy or even trivial to extremely complex and complicated results. Of course, we will discuss just the first type of problem.

Example 3. Prove that every graph on $n \geq 4$ vertices and $m > \frac{n+n\sqrt{4n-3}}{4}$ edges has at least one 4-cycle.

Solution. Let us count, in two different ways, the number of triples (c, a, b) where a, b, c are vertices such that c is connected to both a and b. For a fixed vertex c, there are $d(c)^2 - d(c)$ possibilities for the pair (a, b), where $d(c)$ denotes the valence of c. It follows that there are at least $\sum_c (d(c)^2 - d(c))$ triples. By the Cauchy-Schwarz inequality, if m represents the number of edges of the graph, then

$$\sum_c d(c)^2 - d(c) \geq \frac{4m^2}{n} - 2m \qquad (22.1)$$

Now, if there are no 4-cycles, then for fixed a and b there is at most one vertex c that appears in a triple (a, b, c). Hence we obtain at most $n(n-1)$ triples. It follows that $\frac{4m^2}{n} - 2m \leq n^2 - n$, which implies that $m \leq \frac{n+n\sqrt{4n-3}}{4}$, a contradiction.

Recall that a graph in which every vertex has degree 2 is a disjoint union of cycles. It turns out that this very innocent observation is more than helpful in some quite challenging problems. Here are some examples, taken from different contests:

Example 4. A company wants to build a 2001×2001 building with doors connecting pairs of adjacent rooms (which are 1×1 squares, two rooms being adjacent if they have a common edge). Is it possible for every room to have exactly 2 doors?

[Gabriel Caroll]

Solution. Let us analyze the situation in terms of graphs: suppose such a situation is possible, and consider the graph G with vertices representing the rooms and connecting two rooms if there exists a door between them. Then the hypothesis says that the degree of any vertex is 2. Thus G is a union of disjoint cycles $C_1, C_2, \ldots, C_p$. However, observe that any cycle has even length, because the number of vertical steps is the same in both directions and the same holds for horizontal steps. Therefore the number of vertices of G, which is the sum of lengths of these cycles, is an even number, a contradiction.

Reading the solution to the following problem, one might say that it is extremely easy: there is no tricky idea behind it. But there there are many possible approaches that can fail, and this probably explains its presence on the list of problems proposed for the IMO 1990.

Example 5. Let E be a set of $2n-1$ points on a circle, with $n > 2$. Suppose that precisely k points of E are colored black. We say that this coloring is admissible if there is at least one pair of black points such that the interior of one of the arcs they determine contains exactly n points of E. What is the smallest k such that any coloring of k points of E is admissible?

IMO 1990

Solution. Consider G the graph having vertices the black points of E and join two points x, y by an edge if there are n points of E on one of the two open arcs determined by x and y. Thus the problem becomes: what is the least k such that among any k vertices of this graph at least two are adjacent? The problem becomes much easier with this statement, because of the fact that the degree of any vertex in G is clearly 2, thus G is a union of disjoint cycles. It is clear that for a single cycle of length r, the least value of k is $1 + \left\lfloor \frac{r}{2} \right\rfloor$. Now, observe that if $2n-1$ is not a multiple of 3 then G is actually a cycle (because $(\gcd(n+1, 2n-1) = 1)$, while in the other case G is the union of three disjoint cycles of length $\frac{2n-1}{3}$. Therefore the least k is $n = \left\lfloor \frac{2n-1}{2} \right\rfloor + 1$ if $2n-1$ is not a multiple of 3 and $n-1 = 3\left\lfloor \frac{2n-1}{6} \right\rfloor + 1$ otherwise.

Finally, a more involved example using the same idea, but with some complication which are far from obvious.

Example 6. Consider in the plane the rectangle with vertices $(0,0), (m,0)$ $(0,n), (m,n)$, where m and n are odd positive integers. Partition it rectangle into triangles satisfying the following conditions: 1) Each triangle has at least one side (called the good side; the sides that are not good will be called bad) on a line $x = j$ or $y = k$ for some nonnegative integers j, k, such that the height corresponding to that side has length 1; 2) Each bad side is common for two triangles of the partition. Prove that there are at least two triangles having two good sides each.

IMO 1990 Shortlist

Solution. Let us define a graph G having as vertices the midpoints of the bad sides and as edges the segments connecting the midpoints of two bad sides in a triangle of the partition. Thus, any edge is parallel to one of the sides of the rectangle, being at distance $k + \frac{1}{2}$ from the sides of the rectangle, for a suitable integer k. Also, it is clear that any vertex has degree at most 2, so we have three cases. The easiest is when there exists an isolated vertex. Then the triangles that have the side containing that vertex as common side have two good sides. Another easy case is when there exists a vertex x having degree 1. Then x is the end of a polygonal line formed by edges of the graph, and having the other end a point y, which is the midpoint of a side in a triangle having two good sides. The conclusion follows in this case, too. Thus, it remains to cover the "difficult" case when all vertices have degree 2. Actually, we will show that this case is impossible. Observe that until now we haven't used the hypothesis that m, n are odd. This suggests looking at the cycles of G. Indeed, we know that G is a union of disjoint cycles. If we manage to prove that the number of squares traversed by any cycle is even, it would follow that the table has an even number of unit squares, which is impossible, because mn is odd. Divide first the rectangle by its lattice points into mn unit squares. So, fix a cycle and observe that from the hypothesis it follows that the center of any square is contained in only one cycle. Now, by alternatively coloring the cells of the rectangle with white and black, we obtain a chessboard in which every cycle passes alternatively on white and black squares, so it passes through an even number of squares. This proves the claim and shows that G cannot have all vertices of degree 2.

The next problem is already unobvious, and the solution is not immediate, because it requires two arguments which are completely different: a construction and a proof of optimality. Starting with some special cases is often the best way to proceed, and this is indeed the key here.

Example 7. Let n be a positive integer. Suppose that n airline companies offer trips to citizens of N cities such that for any two cities there exists a direct flight in both directions. Find the least N such that we can always find a company which can offer a trip in a cycle with an odd number of landing points.

Adapted after IMO 1983 Shortlist

Solution. By starting with small values of n, we can guess the answer: $N = 2^n + 1$. But it is not obvious how to prove both that for 2^n the assertion in the problem is not always true and the fact that for $2^n + 1$ cities the conclusion always holds. Let us start with the first claim: the result is not always true if we allow only 2^n cities. Indeed, let the cities be $C_0, C_1, \dots, C_{2^n-1}$. Write every number smaller than 2^n in base 2 with n digits (we allow zeros in the first positions), and let us join two cities C_i and C_j by a flight offered by an airline company A_1 if the first digit of i and j is different, by a flight offered by A_2 if the first digits are identical, but the second digit differs in the two numbers and so on. Because the i-th digit is alternating in the vertices of a cycle for company A_i, it follows that all cycles realized by A_i are even. Therefore $N \geq 2^n + 1$. Now, we prove by induction that the assertion holds for $N \geq 2^n+1$. For $n = 1$ everything is clear, so assume the result for $n - 1$. Suppose that all cycles in the graph of flights offered by company A_n are even (otherwise we have found our odd cycle). Therefore the graph of flights offered by A_n is bipartite, that is there exists a partition $B_1, B_2, \dots, B_m, D_1, D_2, \dots, D_p$ of the cities such that any flight offered by A_n connects one of the cities B_j with one of the cities D_k. Because $m+p = 2^n+1$, we may assume that $m \geq 2^{n-1}+1$. But then the cities $B_1, B_2, \dots, B_m$ are connected only by flights offered by $A_1, A_2, \dots, A_{n-1}$, so by the induction hypothesis one of these companies can offer an odd cycle. This finishes the induction step and shows that $N = 2^n + 1$ is the desired number.

Here comes a very challenging problem with a very beautiful idea:

Example 8. On an infinite checkerboard are placed 111 non-overlapping corners, L-shaped figures made of 3 unit squares. Suppose that for any corner, the 2×2 square containing it is entirely covered by the corners. Prove that one can remove each number between 1 and 110 of the corners so that the property will be preserved.

St. Petersburg 2000

Solution. We will argue by contradiction. Assuming that by removing any 109 corners the property is no longer preserved, it would follow that no 2×3 rectangle is covered by 2 corners. Now, define the following directed graph

with vertices on the corners: for a fixed corner C, draw an edge from it to the corner that helps covering the 2×2 square containing C. It is clear that if in a certain corner there is no entering edge, we may safely remove that corner, contradiction. Therefore, in every corner there exists an entering edge and so the graph constructed has the property that every edge belongs to some cycle. We will prove that the graph cannot be a cycle of 111 vertices. Define the "center" of a corner as the center of the 2×2 square containing it. The first observation, that no two corners can cover a 2×3 rectangle, shows that in a cycle the x coordinate of the centers of the vertices are alternatively even and odd. Thus the cycle must have an even length, which shows that the graph itself cannot be a cycle. Therefore, it has at least two cycles. But then we may safely remove all the corners except those in a cycle of smallest length and the property will be preserved, thus again a contradiction.
The following result is particularly nice:

Example 9. There are n competitors in a table-tennis contest. Any 2 of them play exactly once against each other and no draws are possible. We know that no matter how we divide them into 2 groups A and B, there is some player from A who defeated some player from B. Prove that at the end of the competition, we can sit all the players at a round table such that everyone defeated his or her right neighbor.

Solution. Clearly, the problem refers to a tournament graph, that is, a directed graph in which any two vertices are connected in exactly one direction. We have to prove that this graph contains a Hamiltonian circuit. Take the longest elementary cycle, $v_1, v_2, \ldots, v_m$ with pairwise distinct vertices, and take some other vertex v. Unless all edges come either out of v or into v, there is some i such that $v_i v$ and $v v_{i+1}$ are edges. Then, $v_1, v_2, \ldots, v_i, v, v_{i+1}, \ldots, v_n$ is a longer elementary cycle, contradiction. Therefore, there are only two kinds of vertices $v \in V - \{v_i\}$: (type A) those for which all $v v_i$ are edges; and (type B) those for which all $v_i v$ are edges. If there is some edge ba with a of type A and b of type B, then we can construct once again a longer circuit: $b, a, v_1, \ldots, v_n$. Therefore, for any $a \in A$ and $b \in B$, ab is an edge. Consider the partition $V = B \cup (A \cup \{v_i\})$. Due to the hypothesis, since all edges between the two classes point towards B, we must have $B = \emptyset$. But, once again, $V = A \cup \{v_i\}$ is a forbidden partition, so $A = \emptyset$. Therefore, the circuit is Hamiltonian.

Before discussing the next problem, we need to present a very useful result, which is particularly easy to prove, but has interesting applications. This is why it will be discussed as a separate problem and not as a lemma:

Example 10. Prove that a graph is bipartite if and only if all of its cycles have even length.

Solution. One part of the result is immediate: if the graph is bipartite then obviously it cannot have odd cycles, because there is no internal edge in one of the two classes of the partition. The converse is a little bit trickier. Suppose that a graph G has no odd cycles and start your "journey" with an arbitrary vertex v and color this vertex white. Continue your trip through the vertices of the graph, by coloring all neighbors of the initial vertex in black. Continue in this manner, by considering this time every neighbor of v as an initial point of a new trip and color new vertices by the described rule, avoiding vertices that are already assigned a color. We must prove that you can do your trip with no problem. But the only problem that may occur is to have two paths to a certain vertex (called a problem vertex), each leading to a different color. But this is impossible, since all cycles are even. Indeed, any two paths from v to this problem vertex must have the same parity. Therefore we have a valid coloring of the vertices of the graph, and by construction this proves that G is bipartite. And here is an application:

Example 11. A group consists of n tourists. Among any 3 of them there are 2 who are not familiar. For every partition of the tourists in 2 buses, we can always find 2 tourists that are in the same bus who are familiar with each other. Prove that there is a tourist who is familiar with at most $\frac{2n}{5}$ tourists.

Bulgaria 2004

Solution. Construct a graph G on n vertices corresponding to the n tourists. We construct the edge ab if and only if the tourists a and b are familiar with each other. By the hypothesis, G is not bipartite, so it must have an odd cycle. Let $a_1, a_2, \ldots, a_l$ be the smallest odd cycle. Since l is odd and $l > 3$, we must have $l \geq 5$. It is clear that there are no other edges among the a_i except $a_i a_{i+1}$. If some vertex v is connected to a_i and a_j, it is easy to show that the "distance" between i and j is 2, that is $|i-j|$ equals 2 or $l-2$, since otherwise we would have a smaller odd cycle. Therefore, every vertex which does not belong to the cycle is adjacent to at most 2 a_i's. Even more, every vertex of the cycle is connected to exactly 2 a_i's. Therefore, if $c(v)$ is the number of edges between v and the vertices of the cycle, $c(v) \leq 2$, so

$$\sum_{i=1}^{l} d(a_i) = \sum_{v \in V} c(v) \leq 2n \Rightarrow d(a_k) \leq \frac{2n}{l} \leq \frac{2n}{5} \qquad (22.2)$$

for some k. The solution ends here.

At first glance, the following has nothing to do with graphs and cycles. Well, it does! Here is a beautiful solution by Adrian Zahariuc:

Example 12. In each square of a chessboard is written a positive real number such that the sum of the numbers in each row is exactly 1. It is known that for any 8 squares, no two in the same row or column, the product of the numbers written in these squares does not exceed the product of the numbers on the main diagonal. Prove that the sum of the numbers on the main diagonal is at least 1.

St. Petersburg 2000

Solution. First, let us label the rows and the columns $1, 2, \ldots, 8$, consecutively, in increasing order. Suppose by way of contradiction that the sum of the numbers on the main diagonal is less than 1. Then on row k there is some cell (k, j) such that the number written in it is greater than the number written in cell (j, j), that is, the one in the same column, on the main diagonal. Color (k, j) red and draw an arrow from row k to row j. Some of these arrows must form a loop. From each row belonging to the loop we choose the red cell, and from all other rows we choose the cell on the main diagonal. All these 8 cells lie in different rows and different columns and their product exceeds the product of the numbers on the main diagonal, a contradiction. Therefore our assumption is false, and the sum of the numbers on the main diagonal is at least 1.

And for the die-hards, here are two very difficult problems communicated to us by Adrian Zahariuc:

Example 13. There are two airline companies in Wonderland. Any pair of cities is connected by a one-way flight offered by one of the companies. Prove that there is a city in Wonderland from which any other city can be reached via airplane without changing the company.

Iranian TST 2006

Solution. We would rather reformulate the problem in terms of graph theory: given a bichromatic (say, red and blue) tournament $G(V, E)$ (i.e. a directed graph in which there is precisely one edge between any pair of vertices). We have to prove that there is a vertex v such that, for any other vertex u, there is a monochromatic directed path from v to u. Such a point will be called "strong". Let $|V| = n$. We will prove the claim by induction on n.
The base case is trivial. Suppose it is true for $n - 1$; we will prove it for n. Now suppose by way of contradiction that the claim fails for some G. By the inductive hypothesis we know that for each $v \in V$ there is some $s(v) \in V - \{v\}$ which is a strong point in $G - \{v\}$. Clearly, $s(v) \neq s(v')$ for all $v \neq v'$, since

otherwise $s(v)$ would be strong in G. Let $f = s^{-1}$, i.e. $s(f(v)) = v$ for all v. It is clear that from v we can reach all points through a monochromatic path except $f(v)$. For each v, draw an arrow from v to $f(v)$. These arrows must form a loop. If this loop does not contain all n vertices of the graph, by the inductive hypothesis we must have a strong vertex in this graph, which contradicts the fact that we can't reach $f(v)$ from v. Hence, this loop is a Hamiltonian circuit $v_1, v_2, \dots, v_n$. Let $v_{n+1} = v_1$. From v_i, we can reach all vertices except v_{i+1} because $v_{i+1} = f(v_i)$. We can't reach v from u through paths of both colors since, in that case, from u we could reach all the points we could reach from v, including $f(u)$, which is false.

For $v \neq f(u)$, let $c(uv)$ be the color of all paths from u to v. It is clear that $c(uv) \neq c(vf(u))$. We have $c(uv) \neq c(vf(u)) \neq c(f(u)f(v))$, so $c(uv) = c(f(u)f(v))$ for $u \neq v \neq f(u)$. In other words, $c(v_k v_{k+m}) = c(v_{k+1} v_{k+m+1})$. From here, it is easy to fill in the details. Basically, we just have to take $m > 1$ coprime with n to get that we can travel between any two points through paths of color $c(v_0 v_m)$ and we are done.

Example 14. Does there exist a 3-regular graph (that is, every vertex has degree 3) such that any cycle has length at least 30?

St. Petersburg 2000

<u>**Solution.**</u> Even though the construction will not be easy, the answer is: yes, there does. We construct a 3-regular graph G_n by induction on n such that any cycle has length at least n. Take $G_3 = K_4$, the complete graph on 4 vertices. Now, suppose we have constructed $G_n(V, E)$ and label its edges $1, 2, \dots, m$. Take an integer $N > n2^m$ and let $V' = V \times \mathbb{Z}_N$. If the edge numbered k in G_n is ab, we draw an edge in $G_{n+1}(V', E')$ between (a, x) and $(b, x + 2^k)$ for all $x \in \mathbb{Z}_N$. It is clear that G_{n+1} is 3-regular. We show that G_{n+1} has the desired property, i.e. it contains no cycle of length less than $n + 1$. Suppose $(a_1, x_1), \dots, (a_t, x_t)$ is a cycle with $t \leq n$. Clearly, $a_1, a_2, \dots, a_t$ is a cycle of

G_n. Therefore $t = n$, and all a_i are distinct. We have

$$0 = (x_1 - x_2) + (x_2 - x_3) + \cdots + (x_n - x_1) \equiv \sum_{j=1}^{n} \pm 2^{k_j} \pmod{N}. \quad (22.3)$$

This sum is nonzero since all k_j are distinct, and also it is at most $n2^m < N$ in absolute value, a contradiction. Therefore this graph has all the desired properties, and the inductive construction is complete.

22.2 Practice problems

1. Prove that any graph on $n \geq 3$ vertices having at least $2 + \binom{n-1}{2}$ edges has a Hamiltonian cycle. Does the property remain true if $2 + \binom{n-1}{2}$ is replaced by a smaller number?

2. Some pairs of towns are connected by roads. At least three roads leave each town. Show that there is a cycle containing a number of towns which is not a multiple of three.

   Russia 2000

3. Let n be a positive integer. Can we always assign to each vertex of a 2^n-gon one of the letters a and b such that the sequences of letters obtained by starting at a vertex and reading counterclockwise are all distinct?

   Japan 1997

4. On an $n \times n$ table real numbers are put in the unit squares such that no two rows are identically filled. Prove that one can remove a column of the table such that the new table has no two rows identically filled.

   Bulgarian TST 2004

5. The edges of $K_n (n \geq 3)$ are colored with n colors, and every color is used. Show that there is a triangle whose sides have different colors.

   Hungary-Israel Competition 2001

6. Let G be a simple graph with $2n + 1$ vertices and at least $3n + 1$ edges. Prove that there exists a cycle having an even number of edges. Prove that this is not always true if the graph has only $3n$ edges.

   Miklos Schweitzer Competition

7. The edges of a complete graph with $2^n + 1$ vertices are colored using n colors. Prove that we can find a monochromatic cycle of odd length.

8. For a given $n \geq 2$ find the least k with the following property: any set of k cells of an $n \times n$ table contains a nonempty subset A such that in every row and every column of the table there is an even number of cells belonging to A.

Poland 2000

9. In a society of at least 7 people each member communicates with at least three other members of the society. Prove that we can divide this society in two nonempty groups such that each member communicates with at least 2 members of their own group.

Czech-Slovak Match 1997

10. On the edges of a convex polyhedra we draw arrows such that from each vertex at least one arrow is pointing in and at least one is pointing out. Prove that there exists a face of the polyhedra such that the arrows on its edges form a circuit.

Dan Schwartz, Romanian TST 2005

11. Every street in a city connects two squares and any square may be reached by streets from any other. The governor discovered that if he closed all squares of any route not passing through any square more than once, then any remaining square would be reachable from any other. Prove that there exists a circular route passing through every square of the city exactly once.

S. Berlov, Tuymaada Olympiad 2008

12. A rectangular array of numbers is given. In each row and each column, the sum of all numbers is an integer. Prove that each non-integral number x in the array can be changed into either $\lceil x \rceil$ or $\lfloor x \rfloor$ so that the row-sums and column-sums remain unchanged.

IMO Shortlist 1998

13. There are 25 towns in a country. Find the smallest k for which one can set up bidirectional flight routes connecting these towns so that the following conditions are satisfied:

    (a) from each town there are exactly k direct routes to k other towns;

    (b) if two towns are not connected by a direct route, there is a town which has direct routes to these two towns.

Vietnamese TST 1997

14. There are $2n$ students at a math contest. Each of them submits to the jury a problem. At the end, the jury gives each student one of the problems submitted. Say that the contest is fair if there exist n students who receive their problems from the other n participants. Prove that the number of distributions of problems that end in a fair contest is a perfect square.

Romanian TST 2003

15. In a connected simple graph every vertex has degree at least three. Prove that there exists a cycle such that after removing the edges of this cycle the new graph is still connected.

Komal

16. A simple graph G with vertices $v_1, v_2, \ldots, v_n$ has no isolated vertex. Prove that there exists $I \subset \{v_1, v_2, \ldots, v_n\}$ with the properties

(a)

$$|I| \geq \sum_{i=1}^{n} \frac{2}{\deg v_i + 1}.$$

(b) Every cycle of G contains at least one vertex not belonging to I.

Darij Grinberg

17. Let G be a graph with n vertices and such that the degree of every vertex is strictly greater than $\frac{2n}{5}$. If the graph contains no triangle, then it is bipartite.

Andrasfai, Erdos, Sos

18. A connected graph has 1998 vertices and each vertex has degree 3. Show that one can delete 200 vertices, no two of them joined by an edge, such that the resulting graph is still connected.

Russia 1998

Chapter 23

23.1 Theory and examples

Undoubtedly, polynomials represent a powerful tool in practically any area of mathematics, simply because they manage to create a subtle link between analysis and algebra: on the one hand, considering them as formal series comes handy in arithmetic and combinatorics; on the other hand their analytic properties (location of zeros, complex-analytic properties, etc) are particularly interesting for effective estimations. The purpose of this chapter is to present some striking applications of these ideas in number theory and combinatorics. We will merely scratch the surface, but we are convinced that even this small amount will show the reader what profound mathematical objects polynomials are. A particularly important result to be discussed is the revolutionary "Combinatorial Nullstellensatz" of Noga Alon, which shows perfectly well the power of algebraic methods in combinatorics.

We begin, as usual, with a very easy problem. However, it is not entirely trivial because there are many approaches that can fail. A purely algebraic solution is both easy and insightful.

Example 1. Is there a set of points in space which cuts any plane in a finite, nonzero number of points?

IMO 1987 Shortlist

Solution. The idea is very simple: by taking such a set A to be the set of points of the form $(f(t), g(t), h(t))$, we need to find functions f, g, h such that for any a, b, c not all zero and any d, the equation $af(t) + bg(t) + ch(t) + d = 0$ has a finite nonzero number of solutions. This suggests taking polynomials f, g, h. One of the many choices is $f(t) = t^5$, $g(t) = t^3$ and $h(t) = t$. Indeed, the equation $at^5 + bt^3 + ct + d = 0$ clearly has a finite number of solutions and has at least one, since any polynomial of odd degree has at least one real root. This shows that such a set exists.

You may very well know the classical problem stating that if 2^n+1 is a prime number, then n is a power of 2 (the reader who does not know it is urged to give it some thought before passing to the next problem!). The following example is an adaptation of this classical result, but it is not as immediate as the cited problem.

Example 2. Prove that if $4^m - 2^m + 1$ is a prime number, then all prime divisors of m are smaller than 5.

[S. Golomb] AMM

Solution. Suppose that p is a prime divisor of m, with $p > 3$. Write $m = np$. Then $4^m - 2^m + 1 = P(-2^n)$, where $P(X) = X^{2p} + X^p + 1$. We claim that P is a multiple of $X^2 + X + 1$. Indeed, $X^2 + X + 1$ has distinct complex roots and any of its roots is clearly a root of P. Therefore $X^2 + X + 1$ divides P in $\mathbb{C}[X]$, thus in $\mathbb{Q}[X]$ too. Because $X^2 + X + 1$ is monic, Gauss's lemma implies that P is divisible by $X^2 + X + 1$ in $\mathbb{Z}[X]$. Therefore, $P(-2^n)$ is a multiple of $4^n - 2^n + 1 > 1$, so $4^m - 2^m + 1$ is not a prime number.

We continue with a fairly tricky problem, whose beautiful solution was communicated by Gheorghe Eckstein. This will be a preparation for the next challenging problem.

Example 3. Prove that the number obtained by multiplying all 2^{100} numbers of the form $\pm 1 \pm \sqrt{2} \pm \cdots \pm \sqrt{100}$ is the square of an integer.

Tournament of the Towns

Solution. The crucial observation is that if $P \in \mathbb{Z}[X]$ is an even polynomial, then for every positive integer k, the polynomial $P(X - \sqrt{k})P(X + \sqrt{k})$ is also an even polynomial with integer coefficients. Now, consider the polynomials

$$P_1(X) = X, \quad P_k(X) = P_{k-1}(X - \sqrt{k})P_{k-1}(X + \sqrt{k})$$

for $k \geq 2$. By the first observation, P_{100} is an even polynomial with integer coefficients. But it is clear that the desired product is just $P_{100}(1)P_{100}(-1)$, so it is a perfect square. This finishes the solution.

As we said, the next problem is very challenging. The solution presented here is due to Pierre Bornsztein, and can be adapted to prove much more: the square roots of the squarefree positive integers are linearly independent over the set of rational numbers. There are also elementary proofs of this deep result, but the following argument is simply stunning. Interested readers will find in the exercise section a much more general (and difficult) statement that can be proved using polynomial techniques, and which we strongly recommend.

Example 4. Let $a_1, a_2, \ldots, a_n$ be positive rational numbers such that $\sqrt{a_1} + \sqrt{a_2} + \cdots + \sqrt{a_n}$ is a rational number. Prove that the $\sqrt{a_i}$ are all rational numbers.

Solution. If all $x_i = \sqrt{a_i}$, then x_i^2 are rational numbers and the sum S of the x_i's is also rational. Let us assume that x_1 is not rational and consider the polynomial

$$P(X) = \prod_{u_2,\ldots,u_n=\pm 1} (X - x_1 + u_2x_2 + \cdots + u_nx_n) \tag{23.1}$$

Clearly, when we expand this polynomial $x_2, x_3, \ldots, x_n$ appear with even exponents because the polynomial is invariant under the substitutions $x_2 \to -x_2, \ldots, x_n \to -x_n$. After expansion, the polynomial can be written as

$$P(X) = P(X, x_1, x_2, \ldots, x_n) = N(X, x_1^2, x_2^2, \ldots, x_n^2) - x_1D(X, x_1^2, x_2^2, \ldots, x_n^2)$$

for some polynomials with rational coefficients N and D. Because P vanishes at S, we deduce that $x_1D(S, x_1^2, \ldots, x_n^2) = N(S, x_1^2, \ldots, x_n^2)$, and the assumption that x_1 is irrational implies that $D(S, x_1^2, \ldots, x_n^2) = N(S, x_1^2, \ldots, x_n^2) = 0$.

But then we also have $P(S, -x_1, x_2, \dots, x_n) = 0$, which is impossible, since $P(S, -x_1, x_2, \dots, x_n)$ is a product of positive numbers. This contradiction shows that x_1 is rational and, by induction, all x_i are rational.

The following problem became a classical application of polynomial techniques. It was also used in a Balkan Mathematical Olympiad and more recently in a Chinese TST. The following solution is probably a reason for its popularity.

Example 5. A positive integer p is prime if an only if each equiangular polygon with p vertices and rational side-lengths is regular.

Solution. We will first prove that if n is a positive integer, $\epsilon = e^{\frac{2i\pi}{n}}$, and $a_1, a_2, \dots, a_n$ are positive real numbers, then there exists an n-gon with equal angles and side-lengths $a_1, a_2, \dots, a_n$ if and only if $a_1 + a_2\epsilon + \cdots + a_n\epsilon^{n-1} = 0$. This is not difficult: it is enough to consider the edges of the polygon as oriented vectors in clockwise direction. Clearly, their sum is 0. However, one can translate these vectors so that all of them have origin at O, the origin of the plane. By choosing the positive semiaxis a_1, the complex numbers corresponding to the extremities of the vectors are $a_1, a_2\epsilon, \dots, a_n\epsilon^{n-1}$, from where we find $a_1 + a_2\epsilon + \cdots + a_n\epsilon^{n-1} = 0$. The converse is easy, because the construction follows from the previous argument.

Now assume that p is a prime number, and consider a polygon with side-lengths $a_1, a_2, \dots, a_p$, all rational numbers, and whose angles are equal. It follows that $a_1 + a_2\epsilon + \cdots + a_p\epsilon^{p-1} = 0$ and the irreducibility of the polynomial $1 + X + \cdots + X^{p-1}$ over the field of rational numbers shows that $a_1 = a_2 = \cdots = a_p$, so the polygon is regular (the argument is identical to the proof of the first lemma in chapter **Complex Combinatorics**). For the converse, let us assume that p is not a prime and prove that there exists a non-regular polygon with rational side-lengths and equal angles. Let us write $p = mn$ for some $m, n > 1$. Then $\epsilon = e^{\frac{2i\pi}{p}}$ satisfies the equation $1 + \epsilon^n + \epsilon^{2n} + \cdots + \epsilon^{(m-1)n} = 0$ and also the equation $1 + \epsilon + \cdots + \epsilon^{p-1} = 0$. By adding these two equations, we obtain a relation of the form $a_1 + a_2\epsilon + \cdots + a_p\epsilon^{p-1} = 0$, where all a_i are equal to 1 or 2 and not all of them equal. The observation in the beginning of the solution shows that there exists a polygon with equal angles and side-lengths $a_1, a_2, \dots, a_p$. Clearly, this polygon is not regular.

We continue with two difficult problems. The first one is classical, but very difficult. It belongs to a large class of additive problems in number theory and it is quite remarkable that it has a purely algebraic solution. A similar statement is the famous four-squares theorem, stating that any positive integer is the sum of four squares of integers, or the notoriously difficult Waring problem, stating that for any k there is m such that any sufficiently large integer is a sum of at most m powers of exponent k. We leave it to the interested reader to deduce from the four squares theorem that any positive integer is the sum of 53 fourth powers!

Example 6. Prove that any rational number can be written as the sum of the cubes of three rational numbers.

Solution. If someone really wants to be cruel, they will just write the following identity:

$$\left(\frac{x^3-3^6}{9x^2+81x+3^6}\right)^3+\left(\frac{-x^3+3^5x+3^6}{9x^2+81x+3^6}\right)^3+\left(\frac{9x^2+3^5x}{9x^2+81x+3^6}\right)^3=x.$$

Well, how on earth can we come with such a thing? A natural idea would be to look for a representation of x as a sum of cubes of three rational functions. So let us try to find first two polynomials f, g such that f^3+g^3 has a cubic factor. On the other hand, the factorization

$$f^3+g^3=(f+g)(f+zg)(f+z^2g),$$

where $z=e^{\frac{2i\pi}{3}}$ suggests a smart choice: $f+zg=(X-z)^3$ and $f+z^2g=(X-z^2)^3$. A small computation shows that $f=X^3-3X-1$ and $g=-3X^2-3X$. This already gives us the identity

$$(x^3-3x-1)^3+(-3x^2-3x)^3=(x^2+x+1)^3((x-1)^3-9x),$$

which easily implies the relation presented in the very beginning of the solution, after changing $9x$ to x.

The next problem has a particular flavor, because of the nice idea really well-hidden and of the technical difficulties that appear at all steps of the solution. Definitely not a friendly problem in a mathematical competition, but excellent spiritual food!

Example 7. On an $m \times n$ sheet of paper a grid dividing the sheet into unit squares is drawn. Then, the two sides of length n are taped together to form a cylinder. Prove that one can write a real number in each square, not all numbers being zero, such that each number is the sum of the numbers in the neighboring squares, if and only if there are integers k, l such that $n+1$ does not divide k and

$$\cos\left(\frac{2l\pi}{m}\right) + \cos\left(\frac{k\pi}{n+1}\right) = \frac{1}{2}.$$

[Ciprian Manolescu] Romanian TST 1998

Solution. Number the rings $1, 2, \ldots, n$ going downwards and the columns $1, 2, \ldots, m$, anticlockwise. The idea is to associate to each ring a polynomial $P_i(X) = a_{i1} + a_{i2}X + \cdots + a_{im}X^{m-1}$ and to study how the condition imposed on the numbers translates in terms of these polynomials. This is not difficult, because such numbers exist if and only if

$$P_i(X) \equiv P_{i-1}(X) + P_{i+1}(X) + (X^{m-1} + X)P_i(X) \pmod{X^m - 1},$$

where $P_0 = P_{n+1} = 0$. This can be also written as

$$P_{i+1}(X) \equiv (1 - X - X^{m-1})P_i(X) - P_{i-1}(X) \pmod{X^m - 1}$$

and so $P_i(X) = Q_i(X)P_1(X)$, where Q_i is the sequence defined by $Q_0 = 0, Q_1 = 1$ and

$$Q_{i+1}(X) \equiv (1 - X - X^{m-1})Q_i(X) - Q_{i-1}(X) \pmod{X^m - 1}.$$

The condition for all numbers to be zero becomes $P_1 \equiv 0$. So, the condition of the problem is satisfied if and only if we can find a nonzero polynomial P_1 (of course, $\pmod{X^m - 1}$) such that $P_1 Q_{n+1} \equiv 0 \pmod{X^m - 1}$, which means that Q_{n+1} and $X^m - 1$ are not relatively prime. This is also equivalent to the existence of a number z such that $z^m = 1$ and $Q_{n+1}(z) = 0$. If $x_k = Q_k(z)$, the identity satisfied by Q_i becomes $x_0 = 0, x_1 = 1$ and

$$x_{i+1} = x_i(1 - z - z^{-1}) - x_{i-1}.$$

Now, if $a = 1 - z - z^{-1}$, the relation becomes $x_{i+1} - ax_i + x_{i-1} = 0$. Let r_1, r_2 be the roots of the equation $t^2 - at + 1 = 0$. Then r_1, r_2 are nonzero, so if $x_{n+1} = 0$, then we surely have $r_1 \neq r_2$ and also $x_{n+1} = \frac{r_1^{n+1} - r_2^{n+1}}{r_1 - r_2}$. Thus the condition on m, n is to have $r_1^{n+1} = r_2^{n+1}$, that is there exists x such that $x^{n+1} = 1$, $x \neq 1$ and also $r_2 = xr_1$. Using Viéte's formula, this becomes equivalent to the existence of a nontrivial root of order $n+1$ of 1, say x, such that $a^2 x = (1+x)^2$, that is $2 + 2Re(x) = (1 - 2Re(z))^2$. Of course, this is equivalent to the condition of the problem. This finishes the solution.

Let us now turn to some combinatorial problems. We begin with a very beautiful result. Do not underestimate it because of its short proof – it is far from being trivial. Actually, this old conjecture of Artin plays a very important role in additive number theory and has given birth to some important theorems of Ax and Katz, which are unfortunately well beyond the scope of this book.

Example 8. Let $f_1, f_2, \ldots, f_k$ be polynomials in $\mathbb{Z}/p\mathbb{Z}[X_1, X_2, \ldots, X_n]$ such that $\sum_{i=1}^{k} \deg(f_i) < n$. Then the cardinality of the set of vectors $(x_1, x_2, \ldots, x_n) \in (\mathbb{Z}/p\mathbb{Z})^n$ such that $f_i(x) = 0$ for all $i = 1, 2, \ldots, k$ is a multiple of p.

Chevalley-Warning theorem

Solution. The idea is that the cardinality of the set of common zeros of f_i

can be expressed more conveniently as

$$\sum_{x=(x_1,\dots,x_n)\in(\mathbb{Z}/p\mathbb{Z})^n} (1-f_1(x)^{p-1})(1-f_2(x)^{p-1})\cdots(1-f_k(x)^{p-1}),$$

where we understand by $f_i(x)$ the element $f_i(x_1, x_2, \dots, x_n)$. Indeed, this follows easily from Fermat's little theorem, because the polynomial

$$P(X) = (1-f_1(X)^{p-1})(1-f_2(X)^{p-1})\cdots(1-f_k(X)^{p-1})$$

(here $X = (X_1, X_2, \dots, X_n)$) has the property that $P(x_1, x_2, \dots, x_n) = 0$ if and only if at least one of $f_i(x_1, x_2, \dots, x_n)$ is nonzero and 1 otherwise.

Now, let us prove that $\sum\limits_{x\in(\mathbb{Z}/p\mathbb{Z})^n} P(x) = 0$. In order to do this, it is enough to prove it for any monomial of P, of the form $X_1^{a_1}X_2^{a_2}\cdots X_n^{a_n}$. Observe that in any such monomial we have $a_1 + a_2 + \cdots + a_n < n(p-1)$, because of the condition $\sum\limits_{i=1}^{k} \deg(f_i) < n$. This means that there exists an i such that $a_i < p-1$. Observe that

$$\sum_{x\in(\mathbb{Z}/p\mathbb{Z})^n} x_1^{a_1}x_2^{a_2}\cdots x_n^{a_n} = \prod_{j=1}^{n}\Big(\sum_{x_j\in\mathbb{Z}/p\mathbb{Z}} x_j^{a_j}\Big),$$

and because $a_i < p-1$, by a result proved in the chapter **The Smaller, the Better**, $\sum\limits_{x_i\in\mathbb{Z}/p\mathbb{Z}} x_i^{a_i} = 0$, which shows that $\sum\limits_{x\in(\mathbb{Z}/p\mathbb{Z})^n} P(x) = 0$ in $\mathbb{Z}/p\mathbb{Z}$. This finishes the proof, because it follows that the cardinality of the set is a multiple of p. Finally, observe that if we assume that $f_i(0) = 0$ for all i, it follows that f_i have at least one nonzero common root in the field with p elements, which is anything but trivial!

We continue with an apparently immediate application of Chevalley-Warning theorem: the famous Erdős-Ginzburg-Ziv theorem. There are many other approaches to this beautiful result, but the way in which it follows from Chevalley-Warning's theorem had to be presented.

Example 9. Prove that from any $2n-1$ integers one can choose n whose sum is divisible by n.

Erdős-Ginzburg-Ziv theorem

Solution. Let us suppose first that $n = p$ is a prime number. As we will see, this is actually the hard part of the theorem. Consider the polynomials over $\mathbb{Z}/p\mathbb{Z}$:

$$f_1(X_1, X_2, \dots, X_{2p-1}) = X_1^{p-1} + X_2^{p-1} + \dots + X_{2p-1}^{p-1},$$

$$f_2(X_1, X_2, \dots, X_{2p-1}) = a_1 X_1^{p-1} + a_2 X_2^{p-1} + \dots + a_{2p-1} X_{2p-1}^{p-1},$$

where $a_1, a_2, \dots, a_{2p-1}$ are the $2p-1$ numbers. Clearly, the conditions of Chevalley-Warning's theorem are satisfied and so the system $f_1(X) = f_2(X) = 0$ has a nontrivial solution $(x_i)_{i=1,\dots,2p-1}$. Let I be the set of those $1 \le i \le 2p-1$ such that $x_i \neq 0$. Then from Fermat's little theorem $f_1(x_1, x_2, \dots, x_{2p-1}) = |I| \pmod{p}$ and $f_2(x_1, x_2, \dots, x_{2p-1}) = \sum_{i \in I} a_i \pmod{p}$ and so p divides $|I|$ and $\sum_{i\in I} a_i$. Because I has at least 1 and at most $2p-1$ elements, it follows that it has exactly p elements, and the theorem is proved in this case.

In order to finish the proof of the theorem, it is enough to prove that if it holds for a and b integers greater than 1, it also holds for ab. So, take $2ab-1$ integers look at the first $2a-1$ among them. There are some a whose sum is a multiple of a. Put them in a box labelled 1 and look at the remaining numbers. You have at least $2a(b-1)-1 \ge 2a-1$, so you can find some other a numbers whose sum is a multiple of a. Put them in a box labelled 2. At each stage, as long as you still have at least $2a-1$ numbers which are not yet in a box, you can create another box with a numbers, the sum of which is a multiple of a. So, you can create at least $2b-1$ such boxes. Now, apply the induction hypothesis for the sums of the numbers in the first $2b-1$ boxes divided by a and you will obtain a collection of ab numbers the sum of which is a multiple of ab. This shows that the theorem holds for ab and finishes the proof.

The next example presents a truly amazing theorem, appeared in the revolutionary article "Combinatorial Nullstellensatz" by Noga Alon and which is now a must in algebraic combinatorics. The reader with background in commutative algebra will immediately understand the title of the article: yes, it is related to the even more famous Nullstellensatz of Hilbert, one of the basic results of algebraic geometry and probably one of the most important theorems in mathematics. What does the latter say? Well, the strong form says that if $f_1, f_2, \ldots, f_k$ are polynomials with complex coefficients in n variables and if f is another such polynomial which vanishes at all common zeros of the polynomials $f_1, f_2, \ldots, f_k$, then some power of f can be written in the form $f_1g_1 + f_2g_2 + \cdots + f_kg_k$ for some polynomials $g_1, g_2, \ldots, g_k$. Note for instance that if $f_1, f_2, \ldots, f_k$ have no common zeros then there will be $g_1, g_2, \ldots, g_k$ such that $f_1g_1 + f_2g_2 + \cdots + f_kg_k = 1$, a fact far from being obvious! Actually, the proof of Hilbert's Nullstellensatz is difficult and really needs a fair amount of commutative algebra, so we will not present it here. The reader can find a proof in practically any book of algebraic geometry. Note however that there are substantial differences between this statement and the "Combinatorial Nullstellensatz", and they probably explain why the latter is so well-suited for combinatorial applications.

Example 10. Let F be a field, $f \in F[X_1, X_2, \ldots, X_n]$ a polynomial, and let $S_1, S_2, \ldots, S_n$ be nonempty subsets of F.
a) If $f(s_1, s_2, \ldots, s_n) = 0$ for all $(s_1, s_2, \ldots, s_n) \in S_1 \times S_2 \times \cdots \times S_n$, then f lies in the ideal generated by the polynomials $g_i(X_i) = \prod_{s \in S_i} (X_i - s)$. Moreover, the polynomials $h_1, h_2, \ldots, h_n$ satisfying $f = g_1h_1 + g_2h_2 + \cdots + g_nh_n$ can be chosen such that $\deg(h_i) \leq \deg(f) - \deg(g_i)$ for all i. Finally, if $g_1, g_2, \ldots, g_n \in R[X_1, X_2, \ldots, X_n]$ for some subring R of F, one can choose h_i with coefficients in R.
b) If $\deg(f) = t_1 + t_2 + \cdots + t_n$, where t_i are nonnegative integers such that $t_i < |S_i|$ and if the coefficient of $X_1^{t_1} X_2^{t_2} \cdots X_n^{t_n}$ is not zero, then there exist $s_i \in S_i$ such that $f(s_1, s_2, \ldots, s_n) \neq 0$.

[Noga Alon] Combinatorial Nullstellensatz

Solution. a) The idea is that any element s_i of S_i satisfies an algebraic equation of degree $|S_i|$, so any power of s_i is a linear combination of $1, s_i, \ldots, s_i^{|S_i|-1}$ with coefficients independent of the choice of $s_i \in S_i$. Indeed, if

$$g_i(X_i) = X_i^{|S_i|} - \sum_{j=0}^{|S_i|-1} g_{ij} X_i^j,$$

then $s_i^{|S_i|} = \sum_{j=0}^{|S_i|-1} g_{ij} s_i^j$. This allows us to "reduce" the polynomial f by replacing every X_i^k with a linear combination of $1, X_i, \ldots, X_i^{|S_i|-1}$. This corresponds to subtractions from f of polynomials of the form $g_i h_i$, with $\deg(h_i) \leq \deg(f) - \deg(g_i)$. So we see that by subtracting a linear combination $\sum_{i=1}^{n} g_i h_i$ from f we obtain a polynomial f_1 whose degree in X_i is at most $|S_i| - 1$ and such that $0 = f(s_1, s_2, \ldots, s_n) = f_1(s_1, s_2, \ldots, s_n)$ for all $s_i \in S_i$. But this immediately implies $f_1 = 0$. Indeed, f_1 can be written as $F_0 + F_1 X_1 + \cdots + F_{|S_1|-1} X_1^{|S_1|-1}$ for some polynomials $F_i \in F[X_2, \ldots, X_n]$ such that F_j has degree in X_j at most $|S_j| - 1$. Now, for all $s_2 \in S_2, \ldots, s_n \in S_n$, the polynomial

$$F_0(s_2, \ldots, s_n) + \cdots + F_{|S_1|-1}(s_2, \ldots, s_n) X_1^{|S_1|-1}$$

has at least $|S_1|$ zeros in the field F, so it is identically zero, that is

$$F_0(s_2, \ldots, s_n) = \cdots = F_{|S_1|-1}(s_2, \ldots, s_n) = 0$$

for all $(s_2, \ldots, s_n) \in S_2 \cdots S_n$. An inductive reasoning shows that $F_0 = \cdots = F_{|S_1|-1} = 0$ and so $f_1 = 0$. This finishes the proof of a). b) This is a direct consequence of a). Suppose by contradiction that f vanishes on $S_1 \times S_2 \times \cdots \times S_n$. By taking subsets of S_i with $t_i + 1$ elements, we can assume that $|S_i| = t_i + 1$. Let h_i and g_i be defined as in a). It follows that the coefficient of $X_1^{t_1} X_2^{t_2} \cdots X_n^{t_n}$ in $g_1 h_1 + g_2 h_2 + \cdots + g_n h_n$ is not zero. Because $\deg(h_i) \leq \deg(f) - \deg(g_i)$, the coefficient of $X_1^{t_1} X_2^{t_2} \cdots X_n^{t_n}$ in $g_i h_i$ is zero: any monomial appearing in this polynomial and having degree $\deg(f)$ is a multiple of $X_i^{t_i+1}$, contradiction.

Let us see now some applications of this result. First, some direct consequences which already show the power of the method: try to find other solutions to these problems and you will see that they are far from being trivial. This is probably also a reason for selecting the next problem as problem 6 at the International Mathematical Olympiad in 2007.

Example 11. Let n be a positive integer and consider the set

$$S = \{(x, y, z) | x, y, z \in \{0, 1, \ldots, n\}, x + y + z > 0\}$$

as a set of points in space. Find the minimum number of planes, the union of which contains S but does not contain $(0, 0, 0)$.

IMO 2007

Solution. Let $a_i x + b_i y + c_i z = d_i$ be the equations of these planes and consider the polynomial

$$f(X, Y, Z) = \prod_{i=1}^{k} (a_i X + b_i Y + c_i Z - d_i) - m \cdot \prod_{i=1}^{n} (X - i)(Y - i)(Z - i),$$

where m is chosen such that $f(0, 0, 0) = 0$. If $k < 3n$, then clearly the coefficient of $X^n Y^n Z^n$ in f is nonzero. Thus, by combinatorial Nullstellensatz there are integers $x, y, z \in \{0, 1, \ldots, n\}$ such that $f(x, y, z) \neq 0$. If at least one of x, y, z is nonzero, then clearly both terms defining f are zero, a contradiction. Thus $(x, y, z) = (0, 0, 0)$, which contradicts the fact that $f(0, 0, 0) = 0$. Therefore $k \geq 3n$ and since for $k = 3n$ an example is immediate, we deduce that this is the answer to the problem.

And now a very similar statement:

Example 12. Let p be a prime and let $S_1, S_2, \dots, S_k$ be sets of non-negative integers, each containing 0 and having pairwise distinct elements modulo p. Suppose that $\sum_i(|S_i| - 1) \geq p$. Prove that for any elements $a_1, \dots, a_k \in \mathbb{Z}/p\mathbb{Z}$, the equation $x_1a_1 + x_2a_2 + \cdots + x_ka_k = 0$ has a solution $(x_1, \dots, x_k) \in S_1 \times \cdots \times S_k$ other than the trivial one $(0, \dots, 0)$.

Troi-Zannier's theorem

Solution. [Peter Scholze] Consider the polynomial

$$P(X_1, \dots, X_k) = (a_1X_1 + a_2X_2 + \cdots + a_kX_k)^{p-1} - 1$$

$$+C \prod_{0 \neq s_1 \in S_1} (X_1 - s_1) \prod_{0 \neq s_2 \in S_2} (X_2 - s_2) \cdots \prod_{0 \neq s_k \in S_k} (X_k - s_k)$$

where C is chosen such that $P(0, \dots, 0) = 0$.

Because of the third condition, the coefficient of $x_1^{|S_1|-1} \cdots x_k^{|S_k|-1}$ is non-zero. Therefore there are $t_1 \in S_1, \dots, t_k \in S_k$ with $P(t_1, \dots, t_k) \neq 0$. Since $P(0, \dots, 0) = 0$, it is clearly not the zero solution. Thus,

$$C \prod_{0 \neq s_1 \in S_1} (t_1 - s_1) \prod_{0 \neq s_2 \in S_2} (t_2 - s_2) \cdots \prod_{0 \neq s_k \in S_k} (t_k - s_k)$$

must be zero, which implies that $(a_1t_1 + \cdots + a_kt_k)^{p-1} \neq 1$. It remains only to note that Fermat's little theorem gives $a_1t_1 + \cdots + a_kt_k = 0$.

The category of deep results with short proofs is going to be represented once again, this time with a really important result of additive combinatorics, one of those mathematical fields which exploded in the twentieth century. Of course, there are many other proofs of this result, all of them very ingenious. The result itself is important: as an exercise (solved by Cauchy about two hundred years ago...), try to prove this using this Lagrange's famous theorem stating that any positive integer can be written as a sum of four squares of integers.

There are very elementary arguments, as we will see, but the combinatorial Nullstellensatz also implies this result and actually much more.

Example 13. For any subsets A, B of $\mathbb{Z}/p\mathbb{Z}$ the following inequality holds

$$|A+B| \geq \min(p, |A|+|B|-1).$$

Cauchy-Davenport theorem

Solution. There is one very simple case: $|A|+|B| \geq p+1$. In this case, $A+B = \mathbb{Z}/p\mathbb{Z}$, since for any $x \in \mathbb{Z}/p\mathbb{Z}$, the function $f(a) = x-a$ defined on A cannot take all its values outside B, because it is injective. The difficult case is when $|A|+|B| \leq p$. Let us suppose that $|A+B| \leq |A|+|B|-2$ and let us choose a subset C of $\mathbb{Z}/p\mathbb{Z}$ containing $A+B$ and having $|A|+|B|-2$ elements. The polynomial $f(X_1, X_2) = \prod\limits_{c \in C} (X_1+X_2-c) \in \mathbb{Z}/p\mathbb{Z}[X]$ has degree $|A|+|B|-2$ and vanishes on $A \times B$. In order to obtain a contradiction, it is thus sufficient to prove that $X_1^{|A|-1} X_2^{|B|-1}$ appears with a nonzero exponent in f. However, it is clear that this exponent equals $\binom{|A|+|B|-2}{|A|-1} \pmod{p}$, which is not zero, because $|A|+|B|-2 \leq p-2$. Using the previous theorem, we obtain the desired contradiction.

Before passing to the next example, let us present a truly magnificient (for its simplicity!) proof of the previous result, which is probably more natural when seeing the statement for the first time, but which is by no means as obvious as it looks! We shall prove the result by induction on $|A|$, the case when $|A| = 1$ being obvious. Clearly, we may assume that $|A| > 1$ and also that $|B| < p$. Now, A having more than one element, by shifting it we may assume that it contains 0 and some $x \neq 0$. Now, B is nonempty and $B \neq \mathbb{Z}/p\mathbb{Z}$, so there must be an integer n such that $nx \in B$ but $(n+1)x$ does not belong to B. By shifting B this time we may suppose that $0 \in B$, but x is not in B. Thus, $A \cap B$ is a proper nonempty subset of A and we may use the induction hypothesis for it and $A \cup B$. Because $A+B$ contains $(A \cap B)+(A \cup B)$ and

$$|A \cap B| + |A \cup B| = |A| + |B|,$$

the conclusion follows. Even though this proof is very beautiful and short, it should be noted that Alon's technique is much more powerful. Indeed, Alon shows in his seminal paper that his theorem implies a famous conjecture of Erdős-Heilbronn, with a very similar statement, but with no elementary proof (exercise for the reader: check that the above elementary solution does not work for the following result): for any nonempty subset A of $\mathbb{Z}/p\mathbb{Z}$ one has

$$|\{a+b|a, b \in A, a \neq b\}| \geq \min(p, 2|A|-3).$$

The next problem uses the proof given by Noga Alon for a special case of a difficult conjecture of Snevily. Again, the combinatorial Nullstellensatz is well-suited, but this time it is not so clear that its hypotheses are satisfied. Actually, the most difficult part in using this powerful tool is finding the good polynomial, but there are situations when it is even more difficult to check the hypothesis, because the polynomial can have a quite complicated expression.

Example 14. Let p be a prime number, and let $a_1, a_2, \ldots, a_k \in \mathbb{Z}/p\mathbb{Z}$, not necessarily distinct. Prove that for any distinct elements $b_1, b_2, \ldots, b_k$ of $\mathbb{Z}/p\mathbb{Z}$ there exists a permutation σ such that the elements $a_1 + b_{\sigma(1)}, a_2 + b_{\sigma(2)}, \ldots, a_k + b_{\sigma(k)}$ are pairwise distinct.

Alon's theorem

Solution. Let $B = \{b_1, b_2, \ldots, b_k\}$ and suppose the contrary, that is for all choices of distinct elements $x_1, x_2, \ldots, x_k$ at least two of the elements $x_1 + a_1, \ldots, x_k + a_k$ are identical. That is, if $x_1, x_2, \ldots, x_k$ are distinct elements of B, we have $\prod_{1 \leq i < j \leq k} (x_i + a_i - x_j - a_j) = 0$ in $\mathbb{Z}/p\mathbb{Z}$. We can relax the restriction of $x_1, x_2, \ldots, x_k$ being pairwise distinct by considering the polynomial

$$f(X_1, X_2, \ldots, X_k) = \prod_{1 \leq i < j < k} (X_i - X_j)(X_i + a_i - X_j - a_j).$$

The previous remark shows that f vanishes on B^k. Clearly, f can be written as

$$\prod_{1\le i<j\le k}(X_i - X_j)^2 + g(X_1, X_2, \dots, X_k)$$

for some polynomial g of smaller degree. Also, $\deg(f) = k(k-1)$. We will try to find $t_1, t_2, \dots, t_k$ such that $t_1 + t_2 + \dots + t_k = k(k-1)$, $t_i < k$ and the coefficient of $X_1^{t_1}X_2^{t_2}\cdots X_k^{t_k}$ is nonzero in f. The first two conditions impose $t_1 = t_2 = \dots = t_k = k-1$, so the question is whether the coefficient of $(X_1X_2\cdots X_k)^{k-1}$ in f is zero. Of course, if we manage to prove that $(X_1X_2\cdots X_k)^{k-1}$ appears with a nonzero coefficient in $\prod_{1\le i<j\le k}(X_i - X_j)^2$ we can apply Alon's theorem and obtain the desired contradiction. However, using the result in example 8 of the **Lagrange Interpolation Formula** chapter, we deduce that the free term of $\prod_{1\le i\neq j\le k}\left(1 - \frac{X_i}{X_j}\right)$ is $k!$. But note that

$$\prod_{1\le i\neq j\le k}\left(1 - \frac{X_i}{X_j}\right) = (-1)^{\frac{k(k-1)}{2}}\frac{\prod_{1\le i<j\le k}(X_i - X_j)^2}{(X_1X_2\cdots X_k)^{k-1}}, \qquad (23.2)$$

so the coefficient of $(X_1X_2\cdots X_{k-1})$ is nonzero in $\mathbb{Z}/p\mathbb{Z}$ because of the assumption $k < p$. This finishes the proof of the result.

We have already seen examples of combinatorial problems for which it is almost impossible to find combinatorial solutions. We continue with an example, which is a quite deep result of Alon, Friedland and Katai. Needless to say, the solution using combinatorial Nullstellensatz is practically straightforward. There are, however, limits of the method, for instance one does not know if in the next result one can replace p prime by any positive integer.

Example 15. Let G be a graph with no loops (yet, multiple edges are allowed) and let p be a prime number. Assume that all vertices have degree at most $2p-1$ and the average degree of the graph is greater than $2p-2$. Prove that G has a p-regular subgraph (a subgraph in which every vertex has degree p).

Solution. Let us consider the incidence matrix $(a_{v,e})$, where v denotes a vertex and e an edge and $a_{v,e} = 1$ if $v \in e$ and 0 otherwise. Let x_e be a variable associated with each edge e and consider the polynomial

$$f = \prod_v \left(1 - \left(\sum_{e \in E} a_{v,e} x_e\right)^{p-1}\right) - \prod_e (1 - x_e)$$

The hypothesis implies that f has degree $|E|$ and because the coefficient of $\prod_e x_e$ is nonzero, Alon's Nullstellensatz implies the existence of values $x_e \in \{0,1\}$ such that the evaluation of f at these x_e is not zero. Now, clearly this is not the zero vector, thus using Fermat's little theorem we deduce that all $\sum_{e \in E} a_{v,e} x_e$ are 0 in $\mathbb{Z}/p\mathbb{Z}$, that is if we look at the subgraph of those edges e such that $x_e = 1$, all vertices have degrees multiples of p, smaller than $2p$. Thus this subgraph is p-regular.

The reader is urged to take a look at the problems for training for many other applications of combinatorial Nullstellensatz, a theme that will surely become recurrent in algebraic combinatorics and additive number theory. We will now present a quite subtle result, based on algebraic properties of polynomials. We have already encountered this type of argument in a previous chapter, but the result and the method are too important not be presented.

Example 16. Let F be a family of subsets of a set X with n elements. Suppose that there is a set L with s elements such that $|A \cap B| \in L$ for all distinct members $A, B \in F$. Prove that

$$|F| \leq \binom{n}{s} + \binom{n}{s-1} + \cdots + \binom{n}{1} + \binom{n}{0}.$$

Frankl-Wilson theorem

Solution. Let $L = \{l_1, l_2, \ldots, l_s\}$ and assume without loss of generality that $X = \{1, 2, \ldots, n\}$. Finally, call $A_1, A_2, \ldots, A_m$ the elements of F, such that $|A_1| \le |A_2| \le \cdots \le |A_m|$. We will associate with each set A_i its characteristic vector $v_i = (v_{ij})_{1\le j\le n}$ defined by: $v_{ij} = 1$ if $j \in A_i$ and 0 otherwise. Observe that if $\langle x, y\rangle = x_1y_1 + x_2y_2 + \cdots + x_ny_n$ is the standard euclidean inner product, then $|A_i \cap A_j| = \langle v_i, v_j\rangle$. Now, let us define the polynomials

$$f_i(X) = \prod_{k, l_k < |A_i|} (\langle x, v_i\rangle - l_k)$$

for $i = 1, 2, \ldots, m$. The main idea is to consider the restrictions of these polynomials to the vertices of the unit cube, that is the set $Y = \{0, 1\}^n$. Because $x_i^2 = x_i$ if $x_i \in \{0, 1\}$, it is clear that these restrictions can be written in the form $g_i(x_1, \ldots, x_n)$, where g_i are polynomials of degree at most s and have degree at most 1 in each variable. What is remarkable is that these functions $f_i : Y \to \mathbb{R}$ are linearly independent. This is not difficult: if $\lambda_1 f_1(x) + \lambda_2 f_2(x) + \cdots + \lambda_m f_m(x) = 0$ for $x \in Y$, then by taking $x = v_j$ for all j and using the fact that $f_i(v_j) = 0$ if $j < i$ and $f_i(v_i) \neq 0$ (which is obvious), we immediately deduce by induction that all λ_i are 0. The result follows from the fact that the vector space generated by these functions has dimension m and is a subspace of the vector space of polynomials of maximum degree at most s and partial degrees at most 1, which has dimension

$$\binom{n}{s} + \binom{n}{s-1} + \cdots + \binom{n}{1} + \binom{n}{0}.$$

23.2 Practice problems

1. Let $a_1, a_2, \ldots, a_{100}$ and $b_1, b_2, \ldots, b_{100}$ be 200 distinct real numbers. Consider an 100×100 table and put the number $a_i + b_j$ in the (i, j) position. Suppose that the product of the entries in each column is 1. Prove that the product of the entries in each row is -1.

   Russian Olympiad

2. Let n, m be positive integers with $n < m - 1$ and let $a_1, a_2, \ldots, a_m$ be nonzero integers such that for all $0 \leq k \leq n$ we have $a_1 + a_2 \cdot 2^k + \cdots + a_m \cdot m^k = 0$. Prove that there are at least $n + 1$ pairs of consecutive terms having opposite signs in the sequence $a_1, a_2, \ldots, a_m$.

   Russia 1996

3. The finite sequence $\{a_k\}_{1 \leq k \leq n}$ is called p-balanced if the sums

$$s(k, p) = a_k + a_{k+p} + a_{k+2p} + \cdots$$

   are all equal for $k = 1, 2, \ldots, p$. Prove that if a sequence of 50 real numbers is 3, 5, 7, 11, 13 and 17-balanced, then all its terms are equal to 0.

   St. Petersburg 1991

4. Two numbers are written on each vertex of a convex 100-gon. Prove that it is possible to remove a number from each vertex so that remaining numbers in any two adjacent vertices are different.

   Fedor Petrov, Russia 2007

5. Let A be an $n \times n$ matrix over a field F and define its permanent as

$$Per(A) = \sum_{\sigma \in S_n} a_{1\sigma(1)} a_{2\sigma(2)} \cdots a_{n\sigma(n)}.$$

If $Per(A) \neq 0$, prove that for each $b = (b_1, b_2, \ldots, b_n) \in \mathbb{F}^n$ and for every family of two-element sets $S_1, S_2, \ldots, S_n$ of $\mathbb{F}$, there is a vector $X \in S_1 \times S_2 \times \cdots \times S_n$ such that for each i the i-th coordinate of AX differs from b_i.

Alon's Permanent Lemma

6. Let p be a prime and let $a_1, a_2, \ldots, a_{2p-1}$ be elements of $\mathbb{Z}/p\mathbb{Z}$. Prove that the number of subsets I of $\{1, 2, \ldots, 2p-1\}$ with p elements such that $\sum_{i \in I} a_i = b$ in $\mathbb{Z}/p\mathbb{Z}$ is congruent to 0 or 1 modulo p, for all $b \in \mathbb{Z}/p\mathbb{Z}$.

W. Gao

7. Let p be a prime and let A be a set of positive integers such that:

   (a) the set of prime divisors of the elements in A consists of $p - 1$ elements;

   (b) for any nonempty subset of A, the product of its elements is not a perfect p-th power.

   What is the largest possible number of elements in A ?

IMO Shortlist 2003

8. Let p be a prime and d a positive integer. Prove that for any integer k there are integers $x_1, x_2, \ldots, x_d$ such that $k = x_1^d + x_2^d + \cdots + x_d^d \pmod{p}$.

Gabriel Carrol

9. Let $S_1, S_2, \ldots, S_n$ be subsets of $\mathbb{Z}/p\mathbb{Z}$ and let $S = S_1 \times S_2 \times \cdots \times S_n$. Consider polynomials $f_1, f_2, \ldots, f_k$ in n variables over $\mathbb{Z}/p\mathbb{Z}$ such that

$$(p-1) \cdot \sum_{i=1}^{k} \deg(f_i) < \sum_{i=1}^{n} (|S_i| - 1).$$

Prove that if the system $f_1(x) = f_2(x) = \cdots = f_k(x) = 0$ has a solution $a \in S$, then it has another solution $b \in S$.

David Brink

10. Let $H_1, \ldots, H_m$ be a family of hyperplanes in $\mathbb{R}^n$ that cover all vertices of unit cube $\{0,1\}^n$ but one. Prove that $m \geq n$.

Noga Alon

11. If p is a prime, n is an integer, and $x_1, x_2, \ldots, x_{(p-1)n+1}$ are $(p-1)\,n+1$ elements of the vector space $\mathbb{F}_p^n$, then there exists a non-empty subset $I \subseteq \{1, 2, \ldots, (p-1)\,n+1\}$ such that $\sum_{i \in I} x_i = 0$.

12. Let A be a subset of $\mathbb{Z}/p\mathbb{Z}$, where p is a prime number. Prove that among the elements $a + b$ where $a \neq b \in A$ there are at least $\min(p, 2|A| - 3)$ distinct elements.

Erdos-Heilbronn conjecture

13. Let p be a prime, $n = 4p$ and let $A \subset \{-1, 1\}^n$ be a family of vectors, no two of which are orthogonal. Prove that A has at most $4\sum_{i=0}^{p-1} \binom{n}{i}$ vectors.

Frankl-Wilson

14. Let F be a family of subsets of $\{1, 2, \ldots, n\}$ and let L be a set of nonnegative integers. Say F is k-uniform if $|A| = k$ for all $A \in F$ and say F is L-intersecting if $|A \cap B| \in L$ for all $A \neq B \in F$. If p is a prime, say F is L-intersecting mod p if $|A| \notin L \pmod p$ for all $A \in F$, but $|A \cap B| \in L \pmod p$ for all $A \neq B \in F$.

(a) Prove that an L-intersecting or L-intersecting mod p family has at most $\sum_{i=0}^{|L|} \binom{n}{i}$ elements.

(b) Prove that a k-uniform L-intersecting or k-uniform L-intersecting mod p family has at most $\binom{n}{|L|}$ elements.

Chaudhuri, Frankl, Ray, Wilson

15. Prove that there exists a positive integer n such that any prime divisor of $2^n - 1$ is smaller than $2^{\frac{n}{1993}} - 1$.

Komal

16. A k-forest is a family F of subsets of $\{1, 2, \dots, n\}$ such that

(a) For any $f \in F$, f has k elements;

(b) For any $f \in F$ there exists a partition $\{1, 2, \dots, n\} = V_{1,f} \cup \dots \cup V_{k,f}$ such that f is the only member of F intersecting every $V_{i,F}$.

Prove that a k-forest has at most $\binom{n-1}{k-1}$ members.

Lovasz's theorem

17. Prove that the minimal cardinality of a subset of $(\mathbb{Z}/p\mathbb{Z})^d$ that intersects all hyperplanes is $d(p-1)+1$.

Brouwer-Schrijver's theorem

18. Let n be an even number and let $v_1, v_2, \dots, v_k \in \{-1, 1\}^n$ be vectors of length n such that any $v \in \{-1, 1\}^n$ is orthogonal to at least one of them. Prove that $k \geq n$ and that for all even n this estimate is sharp.

Alon, Knuth

19. Let $S \subset \mathbb{R}^n$ be a set of unit vectors such that there exist real numbers a, b with $a + b \geq 0$ and $< x, y > \in \{a, b\}$ for all $x \neq y \in S$. Here $<>$ is the standard scalar product on $\mathbb{R}^n$. Prove that S has at most $\frac{n(n+1)}{2}$ elements and that this is sharp for $n \geq 7$.

Oleg R.Musin

20. Let $f(n)$ denote the largest cardinality of a subset A of $\mathbb{R}^n$ such that the points in A determine at most two distances. Show that

$$\frac{n(n+1)}{2} \leq f(n, 2) \leq \frac{(n+1)(n+2)}{2}.$$

Larman, Rogers, Seidel, Blokhius

21. A subset E of $(\mathbb{Z}/p\mathbb{Z})^n$ is called a Kakeya set if it contains a line in every direction, ie for all $v \in (\mathbb{Z}/p\mathbb{Z})^n$ with $v \neq 0$ there exists $x \in (\mathbb{Z}/p\mathbb{Z})^n$ such that $x + tv \in E$ for all $t \in \mathbb{Z}/p\mathbb{Z}$.

   (a) Prove that if $P \in \mathbb{Z}/p\mathbb{Z}[X_1, X_2, \ldots, X_n]$ has degree at most $p - 1$ and vanishes on E, then $P = 0$.

   (b) Deduce that any Kakeya set E has at least $\binom{p+n-1}{n}$ elements.

Zeev Dvir's theorem

Bibliography

[1] Aassila M., 300 Defis Mathematiques, Ellipses, 2001.

[2] Aigner M., Ziegler G.M., Proofs from the Book, Springer-Verlag 3rd edition, 2003.

[3] Alon N., Nathanson M.B., Rusza I.Z., Adding distinct congruence classes modulo a prime, Amer. Math. Monthly 102 (1995), 250-255.

[4] Alon N., Nathanson M.B., Rusza I.Z., The polynomial method and restricted sums of congruence classes, J. Number Theory 56 (1996), 404-417.

[5] Andreescu T., Feng Z., Mathematical Olympiads 1998-1999: Problems and Solutions from Around the World, MAA Problem Book Series.

[6] Andreescu T., Feng Z., Mathematical Olympiads 1999-2000: Problems and Solutions from Around the World, MAA Problem Book Series.

[7] Andreescu T., Feng Z., Lee George Jr., Mathematical Olympiads 2000-2001: Problems and Solutions from Around the World, MAA Problem Book Series.

[8] Andreescu T., Feng Z., USA and International Mathematical Olympiads 2000, MAA Problem Book Series.

[9] Andreescu T., Feng Z., USA and International Mathematical Olympiads 2001, MAA Problem Book Series.

[10] Andreescu T., Feng Z., USA and International Mathematical Olympiads 2002, MAA Problem Book Series.

[11] Andreescu T., Andrica D., 360 Problems for Mathematical Contests, GIL Publishing House, Zalau, 2003.

[12] Andreescu T., Andrica D., Complex Numbers from A to... Z, Birkhauser, Boston, 2005.

[13] Andreescu T., Andrica D., Number Theory: Structures, Examples, and Problems, Birkhauser Boston, 2008.

[14] Andreescu T., Gelca R., Mathematical Olympiad Challenges, 2nd edition, Birkhauser, Boston, 2008.

[15] Andreescu T., Kedlaya K., Zeitz P., Mathematical Contests, 1995-1996.

[16] Andreescu T., Kedlaya K., Mathematical Contests 1996-1997, Problems and Solutions from Around the World, American Mathematics Competitions, 1998.

[17] Andreescu T., Kedlaya K., Mathematical Contests 1997-1998, Problems and Solutions from Around the World, American Mathematics Competitions, 1999.

[18] Andreescu T., Cartoaje. V, Dospinescu G., Lascu M., Old and New Inequalities, GIL Publishing House, 2004.

[19] Ankeny N.C., Sums of three squares, Proceedings of the Amer. Math. Soc, 8 (1957), Nr 2, 316-319.

[20] Baker A., Transcendental Number Theory, Cambridge University Press, 1975.

[21] Barbeau E.J., Klamkin M.S., Moser W.O.J., Five Hundred Mathematical Challenges, The Mathematical Association of America, 1995.

[22] Becheanu M., International Mathematical Olympiads 1959-2000. Problems. Solutions. Results, Academic Distribution Center, Freeland, USA, 2001.

[23] Berend D., Bilu Y., Polynomials with roots modulo every integer, Proceedings of the Amer. Math. Soc, 124 (1996), Nr 6, 1663-1671.

[24] Bhargava M., The Factorial Function and Generalizations, Amer. Math. Monthly 107 (2000), 783-799.

[25] Blichfeldt H., A new principle in the geometry of numbers, with some applications, Trans. Amer. Math. Soc 15, 1914, 227-235.

[26] Boju V., Funar L., The Math Problems Notebook, Birkhauser, 2007

[27] Bonavero L., Sur le nombre de sommets des polytopes entiers, Images des Mathmatiques, 33-40, C.N.R.S, 2004.

[28] Bonciocat A.I., Zaharescu A., Irreducibility Results for Compositions of Polynomials with Integer Coefficients, Monatsh. Math. 149, 31-41 (2006).

[29] Bornsztein P., Caruso X., Des formes bilineaires en combinatoire, Revue des Mathematiques Speciales.

[30] Cassels J.W.S, An Introduction to Diophantine Approximation, Cambridge Tracts in Mathematics, Vol 45, 1957.

[31] Cassels J.W.S., Frohlich A., Algebraic number theory, Academic Press, 1967.

[32] Cassels J.W.S., An Introduction to the Geometry of Numbers, Springer-Verlag, Berlin, 1959.

[33] Cuculescu I., International Mathematical Olympiads for Students, Editura Tehnica, Bucharest, 1984.

[34] Davenport H., Multiplicative number theory, Markham Publ. Co., 1967.

[35] Davenport H., Lewis D.J., Schinzel A., Polynomials of certain special types, Acta Arithm. 9, 1964, 108-116.

[36] Davenport H., The geometry of numbers, Q. J. Math, 10:119-121, 1939.

[37] Dorwart H.L., Ore O., Criteria for the Irreducibility of Polynomials, The Annals of Mathematics, Vol 34, No 1(1993), 81-94.

[38] Dorwart H.L., Irreducibility of Polynomials, Amer. Math. Monthly, No. 6, 1935, 369-381.

[39] Ehrhart E., Demonstration de la loi de reciprocite pour un polyedre entier, C. R. Acad. Sci. Paris 265, 1967, 5-7.

[40] Engel A., Problem Solving Strategies, Springer, 1999.

[41] Erdős P., Ginzburg A., Ziv A., Theorem in the Additive Number Theory, Bull. Research Counsil Israel, 1961, 41-43.

[42] Fomin A.A., Kuznetsova G.M., International Mathematical Olympiads, Drofa, Moskva, 1998.

[43] Forman R., Sequences with many Primes, Amer. Math. Monthly, 99 (1992), 548-557.

[44] Freiling C., Rinne D., Tiling a square with similar rectangles, Mathematical Research Letters 1, 547-558, 1994.

[45] Gelca R., Andreescu T., Putnam and Beyond, Springer, 2007

[46] Gerst I., Brillhart J., On the Prime Divisors of Polynomials, Amer. Math. Monthly, 78 (1971), 250-266.

[47] Godsil C., Tools from linear algebra, Handbook of Combinatorics, edited by R. Graham, M. Grotschel and L. Lovasz, Elsevier and M.I.T Press, 1995, 1705-1748.

[48] Graham R.L., Knuth D.E., Patashnik O., Concrete Mathematics, 2nd edition, Addison-Wesley, 1989.

[49] Greitzer S.L., International Mathematical Olympiads 1959-1977, M.A.A., Washington, D.C., 1978.

[50] Guy, R.K., Unsolved Problems in Number Theory, Springer, 3rd edition, 2004.

[51] O'Hara P.J., Another proof of Bernstein's theorem, Amer. Math. Monthly 80, 1973, 673-674.

[52] Hardy G.H, Wright E.M., An introduction to the Theory of Numbers, Oxford 1979.

[53] Hlawka E., Schoibengeier J., Taschner R., Geometric and Analytic Number Theory, Springer-Verlag, 1991.

[54] Huneke C., The friendship theorem, Amer. Math. Monthly, No. 2, 2002, 192-194.

[55] Oleszkiewicz K., An elementary proof of Hilbert's inequality, Amer. Math. Monthly, 100, 1993, 276-280.

[56] Mignotte M., An Inequality about Factors of Polynomials, Mathematics of Computation, 128 (1974), 1153-1157.

[57] Mitrinovic D.S, Vasic P.M, Analytic inequalities, Springer-Verlag, 1970.

[58] Murty M.R., Prime Numbers and Irreducible Polynomials, Amer. Math. Monthly, No. 5, 2002, 452-458.

[59] Nathanson M.B., Additive Number Theory, Springer 1996.

[60] Polya G., Szego G., Problems and theorems in analysis, Springer-Verlag, 1976.

[61] Prasolov V.V, Polynomials, Algorithms and Computation in Mathematics, Volume 11 Springer-Verlag, 2003.

[62] Radulescu T., Radulescu V., Andreescu T., Problems in Real Analysis: Advanced Calculus on the Real Axis, Springer New York, 2009.

[63] Rogosinski W.W., Some elementary inequalities for polynomials, Math. Gaz, 39, No. 327, 1955, 7-12.

[64] Roitman M., On Zgismondy Primes, Proceedings of the Amer. Math. Soc, 125 (1997), No. 7, 1913-1919.

[65] Savchev S., Andreescu T., Mathematical Miniatures, New Mathematical Library, MAA 2002.

[66] Seres I., Irreducibility of Polynomials, Journal of Algebra 2, 283-286, 1963.

[67] Serre J.-P., A Course in Arithmetic, Springer-Verlag 1973.

[68] Siegel, C.L., Lectures on the Geometry of Numbers, Springer-Verlag, 1989 (notes by B.Friedman rewritten by K.Chandrasekharan with assistance of R.Suter).

[69] Sierpinski W., Elementary Theory of Numbers, Polski Academic Nauk, Warsaw, 1964.

[70] Sierpinski W., 250 Problems in Elementary Number Theory, American Elsevier Publishing Company, Inc., New York, Warsaw, 1970.

[71] Stanley R.P., Enumerative Combinatorics, Cambridge University Press, 2nd edition, 2000.

[72] Steele Michael J., The Cauchy-Schwarz Master Class, Cambridge University Press, 2004.

[73] Sun Z.W., Covering the Integers by Arithmetic Sequences, Trans. Amer. Math. Soc, Vol 348, Nr 11, 1996, 4279-4320.

[74] Tomescu I., Melter R.A., Problems in Combinatorics and Graph Theory, John Wiley Sons, 1985.

[75] Tomescu I. et al., Balkan Mathematical Olympiads 1984-1994, Gil, Zalau, 1996.

[76] Turk J., The Fixed Divisor of a Polynomial, Amer. Math. Monthly, 93 (1986), 282-286.

[77] Yaglom A.M., Yaglom I.M., Challenging Mathematical Problems with Elementary Solutions, Dover Publications, 1987.

[78] Zannier U., A note on securrent mod p sequences, Acta Arithmetica XLI, 1982.

Index

Printed by "Combinatul Poligrafic"
Com. nr. 60585